ECOTOXICOLOGY
A Hierarchical Treatment

Edited by
Michael C. Newman
Charles H. Jagoe

LEWIS PUBLISHERS

Boca Raton   New York   London   Tokyo

Library of Congress Cataloging-in-Publication Data

Ecotoxicology : a hierarchical treatment / editors, Michael C. Newman
   and Charles H. Jagoe.
  p.   cm. -- (Savannah River series on environmental sciences)
  Includes bibliographical references (p.) and index.
  ISBN 0-56670-127-9 (alk. paper)
   1. Pollution--Environmental aspects--Congresses. 2. Pollutants-
-Toxicology--Congresses. I. Newman, Michael C. II. Jagoe, Charles
H. III. Series.
QH545.A1E285   1996
574.5′222--dc20   93-31128
                                                    CIP

© 1996 by CRC Press, Inc.
Lewis Publishers is an imprint of CRC Press

No claim to original U.S. Government works
International Standard Book Number 0-56670-127-9
Library of Congress Card Number 95-31128
Printed in the United States of America 1 2 3 4 5 6 7 8 9 0
Printed on acid-free paper

Foreword

It would seem that we are entering an age of interfaces that strive to build bridges between traditional disciplines. New interface societies, symposium volumes and journals are appearing at a rate that must be driving librarians crazy! I believe that a major reason for this trend is the mismatch between traditional disciplines and real world problems which by and large require an interdisciplinary approach. Also contributing to the need for interfaces is the tendency for traditional disciplines, such as biology and chemistry, to become increasingly reductionistic while real world problems are becoming more holistic.

Especially vigorous are interfaces between ecology and various traditional disciplines such as biology, chemistry, physics, economics, agriculture, medicine, etc. Ecotoxicology is one of the first of these to attract a large following because of the growing concern about pollution effects at the ecosystem level. It is obvious that dealing with this concern requires the attention of more than one traditional discipline: chemistry, toxicology, physiology, microbiology, and ecology have to be involved at the very least.

At the beginning, an interface effort enriches the disciplines being interfaced. In the case of ecotoxicology, chemists learn about community metabolism, bioaccumulation, microbial decomposition, and diversity; ecologists learn about transport mechanisms and the importance of distinguishing between the different forms of mercury, sulfur, and other potential pollutants. For an interface to become truly a new discipline, something new has to emerge, that is, new theories or new procedures; otherwise, the interface should remain an interface. If every possible interface, and there are hundreds of possibilities, should become a new discipline with separate societies, journals, college departments, and degrees then science would become so fragmented as to not only completely bewilder the public, but also make any kind of political action even more difficult than it already is.

Let me cite a case where an interface has become an important new discipline, and one, although important, that has not. Nonmarket goods and services, natural capital, and H.T. Odum's Energy are new concepts that have emerged in the ecology-economics interface. Neither the traditional ecologist nor the traditional economist had the slightest interest in these concepts until their importance was highlighted by the interface. Therefore, ecological economics has become a new discipline according to "emergent" criterion.

In contrast, radiation-ecology is an interface that experienced an intense period of research in the 1960s and 1970s, but today is not considered to be an independent discipline, although important research continues. There were three radio-ecology international conferences (1961-1963) in which Oak Ridge and SREL-University of Georgia contributed more papers than any other institutions. Vince Schultz compiled a bibliography of literature in this interface that ended up in a pretty big book. The enrichment phase was very effective in that ecologists learned how to use radionuclide tracers, and physicists learned that white rats and humans were not the only organisms to be considered in assessing radiation effects. Because nothing really new came out of the interaction that justified a major society or journal, radiation ecology remains an important interface but not a new discipline. In other words, we should be cautious in the use of the word "discipline" when it comes to interfaces.

One thing is certain. Any interface effort will involve an increase in both spatial and temporal scales. Thus, in ecotoxicology we go from experimental determinations of LC50 of a toxicant in *Daphnia* and minnows to assessing the impact of a toxicant at the watershed and landscape levels of organization. We go from worrying about the short-term temperature effects in a sewage disposal system to concern about long-term global warming. Traditional scientific methodology involving controlled experiments and hypothesis testing is less appropriate at these higher levels of organization. Accuracy of prediction may be a better test of scientific worth.

Ecotoxicology can be a science that is very important in our efforts to maintain the quality of human life and environmental life-support systems regardless of whether it emerges as an independent new discipline or continues to be an important interface. As you read over the papers in this workshop volume it would be interesting to ask "What is new here, other than a shift in scale, that would justify this interface being considered a new discipline?" As discussed in Chapter 1 by Michael Newman, we are looking for "creative" as contrasted with "normal" science.

Eugene P. Odum
Professor of Ecology, Emeritus
University of Georgia

Preface

Recognized experts working at different levels of ecological organization contributed the chapters contained in this second volume of the Savannah River Symposia on Environmental Sciences. Authors were selected based on the inferential strength of their work and their promise of providing insightful guidance during the current transition of ecotoxicology to a mature science. Additionally, we sought scientists with interests in applying quantitative approaches to ecotoxicological problems. Regulatory concepts and needs were not emphasized; and descriptive and historical themes were avoided. Instead, each chapter author was asked to identify and present the fundamental concepts of ecotoxicology at his or her level of organization. Each was asked to identify important paradigms, "false" paradigms to be reassessed, hypotheses to be tested, or important new techiques to be applied at their level of ecological organization. If a theory or paradigm at the next highest level of organization could be explained by another at their level of organization, they were asked to identify this opportunity for theory reduction.

As a result, this book is a progressive treatment of ecotoxicology at all levels of organization that draws mechanistic interpretation from the next lower level and attempts to predict effects at the next higher level. To quote the succinct insights of Caswell (Chapter 9), "...processes at one level take their mechanisms from the level below and find their consequences at the level above ... Recognizing this principle makes it clear that there are no truly 'fundamental' explanations, and make it possible to move smoothly up and down the levels of a hierarchical system without falling into the traps of naive reductionism or pseudo-scientific holism." The results of this hierarchical "relay race" are described herein.

The Editors

Michael C. Newman is a Research Associate Ecologist at the University of Georgia's Savannah River Ecology Laboratory. After receiving B.A. (Biological Sciences) and M.S. (Zoology) degrees from the University of Connecticut, he went on to earn M.S. and Ph.D. degrees (Environmental Sciences) from Rutgers University. He joined the faculty at the University of Georgia in 1983. His research interests include population level effects of toxicants, toxicity models, factors modifying bioaccumulation kinetics, quantitative methods for ecological risk assessment, and inorganic water chemistry. He has published more than 50 articles on these topics. He directed the development of UNCENSOR, a PC-based program for analyzing data sets with "below detection limit" observations. This program is used by more than 400 professionals. In 1991, he co-edited the book, *Metal Ecotoxicology: Concepts and Applications* with Alan W. McIntosh. In 1995, he authored the book, *Quantitative Methods in Aquatic Ecotoxicology*.

He is also active in professional societies and teaching. He was founder and was first president of the Carolinas Chapter of the Society of Environmental Toxicology and Chemistry (SETAC) and serves on the SETAC awards committee He is a member of the board of editors for the journal, *Archives of Environmental Contamination and Toxicology*, and the two series, *Advances in Trace Substances Research* and *Current Topics in Ecotoxicology and Environmental Chemistry*. He is an editor (aquatic toxicology) for the journal, *Environmental Toxicology and Chemistry*. He has taught at the University of Connecticut, Rutgers University, the University of California - San Diego, the University of South Carolina, and the University of Georgia.

Charles H. Jagoe is a Research Assistant Ecologist at the University of Georgia's Savannah River Ecology Laboratory. He received his B.S. degree (Biology) from Clarkson University in New York, and his M.S. and Ph.D. degrees (Zoology) from the University of Maine. Following a postdoctoral fellowship with the Division of Pinelands Research, Center for Coastal and Environmental Studies (presently the Institute of Marine and Coastal Sciences) at Rutgers University, he joined the faculty at the Savannah River Ecology Laboratory in 1990. He is also an Adjunct Assistant Professor in both the Department of Pharmacology and Toxicology, College of Pharmacy, and the Institute of Ecology at the University of Georgia. His research interests include the physiology and histology of lower vertebrates, the application of physiological, histological, and biochemical responses as biomarkers of pollutant exposure, the toxic effects of metals and radionuclides in the environment, and genotoxicology. He is a member of the American Fisheries Society, and the Society of Environmental Toxicology and Chemistry, and serves on the Board of Directors of the Carolinas Chapter of SETAC. He is the author of over 20 scientific publications.

Acknowledgments

This book contains chapters developed during the Second Savannah River Symposium on Environmental Sciences. This symposium was held in Aiken, South Carolina on June 20 to 22, 1994. Its goal was to identify and present the fundamental concepts of ecotoxicology at all levels of biological organization. Support for the workshop and for preparation of this volume was provided by contract DE-AC09-76SROO-819 between the U.S. Department of Energy and the University of Georgia.

The editors are grateful for the extraordinary efforts of Edith Towns during organization and preparation for the workshop, and compilation and review of this volume. The following individuals contributed significantly to this effort by providing insightful and thorough reviews of manuscripts:

Ankley, G.T.	U.S. EPA, ERL-Duluth
Atchison, G.J.	Iowa State University
Barron, M.G.	RCG/Hagler Bailly, Inc.
Boudou, A.	Université de Bordeaux I
Bradley, B.	University of Maryland
Burton, G.A.	Wright State University
Campbell, P.G.C.	Université du Québec, INRS-Eau
Diamond, J.	Tetra Tech, Inc.
Dixon, P.M.	University of Georgia, SREL
Garrett, K.	University of Georgia, SREL
Gibbons, J.W.	University of Georgia, SREL
Gillespie, R.B.	Indiana University-Purdue University
Evans, D.H.	University of Florida
Forbes, T.	National Environmental Research Institute - Roskilde
Fox, G.A.	San Diego State University
Heath, A.G.	Virginia Polytechnic Institute and State University
Hinton, D.E.	University of California - Davis
Ingersoll, C.	U.S. National Biological Survey
Karr, J.R.	University of Washington
Klaverkamp, J.F.	Fisheries Oceanography - Freshwater Institute
Kramer, V.J.	Michigan State University
Landrum, P.F.	NOAA-Great Lakes Research Laboratories
La Point, T.W.	Clemson University
Laurent, P.	Centre National de la Recherche Scientifique
Leino, R.L.	University of Minnesota
Little, E.E.	U.S. National Biological Survey
Liu, J.	Harvard University
Loehle, C.	Argonne National Laboratory
Luoma, S.N.	U.S. Geological Survey, Menlo Park
Mayer, Jr., F.L.	U.S. EPA, ERL-Gulf Breeze

McIntosh, A.W.   University of Vermont
Mulvey, M.   University of Georgia, SREL
Novak, J.M.   University of Georgia, SREL
Pratt, J.R.   Portland State University
Roesijadi, G.   University of Maryland, CBL
Sander, B.   California State University - Long Beach
Scherer, E.   Fisheries Oceanography - Freshwater Institute
Schmeider, P.K.   U.S. EPA, ERL-Duluth
Schultz, I.   University of Georgia, SREL
Shugart, L.R.   Oak Ridge National Laboratory
Sibly, R.M.   University of Reading
Simkiss, K.   University of Reading
Strojan, C.   University of Georgia, SREL
Sugg, D.W.   University of Georgia, SREL
Suter, II., G.W.   Oak Ridge National Laboratory
Taub, F.B.   University of Washington
van Straalen, N.M.   Free University - Amsterdam

Contributors

Gary J. Atchison
Department of Animal Ecology
Iowa State University
Ames, Iowa 50011

Michael D. Bryan
Beak Consultants, Inc.
Sacramento, California 95834

John Cairns, Jr.
Department of Biology
Virginia Polytechnic Institute
  and State University
Blacksburg, Virginia 24061

Peter G.C. Campbell
INRS-Eau
Université du Québec
Ste-Foy, Quebéc, Canada
G1V 4C7

Hal Caswell
Biology Department
Woods Hole Oceanographic Institution
Woods Hole, Massachusetts 02543

Ronald K. Chesser
Savannah River Ecology Laboratory
University of Georgia
Aiken, South Carolina 29802

Philip M. Dixon
Savannah River Ecology Laboratory
University of Georgia
Aiken, South Carolina 29802

Gail A. Harkey
Cooperative Institute for Limnology
  and Ecosystems Research
University of Michigan
Ann Arbor, Michigan 48105

Charles H. Jagoe
Savannah River Ecology Laboratory
University of Georgia
Aiken, South Carolina 29802

Jussi Kukkonen
Department of Biology
University of Joensuu
Joensuu, Finland

Peter F. Landrum
Great Lakes Environmental
  Research Laboratory
NOAA, U.S. Department of
  Commerce
Ann Arbor, Michigan 48105

Michael C. Newman
Savannah River Ecology Laboratory
University of Georgia
Aiken, South Carolina 29802

James R. Pratt
Portland State University
Environmental Programs
Portland, Oregon 97207

Mark B. Sandheinrich
River Studies Center
Department of Biology and
  Microbiology
University of Wisconsin-LaCrosse
LaCrosse, Wisconsin 54601

David W. Schindler
Department of Biological Sciences
University of Alberta
Edmonton, Alberta, Canada
T6G 2E9

Lee R. Shugart
Environmental Sciences Division
Oak Ridge National Laboratory
Oak Ridge, Tennessee 37831

Richard M. Sibly
School of Animal and Microbial
  Sciences
University of Reading
Reading, England RG6 6AJ

Ken Simkiss
School of Animal and Microbial
  Sciences
University of Reading
Reading, England RG6 6AJ

Carl L. Strojan
Savannah River Ecology
  Laboratory
University of Georgia
Aiken, South Carolina 29802

Derrick W. Sugg
Savannah River Ecology
  Laboratory
University of Georgia
Aiken, South Carolina 29802

André Tessier
University of Quebec
INRS, Eau
Ste-Foy, Quebec, Canada
G1V 4C7

Contents

Chapter 1.
Ecotoxicology as a Science 1
Michael C. Newman

I. Ecotoxicology Assessed from a Scientific Context 1
II. Ecotoxicology and the Qualities of a Scientific Discipline 2
    A. Balance of Normal and Innovative Science 3
    B. The Importance of Strong Inference 3
    C. Theory Maturation and Strength of Inference 5
    D. Model Maturation 6
III. The Emergence of Ecotoxicology as a Science 7
Acknowledgments 8
References 8

Chapter 2.
Ecotoxicology of Metals in the Aquatic Environment:
Geochemical Aspects 11
Peter G.C. Campbell and André Tessier

I. Introduction 11
II. Geochemical Considerations 12
    A. Oxic Sediments 15
    B. Anoxic/Suboxic Sediments 18
    C. Application to *in Situ* Sediments 19
III. Interactions Between Dissolved Trace Metals and
    Aquatic Organisms 20
    A. Formulation of the FIAM 21
    B. Applicability of the FIAM — Studies with Dissolved Metals 23
    C. Applicability of the FIAM — Studies with Sediments 25
        i. Bioassays 26
            a. Laboratory Studies 26
            b. Field Studies 29
            c. Summary 32
        ii. Field Studies on Natural Sediments 34
IV. Interactions Between Particulate Trace Metals
    and Aquatic Organisms 36
    A. Background Information 38
    B. Laboratory Feeding Experiments 39
V. Ecotoxicological Considerations — Metallothionein
    as a Biochemical Indicator of Metal-Induced Stress 43
    A. Background Information 43
    B. Biochemical Indicators 44
    C. Metallothionein as a Monitoring Tool 45
    D. Summary 49

Acknowledgments 49
Appendix 50
References 51

Chapter 3.
Ecotoxicants at the Cell-Membrane Barrier 59
Ken Simkiss

I.   Background 59
II.   Origin of the Concepts on Membrane Permeability 60
III. Entry into the Cell 61
      A. The Lipid Route 61
      B. The Aqueous Route 62
      C. The Endocytotic Route 65
IV.   Complex Interactions 67
      A. Lipid Effects 67
          i.   Membrane Fluidity 67
          ii.   Narcosis 67
          iii. Parabolic Responses 68
      B. Channel Routes 68
          i.   The Permeant Form 68
          ii.   One Ion, One Channel? 69
          iii. Channel Selectivity 69
      C. Vesicular Uptake 70
          i.   Transferrin Specificity 70
          ii.   Particle Assimilation 70
V.   Membrane Ecotoxicology 71
      A. The Lipid Membrane 71
          i.   LC50 of Chlorophenols 71
          ii.   Insecticide Effects 71
          iii. Oil Spills 71
      B. Channel Interference 75
          i. Cd^{2+} Uptake into *O. gammarellus* 75
          ii. Cu^{2+} Toxicity to *Salmo clarki* 75
          iii. Zn^{2+} Effects on Fish Gill Ion Regulation 77
VI.   Membranes in the Ecotoxicology Hierarchy 78
References 80

Chapter 4.
Evaluation of Organic Contaminant Exposure in Aquatic Organisms:
The Significance of Bioconcentration and Bioaccumulation 85
Peter Landrum, Gail A. Harkey, and Jussi Kukkonen

I.   Introduction 85
II.   Aqueous Exposures 86
      A. Bioconcentration of Contaminants 86
      B. Factors That Influence Bioavailability in Aqueous Exposures 89

C. Normalization of Contaminants
    Accumulated from Aqueous Exposures 94
III. Sediment Exposures 95
IV. Food Web Transfer.. 103
V. Estimation Methods 105
VI. Steady-State Models 108
VII. Uses and Limits of Toxicokinetics 110
VIII. Bioavailability 114
IX. Utility and Assessment 116
References 117

Chapter 5.
Molecular Markers to Toxic Agents 133
Lee R. Shugart
I. The Problem 133
II. Biological Markers 134
    A. Concept 134
    B. Approach 135
        i. Introduction 135
        ii. Objectives 136
        iii. General Elements 137
        iv. Research Component.. 137
        v. Limitations 139
    C. Dose-Response and Multiple-Response Paradigms 140
III. Molecular Markers to Toxicants 142
    A. Introduction 142
    B. Protein Induction 146
        i. Cytochrome P450 Enzymes 146
        ii. Stress Proteins 148
        iii. Metallothionein... 149
    C. Metabolites as Indicators 149
    D. DNA Markers 150
        i. DNA Damage 150
        ii. Surrogates to DNA Damage 152
        iii. Consequences of Structural Perturbations 152
IV. Genetic Ecotoxicology 154
V. Conclusions... 156
Acknowledgments 157
References 157

Chapter 6.
Responses at the Tissue Level: Quantitative Methods in Histopathology
Applied to Ecotoxicology 163
Charles H. Jagoe

I. Conceptual Background 163

II.   The Necessity of Quantitation 167
        A. Situations in which Qualitative Observations Are Sufficient 168
        B. Problems with Purely Qualitative Observations 168
        C. Semiquantitative Approaches 169
        D. Rationales For Quantitative Studies 172
III.   Methods for Quantifying Effects at the Cell and Tissue Levels 173
        A. Stereology and Morphometry 174
        B. Sampling Design and Efficiency 179
IV.   An Example: Effects of Low pH and Dissolved
        Metals on Fish Gills 181
V.   Conclusion 189
Acknowledgments 189
References 190

Chapter 7.
Effects of Pollutants on Individual Life Histories
and Population Growth Rates 197
Richard M. Sibly

I.   Introduction 197
II.   Effects of Pollutants on Individual Organisms,
        and the Consequences for Population Growth Rate 198
III.   Population Density and Population Ecology 206
IV.   Evolutionary Analysis 213
V.   Conclusions 217
Acknowledgments 218
References 219

Chapter 8.
Ecologically Meaningful Estimates of Lethal Effect in Individuals 225
Michael C. Newman and Philip M. Dixon

I.   Overview 225
II.   The Dose-Response (Time Endpoint) Approach 226
III.   The Time-Response (Survival Time) Approach 229
        A. Introduction 229
        B. Nonparametric (Product-Limit) Methods 232
        C. Semiparametric Cox Proportional Hazard Methods 233
        D. Parametric Methods 234
           i.   General 234
           ii.   Application of Parametric Models 236
           iii. Multiple Comparisons 241
           iv.   Summary of Parametric Methods 243
IV.   Conclusion 244
Acknowledgments 244
Appendices 245
References 250

Chapter 9.
Demography Meets Ecotoxicology: Untangling
the Population Level Effects of Toxic Substances 255
Hal Caswell

I.   Introduction 255
       A. Diversity, Stage-Specificity, and Population Models 256
II.   Demographic Models 258
       A. Matrix Population Models 259
           i.   Projection and Prediction 260
           ii. Sensitivity Analysis: The Other Piece of the Puzzle 261
III.   Life Table Response Experiments 261
IV.   Decomposing Treatment Effects: Why and How? 264
       A. The General Decomposition Principle 265
       B. Fixed Effects Designs 266
           i.   Alternative Parameterizations of Stage-
                Classified Models 269
                a. Stage-Specific Vital Rates 269
                b. Analyzing Effects on Age at Maturity 273
       C. Regression Designs 276
           i.   Decomposing Regression Effects 278
           ii. An Example 280
                a. Expressing the Vital Rates as a Function
                   of Treatment 280
                b. Effects on the Vital Rates 281
                c. Contributions to Effects on λ 282
V.   Discussion 285
       A. Demographic Bioassay 285
       B. Unresolved Issues 287
Acknowledgments 288
References 289

Chapter 10.
Toxicants as Selective Agents in Population and Community Dynamics 293
Ronald K. Chesser and Derrick W. Sugg

I.   Introduction 293
II.   Methods 295
       A. Constants 295
       B. Initial Values 296
       C. Carrying Capacities 299
       D. Toxicant Flow 300
       E. Selection 301
       F. Population Growth 302
       G. Immigration 302
III.   Ecosystem Stability 303
IV.   Perturbations 308

V.   Species Redundancies 310
VI.   Immigration 312
VII. Generalizations 313
Acknowledgments 316
References 316

Chapter 11.

Effects of Environmental Stressors on Interspecific
Interactions of Aquatic Animals 319
Gary J. Atchison, Mark B. Sandheinrich, and Michael D. Bryan

I.   Overview 319
II.   General Background 320
III. Predation 321
      A. Hunger and Motivation to Feed 322
      B. Searching Behavior 322
      C. Prey Choice 323
      D. Capture 324
      E. Optimal Foraging Theory and Ecotoxicology 325
      F. Risks of Predation 328
      G. Balancing Predation Risk and Foraging 330
IV.   Competition for Resources 331
V.   Conclusions 337
Acknowledgments 337
References 337

Chapter 12.

Ecotoxicology and the Redundancy Problem: Understanding
Effects on Community Structure and Function 347
James R. Pratt and John Cairns, Jr.

I.   Introduction 347
II.   Possible Relationships Between Structure and Function 348
III. Ecosystem Differences 349
IV.   Predictive Approaches 350
      A. What Has Ecotoxicology Learned from Ecology? 350
      B. Toxicological Paradigms 351
      C. Ecological Paradigms 352
V.   Linking Ecology and Ecotoxicology 354
      A. What Could Ecology Learn from Ecotoxicology? 354
      B. Redundancy in Terrestrial Systems 355
      C. Redundancy in Aquatic Systems 356
      D. Species Turnover 360
VI.   Toward a New Ecotoxicology: Slaying Some Myths 362
      A. Ecological Redundancy 362

B. Biological Balance and Species Turnover 363
C. Ecological Resilience 363
D. Communities 364
E. A New Ecotoxicology 366
References 367

Chapter 13.
Ecosystems and Ecotoxicology: A Personal Perspective 371
David W. Schindler

I.   Introduction 371
II.   A Definition of Ecosystem Ecology 372
III. The Conceptual Basis for Evaluating Indicators
      of Stress in Ecosystems 373
      A. Advantages of Lakes 373
      B. Monitoring Approaches 374
         i.   Indicator Populations 375
         ii.   Lacustrine Communities as Indicators
               of Ecosystem Stress 376
         iii. Ecosystem Processes and Food Chain
               Functions as Indicators of Stress 376
         iv.   Functional Redundancy and Early Indicators of Stress ... 377
         v.   How Many Species Do We Really Need? 378
         vi.   Where Are the Reference Ecosystems? 380
         vii. The Element of Surprise 380
      C. If We Cannot Monitor, What Should We Measure? 383
         i.   Toxicants in Lake Sediment and Top Carnivores 383
         ii.   Screening with Stable Isotopes 384
         iii. Sensitive Physiological Measurements 385
         iv.   Interpretation of the Sedimentary Record 386
      D. Scaling Up: Does Ecosystem Size Matter
         in Making Predictions? 390
         i.   Effects of Lake Size 390
         ii.   Monitoring Landscape-Scale Stresses 391
      E. Humans as Ecological Indicators 392
      F. A Preventive Approach to Ecosystem Management 392
Acknowledgments 393
References 393

Chapter 14.
Summary 399
Carl L. Strojan

Index 403

CHAPTER 1

Ecotoxicology as a Science

Michael C. Newman

I. ECOTOXICOLOGY ASSESSED FROM A SCIENTIFIC CONTEXT

Science is concerned with creating an intellectual model of the material world. Technology is concerned with procedures and tools and their general use to gain or use knowledge. Practice is concerned with how to treat individual cases. Confusing the three can be dangerous.

Slobodkin and Dykhuizen (1991)

The goal of science is to organize and classify knowledge based on explanatory principles (Nagel, 1961). It follows that the goal of ecotoxicology as a science is the organization of knowledge about the fate and effects of toxicants in ecosystems based on explanatory principles (Newman, 1995). The consistency of this goal with the definition of ecotoxicology originally given by Truhaut (1977) and more recent definitions (e.g., Cairns and Mount, 1990; Jørgensen, 1990) imparts a comforting unanimity during our initial efforts to describe this emerging scientific discipline. Unfortunately, this appearance of consistency passes quickly when this goal is used to judge present activities in ecotoxicology. Inconsistencies arise from the complex interweaving of various scientific, technological, and practical goals within this socially obligated endeavor.

What are these various goals? The goal given above suffices for scientific ecotoxicology. However, the technological objective of ecotoxicology is development and effective application of tools and procedures to acquire a better understanding of toxicant fate and effects in ecosystems. Practical ecotoxicology applies available knowledge, tools, and procedures to specific problems. For justifiable reasons, most present efforts address crucial issues in technological

1-56670-1127-9/95/$0.00+$.50

and practical ecotoxicology. Taken together, the predominance of technological and practical motivations imparts a distinctly nonscientific structure to the field. Consequently, a contradiction emerges: the accepted (scientific) goal of the field is inconsistent with the activities of most ecotoxicologists. In the resulting confusion, standard methods essential in practical ecotoxicology may be timidly applied to scientific questions, despite the availability of more appropriate methodologies and the absence of any regulatory requirement for using the standard methods. An individual's proficiency may be gauged more from his or her rote application of such methods and regulations than from scientific creativity and problem-solving skills.

Although the confusion is lessened by recognizing the distinct goals of scientific, technological, and practical ecotoxicology, the imbalance in relative effort remains. Scientific ecotoxicology frequently comes as an afterthought as the immediate and crucial technological and practical goals are addressed. Unfortunately, the long-term vitality of the field depends upon the growth of knowledge based on explanatory principles. Students are not routinely trained to effectively address ecotoxicological questions in a scientific manner. After defining ecotoxicology as a scientific discipline, mentors teach by example that technical acumen is more important than circumspective development of hypotheses and formal testing through the falsification process. Students of ecotoxicology are instructed extensively in techniques, specific qualities of important toxicants, and important regulatory practices, while problem solving skills and inferential methods are quietly neglected. At present, measurement is taught as intrinsically valuable, a characteristic of "unnatural science" that Medawar (1982) calls *idola quantitatis*. This process perpetuates itself as students so taught move on to work in the field, assess proposals, and mentor new students. At present, most innovation in ecotoxicology diffuses in from other fields such as chemistry, ecology, epidemiology, statistics, and mammalian toxicology. A degree of cross-fertilization of ideas occurs among all fields but, lamentably, the ability to generate, clearly state, and test novel concepts remains underdeveloped in ecotoxicology.

II. ECOTOXICOLOGY AND THE QUALITIES
OF A SCIENTIFIC DISCIPLINE

[A balance of faculties] *should be cultivated in scientific research. Imaginativeness and a critical temper are both necessary at all times, but neither is sufficient. The most imaginative scientists are by no means the most effective; at their worst, uncensored, they are cranks. Nor are the most critical minded. The man notorious for his dismissive criticisms, strenuous in the pursuit of error, is often unproductive, as if he had scared himself out of his own wits ...*

Medawar (1982)

A. BALANCE OF NORMAL AND INNOVATIVE SCIENCE

Now that we have defined the goal of science and related it to the goals of ecotoxicology, let's examine the means by which such a goal is reached. Kuhn (1970) identified two essential categories of scientific endeavor, normal and innovative. Scientists engaged in normal science do not intend to generate new ideas or discoveries. Instead, they abandon the large picture to their fascination with solving puzzles, regardless of their intrinsic value. The importance of normal science lies in the incremental enrichment of our breadth, depth, and precision of knowledge about established theories and paradigms (Kuhn, 1970). In contrast, innovative science involves rejection or major modification of existing paradigms, and formulation of new paradigms. Innovation occurs after normal science has accumulated sufficient detail to test established paradigms more rigorously. It follows that a balance of normal and innovative science must exist within any discipline for effective progress. For example, a preoccupation with every detail of every instance in a scientific discipline leads to the "tyranny of the particular" (Medawar, 1967). Facts accumulate faster than they can be incorporated into theory, with a consequent inefficiency in developing, organizing, and using knowledge for predictive purposes. In ecotoxicology, the necessity for standardization and the immediate need for action in specific situations, combined with the normal scientist's fascination with particulars, contributes to our present dearth of innovative science. It encourages a preoccupation with methodology, particulars, and *idola quantitatus*.

How should the science of ecotoxicology change to progress more effectively? Remarkably, most of the essential components for rapid advancement are already present. Technological and practical ecotoxicologists have already adopted quality control methods for accurate and precise measurement, a crucial requirement for rapid advancement (Newman, 1995). Normal science has flourished during the present presynthetic stage of ecotoxicology; consequently, facts are plentiful. The present movement of ecotoxicology from a predominantly descriptive discipline to a mature science requires only that the value and qualities of innovative science be taught to students both formally and by example. Possessing sufficient facts in many areas, we now need to focus more on the question, "Why is this so?"

B. THE IMPORTANCE OF STRONG INFERENCE

Advancement in the new science of ecotoxicology can be fostered in several ways. Required at this time is a stronger inferential approach, which must be perceived and taught to be as valuable as the present regulatory approach of most ecotoxicology. The writings of John Platt (1964) are particularly pertinent to this point.

Platt (1964) observed that scientific disciplines progress at very different rates and that these differences appear to arise from the value placed on systematic scientific thinking and rigorous testing of hypotheses. He referred to such systematic application of inductive inference within a field as strong inference. Problems are formally addressed using working hypotheses, alternative hypotheses, critical testing with appropriate accuracy and precision, and repetition of the testing with sequential hypotheses until only one explanation survives the falsification process, i.e., the scientific method. Rigorous testing with high risk of falsification is an essential feature of this process. Chronic application of low-risk tests in any field of study is undesirable, as belief can be falsely increased solely by frequent repetition of a theme (Popper, 1965). Platt (1964) argued simply that a sequence of a few carefully selected hypotheses, rigorously tested, will advance understanding faster than many poorly formulated or weakly tested ones. The cumulative effect within a discipline of each worker using strong inference is accelerated progress. Also critical to Platt's formulation of strong inference is the concept of multiple working hypotheses.

Chamberlin (1897) formulated the concept of multiple working hypotheses nearly a century ago yet, in reality, it is rarely practiced explicitly today. Many aspects of Chamberlin's argument are particularly germane to ecotoxicology and will be discussed in detail. In developing his arguments for multiple working hypotheses, Chamberlin described three historical phases of intellectual evolution. Initially, so little was known in most subject areas that experts were assumed capable of understanding fully any particular subject. An immediate and sufficient answer based on some general theory was given when a question was asked. Such a ruling theory provided unquestioned or weakly questioned explanation. This process of uncritical assertion of a theory (precipitate explanation) reinforces the ruling theory by repeated application alone, not by rigorous testing. Although formally discarded as untrustworthy in modern science, a tendency toward precipitate explanation still exists. It is pervasive in ecotoxicology due to technological advocacy (a reluctance to question or tendency to unobjectively support a particular technique, regulatory approach, or standard method) and a general inconsistency in adhering to a scientific context. Although the goals of technical and practical ecotoxicology justifiably encompass such unscientific behavior, the purpose of science is decidedly not to win or maintain primacy for a particular theory, idea, method, or approach (Cournand, 1977). Scientific progress in ecotoxicology continues to be hindered by precipitate explanation.

In the present phase of intellectual evolution, ruling theories have been replaced by the familiar working hypothesis. Facts are gathered and tests are formulated to falsify the working hypothesis. A working hypothesis should have no favored status except that accrued after surviving repetitive and rigorous testing. Chamberlin argued that many applications of the working hypothesis concept retain elements of precipitate explanation. There is a tendency to give favored status to the central working hypothesis and to consider alternate hypotheses as secondary. Loehle (1987) refers to this tendency to "confirm one's theory, or to not seek out or use disconfirming evidence" as confirmation bias. Chamberlin suggested that the method of multiple working hypotheses can be used to lessen confirmation bias. With this method, equal amounts of effort are spent in testing all reasonable hypotheses simultaneously. The method of multiple working hypotheses also avoids the tendency to stop testing when a single cause is found and to ignore the possibility of additional causes. Consideration of multiple causes is particularly important in ecology and ecotoxicology (Hilborn and Stearns, 1982). Indeed, Quinn and Dunham (1983) point out that the numerous interactive effects in ecological systems must be incorporated into this process. An awareness of interactions must be used to supplement the steps just described and to temper conclusions drawn from hypothesis testing when applying strong inference to ecological questions.

C. THEORY MATURATION AND STRENGTH OF INFERENCE

With the goals and general qualities of modern scientific inquiry summarized, we can address theory maturation within scientific disciplines. The process of theory maturation gives explanation to transitions and contrasting attitudes seen in ecotoxicology today.

Symptomatic of the present evolution of ecology, and specifically ecotoxicology, are the contrasting views regarding the applicability of the classic hypothetico-deductive method. Quinn and Dunham (1983) reject this method and the associated strong inference concept with the remarkable statement that the logic of ecology (and evolution) is different from that of traditional science. Hilborn and Stearns (1982) also reject strong inference and identify ecological science as unique. Slobodkin and Dykhuizen (1991) also urge caution in applying traditional scientific methods to ecotoxicology. In contrast, Cairns (1990) argues that too many of our present practices in ecotoxicology are driven by the history of the field, not their scientific soundness. He advocates rejection of many of our present paradigms and adherence to a more rigorous falsification process (Cairns, 1992). The present author strongly supports Cairns's argument.

Is ecology (and ecotoxicology) unique as a science or can traditional scientific methods be applied profitably to ecotoxicology? A brief discussion of the maturation process exhibited by scientific disciplines will reveal the partial truths in these contrasting views. All sciences pass through a period in which facts accumulate faster than they can be assimilated into theory (Medawar, 1967). Accrual of facts and description is paramount in these early stages, and rigorous falsification is less pertinent. This descriptive stage may end after an uncomfortable transition period characterized by a continued, but now unjustifiable, preoccupation with detail (i.e., tyranny of the particular). Eventually, a mature and healthy science emerges in which normal and innovative scientists work together to carefully examine facts and test hypotheses. If a particular hypothesis withstands the rigors of strong inference, it is incorporated into a unifying set of theories and paradigms.

The contrasting views described above suggest that ecotoxicology is making that confused, yet exciting, transition to a mature scientific discipline. As rightfully suggested (Hilborn and Stearns, 1982; Quinn and Dunham, 1983; Slobodkin and Dykhuizen, 1991), such a deficiency of facts and basic understanding of phenomena existed until recently that insistence on rigorous falsification would have led to premature rejection of many correct theories. Insufficient information was available to formulate concise and discriminating hypotheses, and interpretation of test results would have remained superficial. Loehle (1987) cautions against such dogmatic falsification. Hinderance of progress by dogmatic falsification is exaggerated by the surprising, yet pervasive, resistance of most practicing scientists to innovation (Barber, 1961). The rejections of strong inference described above appear to be aimed at dogmatic or inappropriate falsification. Fact accrual and description remain the most valuable means of advancing our knowledge in several areas of ecology and ecotoxicology. However, there are many more areas in which the tyranny of the particular exists and, as argued by Cairns and the present author, strong inference is essential for further advancement. Without strong inference, progress is hindered by precipitate explanation, *idola quantitatis,* confirmation bias, technical advocacy, and theory tenacity (a resistance to discard a theory, belief, or framework during problem solving despite evidence to the contrary [Loehle, 1987]). Consequently, ecotoxicology remains ripe for pathological science, i.e., science practiced with excess loss of objectivity (Hall, 1989; Rousseau, 1992).

D. MODEL MATURATION

Just as theory maturation has yet to occur fully in ecotoxicology, there has been an epiphenomenal delay in model maturation. Models in any quantitative science function as either exploratory tools, redescriptions, or generative representations (Taylor, 1989). An exploratory tool is formulated to highlight behaviors of the system of interest. For example, the logistic growth model may be used to explore possible behaviors of populations, although it is understood that the model does not accurately reflect the behavior of any specific population. The one-compartment, first-order bioaccumulation model is often used as an exploratory tool in ecotoxicology.

A redescription simply summarizes observations and permits limited prediction under the assumption that the modeled pattern will be repeated. Many examples of this type of model occur in ecotoxicology. A response surface model generated with a polynomial is a redescription model. As implemented in ecotoxicology today, the probit model for toxic response is a redescription of data. A redescription makes the transition to a generative representation if "the model captures the necessary and sufficient conditions to explain the phenomenon observed [so that] we can make confident predictions for situations not yet observed" (Taylor, 1989). Many quantitative structure-activity relationships (QSAR) are presently making this transition from redescription to generative representations.

As our understanding increases, more ecotoxicological models must make the transition from redescriptions to generative representations. For example, the empirical relationship between metal toxicity and water hardness should be replaced by models quantitatively linking metal speciation to toxic action. Another example is the empirical incorporation of a temporal dimension to toxicity data by plotting some endpoint value (e.g., LC50) against time to estimate an incipient lethal level. This approach should be replaced by methods such as those described in Chapter 8. Obviously, model maturation is slowed by legal or regulatory adherence to any specific redescription model. Finally, more effective use must be made of exploratory tools. For example, very few ecotoxicologists have taken advantage of optimal foraging theory, despite a rich literature surrounding it, (e.g., Stephens and Krebs, 1986), the notable exception being papers by Atchison and Sandheinrich (Atchison et al., 1987; Sandheinrich and Atchison, 1990; Chapter 11 of this book). With the transition to a mature science, the emphasis on redescription will be replaced by an emphasis on explanatory principles. Hopefully, the result will be that arrested development at the redescription stage will be remedied and a more effective use of exploratory models will occur.

III. THE EMERGENCE OF ECOTOXICOLOGY AS A SCIENCE

Well, there are two kinds of biologists, those who are looking to see if there is one thing that can be understood, and those who keep saying it is very complicated and that nothing can be understood ... You must study the simplest system you think has the properties you are interested in.

Cy Levinthal as quoted in Platt (1964)

Ecotoxicology is making the transition to a mature science. Whether this occurs quickly or slowly depends on our collective openness to change, dissatisfaction with mediocrity, and sense of urgency regarding current environmental issues. The distinct goals and activites of practical, technological, and scientific ecotoxicologists must be understood and valued during the transition. As scientific ecotoxicologists, it is time to abandon the false paradigm that "ecological systems are too complex to permit any useful level of prediction." This hobbling belief permeates much of ecotoxicology today and prolongs the tyranny of the particular (Newman, 1995). Although true in early stages of our science and still true in some areas of ecotoxicology, this false paradigm is now invoked more to avoid rigor and allow business to comfortably continue as usual. With its rejection, emphasis can be placed on paradigms that function as touchstones, not talismans. In addition to descriptive, analytical, and regulatory training, it is important that students develop strong problem-solving skills and a reflexive tendency to insist on knowing, "Why is this so?".

ACKNOWLEDGMENTS

This work was supported by contract DE-AC09-76SROO-819 between the U.S. Department of Energy and the University of Georgia. Drs. W. Gibbons, C. Loehle, A. McIntosh, M. Mulvey, and C. Strojan provided excellent input on earlier versions of the manuscript.

REFERENCES

Atchison, G.J., M.G. Henry, and M.B. Sandheinrich, 1987. Effects of metals on fish behavior: a review. Environ. Biol. Fishes 18, 11–25.

Barber, B., 1961. Resistance by scientists to scientific discovery. Science 134, 596–602.

Cairns, J., Jr., 1990. The genesis of biomonitoring in aquatic ecosystems. Environ. Prof. 12, 169–176.

Cairns, J., Jr., 1992. Paradigms flossed: the coming of age in environmental toxicology. Environ. Toxicol. Chem. 11, 285–287.

Cairns, J., Jr. and D.I. Mount, 1990. Aquatic toxicology, Part 2 of a four–part series. Environ. Sci. Technol. 24, 154–161.

Chamberlin, T.C., 1897. The method of multiple working hypotheses. J. Geol. 5, 837–848.

Cournand, A., 1977. The code of the scientist and its relationship to ethics. Science 198, 699–705.

Hall, R.N., 1989. Pathological science. Phys. Today (October), 36–48.

Hilborn, R. and S.C. Stearns, 1982. On inference in ecology and evolutionary biology: the problem of multiple causes. Acta Biotheor. 31, 145–164.

Jørgensen, S.E., 1990. Modelling in Ecotoxicology, Elsevier, New York.

Kuhn, T.S., 1970. The Structure of Scientific Revolutions, University of Chicago Press, Chicago.

Loehle, C., 1987. Hypothesis testing in ecology: psychological aspects and the importance of theory maturation. Q. Rev. Biol. 62, 397–409.

Medawar, P.B., 1967. The Art of the Soluble. Methuen & Co., London.

Medawar, P.B., 1982. Pluto's Republic. Oxford University Press, Oxford.

Nagel, E., 1961. The Structure of Science. Problems in the Logic of Scientific Explanation. Harcourt, Brace and World, New York.

Newman, M.C., 1995. Quantitative Methods in Aquatic Ecotoxicology. Lewis Publishers, Chelsea, MI.

Platt, J.R., 1964. Strong inference. Science 146, 347–353.

Popper, K.R., 1965. Conjectures and Refutations. The Growth of Scientific Knowledge. Harper & Row, New York.

Quinn, J.F. and A.E. Dunham, 1983. On hypothesis testing in ecology and evolution. Am. Nat. 122, 602–617.

Rousseau, D.L., 1992. Case studies in pathological science. Am. Sci. 80, 54–63.

Sandheinrich, M.B. and G.J. Atchison, 1990. Sublethal toxicant effects on fish foraging behavior: empirical vs. mechanistic approaches. Environ. Toxicol. Chem. 9, 107–119.

Slobodkin, L.B. and D.E. Dykhuizen, 1991. Applied ecology, its practice and philosophy. In: Integrated Environmental Management, J. Cairns, Jr. and T.V. Crawford (Eds.), Lewis Publishers, Chelsea, MI, pp. 63–70.

Stephens, D.W. and J.R. Krebs, 1986. Foraging Theory. Princeton University Press, Princeton, NJ.

Taylor, P. 1989. Revising models and generating theory. Oikos 54, 121–126.

Truhaut, R. 1977. Ecotoxicology: Objectives, principles and perspectives. Ecotoxicol. Environ. Saf. 1, 151–173.

CHAPTER 2

Ecotoxicology of Metals in the Aquatic Environment: Geochemical Aspects

Peter G.C. Campbell and André Tessier

I. INTRODUCTION

Metals introduced into the aquatic environment, whether from natural sources or industrial origins, tend to accumulate in sediments (Livett, 1988). The fluxes of metals to sediments may involve the settling of metal-laden particles or the downward transport of dissolved metal forms across the sediment-water interface. In both cases, the overall result is often a marked enhancement of metal concentrations in the upper strata of lake sediments — metal concentrations in systems affected by anthropogenic inputs can attain levels 1000 to 5000 times higher (on a microgram per gram wet weight basis) than those in the overlying water column. For a typical circumneutral water body of shallow to medium depth (e.g., 50 m), the surficial mobile sediments constitute the largest trace metal reservoir (metal mass balance calculations, expressed on a weight per square meter basis), see, for example, St-Cyr et al. (1994). In addition, under normal oxic conditions, the sediment-water boundary is a particularly important aquatic habitat, characterized by the presence of a diverse and abundant fauna and flora. Similarly, from a functional point of view, the organisms present at the sediment-water interface are known to play an important role in the overall metabolism of the water body (Schindler et al., 1985; Schindler, 1987).

This juxtaposition in benthic sediments of high contaminant metal concentrations and an important aquatic community has stimulated considerable interest among both research scientists and environmental managers. This chapter focuses on this research, i.e., on possible biological effects of toxic metals associated with aquatic sediments, and attempts to present a balanced review of the current state of understanding in this area. The basic premise is

1-56670-1127-9/95/$0.00+$.50

that metals present in contaminated sediments may affect aquatic life in two ways (Figure 1): indirectly (i.e., by partitioning of the metals into the ambient water, followed by their assimilation from the aqueous phase), and directly (e.g., in macrofauna, by ingestion of the sediments and assimilation of the metals from the gut). Both routes of metal exposure are considered. The metals of interest in this chapter are those that (1) are commonly present in contaminated sediments, (2) are recognized as potentially toxic at low concentrations to aquatic biota, and (3) exist in natural waters as dissolved cations (e.g., Cd, Cu, Ni, Pb, Zn).

The reasoning that guided the review is summarized in Table 1. Considering the indirect route of exposure to sediment-bound metals (via water), we first consider the reactions that are likely to control metal concentrations, $[M]_i$, in the interstitial or pore waters of contaminated sediments (Section II: Geochemical Considerations). Two approaches are described for estimating $[M]_i$ on the basis of the geochemistry of the surficial sediments. Secondly, having considered possible geochemical controls on metal concentrations in pore waters, we address the bioavailability of these dissolved metals (in the pore waters, and in the overlying water column following vertical transfer), using as a convenient paradigm the free-ion activity model (FIAM) for metal-organism interactions (Section III: Interactions Between Dissolved Trace Metals and Aquatic Organisms).

After dealing with dissolved metals (geochemical controls, bioavailability), we focus on those metals that remain in particulate form, i.e., the direct route of exposure (Section IV: Interactions Between Particulate Trace Metals and Aquatic Organisms). This reasoning is designed to apply to organisms for which ingestion of particulate material constitutes an important vector for metal uptake. We explore possible links between digestive processes in benthic organisms and the "bioavailability" of ingested sediment-bound metals, and then review the results of laboratory experiments designed to determine the assimilation efficiencies of metals present in different forms in the "food."

In the final review section (Section V) we adopt a more holistic approach and ask the question, "What can the indigenous biota tell us about the availability of the metals in contaminated sediments?" At the organismal level, we consider the monitoring of metal bioaccumulation per se, as well as the use of biochemical indicators (biomarkers) of metal-induced, sublethal stress.

II. GEOCHEMICAL CONSIDERATIONS

For the purposes of this chapter, we shall take the existence of metal-contaminated sediments as a given (i.e., we neglect how they came to be contaminated and consider only the settled, quiescent sediments). We then explore how the metals present in surficial sediments could become biologically available, how they could affect aquatic biota. This line of reasoning leads

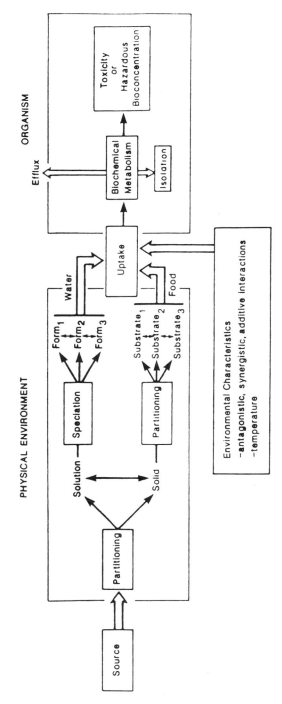

Figure 1 Processes affecting the biological availability of sediment-associated metals before, during, and after uptake. (Adapted from Luoma, 1983.)

**Table 1   Summary of the Reasoning Guiding the Literature Review on the
Biological Effects of Metal-Contaminated Sediments**

Step	Comments/Rationale
1. Ignore how the sediments came to be contaminated, i.e., metal cycling in the water column, the conversion of metals to settleable form, and the deposition of particulate material to the (lake) bottom.	In other words, only consider the contaminated sediments after settling.
2. Concentrate on the upper sediment stratum.	The effective contact zone between the aquatic biota and deposited sediments (Luoma and Davis, 1983; NRCC, 1988).
3. In the case of metal exposure via the dissolved phase, consider possible controls on metal concentrations in the interstitial (pore) water, $[M]_i$.	Two possible approaches: (i) control by sorption reactions on Fe, Mn-oxyhydroxides or on sedimenatary organic matter; (ii) control by substitution reactions with amorphous sulfides.
4. Depending on which approach proves more appropriate for contaminated sediments in their natural setting, estimate $[M]_i$ based on the geochemistry of the surficial sediments after diagenesis.	In real world, the distinction between oxic and anoxic sediments is often blurred; most aerobic benthic organisms survive in sediments that are underlain or even surrounded by completely anaerobic sediments.
5. Consider the bioavailability of these dissolved metals (in the pore waters; in the overlying water, assuming vertical transfer), using the free-ion activity model (FIAM).	FIAM should apply to organisms that do not assimilate particulate material (e.g., rooted aquatic plants), and to organisms that do assimilate particles but for which the dissolved phase remains the primary vector for metal uptake (e.g., molluscs).
6. Having dealt with dissolved metals (geochemical controls, bioavailability), now consider those metals that remain in particulate form. Review results of laboratory feeding experiments designed to determine the assimilation efficiencies of metals present in different forms in the ingested particles.	Applies to organisms for which particles constitute the primary vector for metal uptake. Key factors include the particle size ingested, digestion chemistry in the gut (pH, pE, residence time).
7. To complement the "reductionist" approach outlined in steps 1 to 6, consider a holistic (ecotoxicological) approach.	What can the indigenous biota tell us about the bioavailability of metals in contaminated sediments? (metal burdens; biochemical indicators of metal exposure).

naturally to a consideration of the geochemical controls on metal concentrations in pore waters, $[M]_i$ (Figure 2). These concentrations will reflect the metal's chemical potential in the solid and solution phases at the sediment-water boundary — changes in this chemical potential would be expected to affect the bioavailability of the metal.

Two approaches can be used to estimate $[M]_i$: one applies to oxic conditions and assigns the control of $[M]_i$ to sorption reactions on such sorbents as Fe- or Mn-oxyhydroxides or sedimentary organic matter; the second applies to anoxic conditions and assumes that $[M]_i$ is controlled by precipitation-dissolution reactions with reactive amorphous sulfides (*Acid Volatile Sulfides*, or AVS).

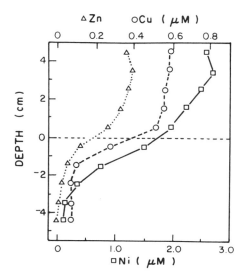

Figure 2   Variations in the concentrations of dissolved Cu, Ni, and Zn close to the sediment-water boundary. Samples were collected by *in situ* dialysis in Clearwater Lake, an acidic and metal-contaminated lake near Sudbury, Ontario. Note that concentrations decrease on passing from the overlying water column into the sediment interstitial water. (Adapted from Tessier et al., 1994.)

A. OXIC SEDIMENTS

The evidence for sorptive control of dissolved trace metal concentrations under oxic conditions has been reviewed by Tessier (1992). With the notable exception of Fe and Mn, measured trace metal concentrations in oxic waters are consistently much lower than those predicted from solubility equilibria involving known solid phases. Reactions other than precipitation must thus be involved in the geochemical control of trace metal concentrations under such conditions. Given the presence in natural sediments of solid phases known to be good sorbents (e.g., Fe-oxyhydroxides, Mn-oxyhydroxides, sedimentary organic matter), sorption reactions have generally been invoked to explain the observed undersaturation (Schindler, 1967). Indeed, in laboratory studies metals added to suspensions of these various solid phases are rapidly removed from solution and the metal concentrations in solution at equilibrium are lower than those predicted from solubility relationships (K_{sp}). Similarly, in experiments where natural sediments have been compared with individual sorbents, the dissolved metal concentrations at equilibrium have been shown to respond to changes in various experimental variables (e.g., ΔpH, $\Delta[M]$, $\Delta\{sorbent\}$) in much the same manner in both systems (e.g., Lion et al., 1982).

Models for the sorptive control of trace metal concentrations generally assume that the individual sorbents present in the surface sediments compete for the trace metal. The partitioning of the trace metal among the various sorbents can then be described by the following overall reactions:

$$M^{z+} + \equiv S_1 - OH_a \stackrel{*K_1}{\rightleftarrows} \equiv S_1OM + aH^+ \tag{1}$$

$$M^{z+} + \equiv S_2 - OH_b \stackrel{*K_2}{\rightleftarrows} \equiv S_2OM + bH^+ \tag{2}$$

$$M^{z+} + \equiv S_3 - OH_c \stackrel{*K_3}{\rightleftarrows} \equiv S_3OM + cH^+ \tag{3}$$

$$*K_i = \frac{\{\equiv S_i - OM\}[H^+]^n}{\{\equiv S_i - OH_n\}[M^{z+}]} \tag{4}$$

where S = sorbent (e.g., "S_1" = Fe(III) oxyhydroxide; "S_2" = Mn(IV) oxyhydroxide; "S_3" = sedimentary organic matter); $\{\equiv S_i-OH_n\}$ = concentration of free binding sites on sorbent "i"; a, b, c, n = average apparent number of protons released when metal M is sorbed; $\{\equiv S_i-OM\}$ = concentration of sites occupied by metal M; $[M^{z+}]$ = concentration of the free metal ion; $*K_i$ = apparent overall equilibrium constant for sorption on substrate "i". Note that here and elsewhere in the text the concentrations of solid phases are indicated by { } parentheses, whereas concentrations of dissolved species are designated by [ ] brackets; the notation "$\equiv$" refers to adsorption sites. Charges on the various solid phases have been omitted for simplicity.

To estimate the sorption constants, we need to be able to determine the variables on the right-hand side of Equation 4, for one or more of the one metal-sorbent combinations present in natural sediments. Since the value of "n" in Equation 4 is generally unknown, it is convenient to define a conditional sorption constant as follows.

$$K_i = \frac{*K_i}{[H^+]^n} = \frac{\{\equiv S_i - OM\}}{\{\equiv S_i - OH_n\}[M^{z+}]} \tag{5}$$

In some cases, chemical extractions can be used to extract a particular sorbent (e.g., amorphous Fe(III) oxyhydroxide) and its associated metals.* With certain simplifying assumptions (described in Appendix 1, and discussed in detail in Tessier, 1992), the ratio $\{\equiv S_i-OM\}/\{\equiv S_i-OH_n\}$ can be replaced by the ratio of analytical concentration of metal M associated with a particular

---

* The use of partial chemical extractions to remove particular sorbents and their associated metals from aquatic sediments is not without its problems. For an exchange of views on this subject, see Nirel and Morel (1990) and Tessier and Campbell (1991).

Table 2   Field-Derived Equilibrium Constants for the Sorption of Trace Metals on Amorphous Fe(III) Oxyhydroxides

Metal	Relation	
Cd	Log K_{Fe-Cd} = 1.03 pH − 2.44	(r^2 = 0.80; n = 26)
Cu	Log K_{Fe-Cu} = 0.64 pH + 0.10	(r^2 = 0.75; n = 39)
Ni	Log K_{Fe-Ni} = 1.04 pH − 2.29	(r^2 = 0.87; n = 29)
Pb	Log K_{Fe-Pb} = 0.81 pH + 0.67	(r^2 = 0.81; n = 7)
Zn	Log K_{Fe-Zn} = 1.21 pH − 2.83	(r^2 = 0.89; n = 41)

From Tessier, A., 1992. In: Environmental Particles — Environmental, Analytical and Physical Chemistry Series, edited by J. Buffle and H. P. Van Leeuwen, Lewis Publishers, Boca Raton, FL, pp. 425–453.

sorbent divided by the sorbent concentration. For example, for sorption of metal M on iron oxyhydroxide, the ratio {Fe–OM}/{Fe–ox} can be substituted in Equation (5) to give

$$K_{M-Fe} = \frac{\{Fe - OM\}}{\{Fe - ox\}[M^{z+}]} \qquad (6)$$

Note that {Fe–ox} corresponds to the analytically determined concentration of amorphous iron oxyhydroxides and {Fe–OM} refers to the concentration of metal "M" coextracted with the Fe–ox sorbent. Alternatively, diagenetic Fe-oxyhydroxides can be collected on Teflon sheets inserted in the sediments and left *in situ* for several months (Belzile et al., 1989; Tessier et al., 1993); dissolution of the oxyhydroxide deposit yields the required ratios of {Fe–OM}/{Fe–ox}. The remaining variable in Equation 6, i.e., [M^{z+}], can be obtained by sampling and analyzing the water immediately overlying the oxic surface sediments. Once again certain simplifying assumptions must be made, notably with respect to the speciation of metal "M" in the oxic pore water (i.e., that there is no appreciable complexation of metal "M" by the dissolved organic matter present in the interstitial water — see Appendix 1).

Using field measurements of this type in lakes of different pH, Tessier (1992) calculated conditional constants, K_{Fe-M}, for the sorption of Cd, Cu, Ni, Pb, and Zn on natural amorphous iron oxyhydroxides and determined their pH dependence (Table 2). As anticipated from Equations 1 to 3, metal cation sorption is favored at high pH — the slopes of the log K_{M-Fe} vs. pH relations shown in Table 2 are all positive. At low pH values, the field-derived K_{M-Fe} values decrease in the sequence Pb > Cu > Zn > Ni ≈ Cd (Tessier, 1992); a similar affinity sequence is observed in laboratory sorption studies on synthetic iron oxyhydroxides in well-defined media. This correspondence between field and laboratory measurements can be taken as an indication that sorption phenomena are indeed responsible for controlling dissolved trace metal concentrations under oxic conditions.

Given the field-derived constants from Table 2, the amount of amorphous iron oxyhydroxide (sorbent), the concentration of sorbed metal, and the ambient pH, one can estimate the concentration of the free metal ion, M^{z+}, in equilibrium with the oxic sediments (Equation 7).

$$[M^{z+}] = \frac{\{Fe-OM\}}{K_{M-Fe}[Fe-ox]} \tag{7}$$

In a recent publication (Tessier et al., 1993), the model for Cd sorption has been generalized to include both Fe–ox and sedimentary organic matter as sorbents:

$$[Cd^{2+}] = \frac{\{Cd\}_T[H^+]^{x+y}}{N_{Fe} \cdot *K_{Fe-Cd}\{Fe-ox\}[H^+]^y + N_{OM} \cdot *K_{OM-Cd}\{OM\}[H^+]^x} \tag{8}$$

where $\{Fe-ox\}$ = concentration of amorphous iron oxyhydroxides (mmol/g); $\{OM\}$ = concentration of organic carbon in the surficial sediments (mmol C/g); $\{Cd\}_T$ = total Cd concentration in the sediments (nmol/g); x and y = apparent average numbers of protons released per Cd^{2+} ion adsorbed on iron oxyhydroxides and organic matter, respectively; N_{Fe} = number of moles of sorption sites on iron oxyhydroxides, per mole Fe–ox; N_{OM} = number of moles of sorption sites on sediment organic matter, per mole organic carbon; $*K_{Fe-Cd}$ and $*K_{OM-Cd}$ = apparent overall equilibrium constants for the sorption of Cd on iron oxyhydroxides and organic matter, respectively. The values of the geochemical constants x (0.82), y (0.96), $N_{Fe} \cdot *K_{Fe-Cd}$ ($10^{-1.22}$) and $N_{OM} \cdot *K_{OM-Cd}$ ($10^{-2.38}$) were determined experimentally from the field geochemical data (Tessier et al., 1993).

B. ANOXIC/SUBOXIC SEDIMENTS

Under anoxic, sulfate-reducing conditions, the partitioning of trace metals between the pore waters and the solid phase is likely to be controlled by precipitation-dissolution reactions involving amorphous sulfides. Evidence supporting this hypothesis has recently been presented by Di Toro and co-workers (1990, 1992). By titrating suspended synthetic iron monosulfide or natural sediments with Cd, these workers demonstrated that the added Cd could react with the amorphous sulfides according to Equation 9 (Figure 3). Since the solubility product for CdS(s) is lower than that of FeS(s), cadmium will tend to displace iron in the solid sulfide phase (Equation 9).

$$Cd^{2+} + FeS(s) \rightleftarrows CdS(s) + Fe^{2+} \tag{9}$$

Di Toro et al. concluded that under their experimental conditions the concentration of the free Cd^{2+} ion was controlled not by sorption reactions, nor by reaction with free S^{2-} in solution, but rather by reaction with solid iron and manganese monosulfides. The reactive sulfide fraction was operationally defined by cold acid extraction of the sediments (1-h treatment with 0.5 M HCl at room temperature; the forms extracted in this manner are defined as "acid-volatile sulfide", or AVS).

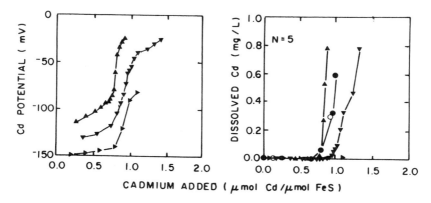

Figure 3 Cadmium titration of amorphous FeS. The added Cd is normalized with respect to the concentration of FeS initially present. The ordinate corresponds to the Cd electrode response (left-hand panel) or total dissolved Cd (right-hand panel). (Adapted from Di Toro et al., 1990.)

Di Toro et al. (1990, 1992) further hypothesized that, since the sulfides formed by Ni, Zn, Pb, Cu, and Hg are also less soluble than FeS or MnS, they too should tend to displace Fe or Mn from the AVS fraction. This hypothesis was tested by Casas and Crecelius (1994), who titrated three anoxic marine sediments with Cu, Pb, and Zn, and monitored the pore water for dissolved metal. For Pb and Zn, metal was not detected in the pore water until the sediment AVS concentration was exceeded, as predicted by the AVS model; for Cu, however, metal levels in pore water remained low even after the AVS concentration was exceeded, suggesting that other sediment phases were important in binding Cu.

Using the solubility products for metal sulfides (K_{sp}; Table 3) and the appropriate field data (pH; total reduced sulfur concentration, ΣH_2S), one could in principle estimate the free-metal ion concentration in pore-water solution on the basis of chemical equilibria. However, given the mixture of metals present in sediments, it is possible that metal solubility will be controlled not by pure sulfide phases such as those represented in Table 3, but rather by a solid solution of the trace metal in one of the major sulfide phases. Under such conditions, the calculation of the free-metal ion concentration would require the stability constants for metal complexes with all the dissolved ligands, including the metal-sulfide complexes, the same field data as above (pH; total reduced sulfur concentration, ΣH_2S), and in addition the total dissolved metal and ligand concentrations.

C. APPLICATION TO *IN SITU* SEDIMENTS

The two approaches described above have as a common goal the prediction of dissolved metal concentrations in the pore waters of natural, *in situ* sediments. In effect, pore-water metal concentrations are considered to reflect the chemical potential of the metal in the solid and solution phases at the sediment-water boundary; changes in this chemical potential should in turn affect the bioavailability of the metal. The two approaches differ, however, in their

Table 3   Metal Sulfide Solubility Products

Metal sulfide		log K_{sp}[a]	log (K_{MS}/K_{FeS-am})[b]	Ref.
FeS	(s, am)	+4.1		Davison, 1991
	(s, mackinawite)	+3.4		Davison, 1991
	(s, greigite)	+2.6		Davison, 1991
NiS	(s, α)	+1.5	−2.6	NIST, 1993
	(s, ß)	−4.0	−8.1	NIST, 1993
	(s, γ)	−5.7	−9.8	NIST, 1993
ZnS	(s, sphalerite)	−4.5	−8.6	Daskalakis & Helz, 1993
CdS	(s, greenockite)	−7.3	−11.4	Daskalakis & Helz, 1992
PbS	(s)	−7.9	−12.0	NIST, 1993
CuS	(s, am)	−11.9	−16.0	Shea & Helz, 1989
	(s, covellite)	−15.3	−19.4	Shea & Helz, 1989
HgS	(s, black)	−31.7	−35.8	NIST, 1993
	(s, red)	−32.1	−36.2	NIST, 1993

[a] K_{sp} given for the reaction MS + 2 H$^+$ $\rightleftarrows$ M^{2+} + H$_2$S; multiple values are given when different sulfide solid phases exist.

[b] Ratio log (K_{MS}/K_{FeS}) calculated for amorphous FeS.

choice of the reactions controlling metal concentrations in sediment pore waters. This divergence stems from a different conception of what constitutes "biologically important sediments" — fully oxidized, surficial sediments (where AVS levels should be vanishingly low and $[M]_i$ should be controlled by sorption reactions) vs. partially oxidized, suboxic sediments (where significant AVS levels would be expected to persist, and exchange reactions with the amorphous sulfides would control metal partitioning between the dissolved and solid phases).*

Choosing between the two approaches is not as straightforward as might appear at first glance. In the real world, due to spatial heterogeneity at the millimeter to centimeter scale, the distinction between oxic and anoxic sediment strata is often blurred. Most aerobic benthic organisms survive in sediments that are underlain or even surrounded by completely anaerobic sediments, which constitute a potential source of AVS. Di Toro et al. (1992) suggest that the presence of AVS in the anaerobic layer may be sufficient to reduce the metal activity to which the aerobic benthic animals are exposed, but admit that field data to support this contention are lacking. The question of field validation is considered in detail in Section III.C.

III. INTERACTIONS BETWEEN DISSOLVED TRACE METALS AND AQUATIC ORGANISMS

In the previous section, we have considered possible geochemical controls on metal concentrations in pore waters in contaminated sediments. This section takes the reasoning one step further (see Table 1) and examines the bioavailability of these dissolved metals.**

* For a thought-provoking discussion of this point, see Luoma and Davis (1983).

** Much of the discussion in this section has been extracted from a recent review chapter (Campbell, 1995).

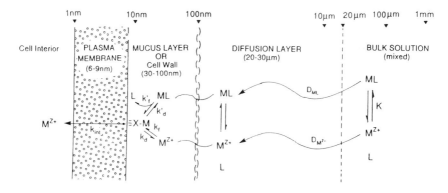

Figure 4 Conceptual model of metal-organism interactions. M^{z+} = free-metal ion; ML = metal complex in solution; M-X-membrane = surface metal complex; k_f, k_f' = rate constants for formation of the surface complex; k_d, k_d' = rate constants for dissociation of the surface complex; k_{int} = rate constant for "internalization" or transport of the metal across the biological membrane. Charges on ligand not shown for simplicity. (Adapted from Campbell, P.G.C., 1995. In: Metal Speciation and Bioavailability in Aquatic Systems, edited by A. Tessier and D.R. Turner John Wiley & Sons, New York, 1995, pp. 45–102.)

Much qualitative evidence exists to the effect that the total aqueous concentration of a metal is not a good predictor of its "bioavailability," i.e., that the metal's speciation will greatly affect its availability to aquatic organisms. Note that the term "speciation" is widely used/misused in the trace metal literature (Bernhard et al., 1986); in the present context, the term will be used in the passive sense to mean "the distribution or partitioning of a metal among various physico-chemical forms in the external medium."

A. FORMULATION OF THE FIAM

Over the past 20 years, a convincing body of evidence has been developed to support the tenet that the biological response elicited by a dissolved metal is usually a function of the free metal ion concentration, $M^{z+}(H_2O)_n$, which in turn is determined not only by the total dissolved metal concentration but also by the concentration and nature of the ligands present. To rationalize these experimental observations and explain what was perceived as "the universal importance of free metal ion activities in determining the uptake, nutrition and toxicity of all cationic trace metals," Morel (1983) formulated the free-ion activity model (FIAM) for metal-organism interactions.

To elicit a biological response from a target organism or to accumulate within this organism, a metal must first interact with and/or traverse a cell membrane. This interaction of the metal with the cell surface, involving either the free metal ion (M^{z+}) or a metal complex (ML^{z+}) as the reactive species, can be represented in terms of the formation of M-X-cell surface complexes, where X-cell is a cellular ligand present at the cell surface (Figure 4).

In the simplest case, where the free metal ion is the species reacting at the cell surface, one can envisage the following reactions (where, for simplicity, charges on the ligand are not shown):

solution equilibria

$$M^{z+} + L \overset{K_c}{\rightleftarrows} ML \tag{10}$$

$$K_c = [ML]/([M^{z+}][L]) \tag{10a}$$

surface reaction of M^{z+}

$$M^{z+} + {}^-X\text{-cell} \overset{K_{ad}}{\rightleftarrows} M^{z+} - X\text{-cell} \tag{11}$$

$$\{M^{z+} - X\text{-cell}\} = K_{ad}\{{}^-X\text{-cell}\}[M^{z+}] \tag{11a}$$

where K_c and K_{ad} are apparent (concentration) equilibrium constants. The biological response is assumed to be proportional to the concentration of the surface complex, $\{M^{z+}\text{-X-cell}\}$ (see below). Provided the concentration of free $^-$X-cell sites remains approximately constant, Equation 11a indicates that the biological response will vary as a function of $[M^{z+}]$. A similar situation prevails even if the metal complex (ML) is the species reacting at the cell surface, provided that the reaction proceeds by ligand exchange (Campbell, 1995).

A number of key assumptions underlie the free-ion activity model, some involving the biological surface and others the kinetics of metal-organism interactions.

Biological surface:
- The key interaction of a metal with a living organism involves the plasma membrane, which is impermeable to the free metal ion, M^{z+}, and to its (hydrophilic) complexes, $ML^{\pm}$.
- The interaction of the metal with the plasma membrane can be described as a surface complexation reaction, forming M^{z+}-X-cell (Equation 11). The biological response, whether it be metal uptake, nutrition, or toxicity, is proportional to the concentration of this surface complex; variations in $\{M^{z+}\text{-X-cell}\}$ follow those of $[M^{z+}]$ in solution (Equation 11a).
- The biological surface does not change during the metal exposure experiment (i.e., the FIAM will be more applicable to short-term experiments than to long-term chronic exposure).

Kinetics:
- Metal transport in solution, towards the membrane, and the subsequent surface complexation reaction occur rapidly, such that a (pseudo-)equilibrium is established between metal species in solution and

those at the biological surface ("rapid" = faster than metal uptake, faster than the expression of the biological response).

• Thus, the identity of the metal form(s) reacting with the plasma membrane is of no biological significance. No one species in solution can be considered more (or less) available than another.

Possible mechanistic links between the formation of the surface complex, M^{z+}-X-cell, and the initiation of a biological effect have been suggested by several workers (Morel, 1983; Pagenkopf, 1983; Sunda, 1991). If ⁻X-cell represents a physiologically active site at the cell surface, then the binding of metal M might induce a direct biological response (e.g., fish gills, Pagenkopf, 1983; algal cell, Steemann Nielsen et al., 1969). Alternatively, if ⁻X-cell corresponds to a transport site that allows metal M to traverse the cell membrane and enter the cytosol, then binding at the surface site would simply precede transport into the cell (i.e., the actual reaction of M with the metabolically sensitive site would occur intracellularly, following transport [Morel, 1983]). In a variation of this scenario, ⁻X-cell might correspond to a transport site normally used by an essential micronutrient; binding at the cell surface site by metal M would then inhibit the supply of the essential element and induce nutrient deficiency (e.g., phytoplankton, Sunda and Huntsman, 1983 [Mn, Cu]; Harrison and Morel, 1983 [Fe, Cd]).

B. APPLICABILITY OF THE FIAM — STUDIES WITH DISSOLVED METALS

In principle, the FIAM should apply to aquatic organisms that do not assimilate particulate material (e.g., algae and rooted aquatic plants), and to organisms that do assimilate particulate material but for which the dissolved phase remains the primary vector for metal uptake (e.g., fish).* A considerable body of experimental evidence has indeed accumulated in support of the FIAM. Examples of such experiments have recently been critically reviewed by Campbell (1995) and several points of general interest can be extracted from this review.

Organisms Studied — Studies on marine and freshwater phytoplankton outnumber all others. This distribution presumably reflects the sensitivity of algae to trace metals (Morel, 1983) and the relative ease with which they can be studied in the laboratory (defined inorganic growth media, rapid response). References to invertebrates are also frequent, with a distinct bias towards marine species. In contrast, the few fish studies were all performed with freshwater species.

Various "endpoints" have been studied in the bioassays, covering a spectrum from highly specific (e.g., metal accumulation per se: surface adsorption; absorption; subcellular distribution) to more holistic (e.g., growth, motility, mortality).

---

* One could also argue that even in the case where ingested particulate material constitutes the primary vector for metal uptake, the response of the consuming organism may still depend indirectly on the free-metal ion concentration in the ambient water, provided the particulate material being ingested is itself in equilibrium with metal concentrations in the ambient water.

Experimental Conditions (Metals/Ligands) — Almost without exception, the experiments designed to test the FIAM have been performed at a fixed pH, with divalent trace metals (Cu, Cd, Ni, Pb, Zn), in artificial (inorganic) media or in filtered seawater, and in the presence of known quantities of synthetic ligands such as ethylene-diamine tetraacetic acid (EDTA) or nitrilotriacetic acid (NTA). These ligands have been employed as metal buffers. By manipulating $[M]_T$ and/or $[L]_T$, one can adjust the free-ion concentration, $[M^{z+}]$, over the desired range. Perhaps fortuitously, these ligands form hydrophilic complexes with the metals of interest, towards which biological membranes are virtually impermeable. Copper has been studied far more extensively than any other metal. Relevant citations decrease in the order Cu > Cd > Zn $\gg$ all others.

Major Conclusions — Studies performed in the absence of natural dissolved organic matter (DOM):

- In experiments performed at constant pH and water hardness, in the presence of synthetic ligands forming hydrophilic metal complexes ($ML^{\pm}$), the biological response consistently varies as a function of the concentration of the free-metal ion, as predicted by the FIAM (52 of 59 cases reviewed). See Figure 5 for a typical example.
- Documented examples of experiments in which changes in metal bioavailability do not conform to the FIAM are relatively few in number, and most of these can be explained as cases in which the metal forms a neutral, lipophilic complex (ML_n^o) to which biological membranes are permeable (e.g., Phinney and Bruland, 1994).
- From studies performed with freshwater organisms under conditions where the pH and/or the water hardness was varied, it is clear that one must consider potential competition for the metal binding site, $^-$X-cell, by the hydrogen ion, H^+, and by the hardness cations, Ca^{2+} and Mg^{2+}. These effects were explicitly included in Morel's original formulation of the FIAM (Morel, 1983), but since most of the early work was done in seawater, variations of $[H^+]$ and $[Ca^{2+}]$ were of relatively little concern. The influence of the hydrogen ion on metal-organism interactions was reviewed by Campbell and Stokes (1985).

Studies performed in the presence of natural DOM:

- Studies suitable for testing the applicability of the FIAM in the presence of DOM are very scarce. There are numerous reports in the literature of the effects of DOM on metal bioavailability, but virtually all of these studies are qualitative in nature (i.e., metal speciation is undefined).
- The few quantitative studies that do exist are more or less evenly divided between examples that conform to the FIAM and others that appear to be in contradiction, i.e., unlike the studies performed in the absence of DOM, no consensus is evident for the experiments run in the presence of DOM. The applicability of the FIAM in natural waters in the presence of natural DOM remains to be demonstrated (Campbell, 1995).

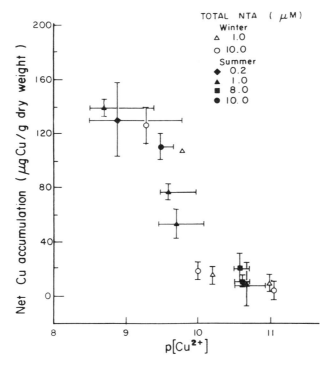

Figure 5 Net copper accumulation by the American oyster, *Crassostrea virginica*, after 14 d exposure, expressed as a function of pCu^{2+} (–log cupric ion activity). (Adapted from Zamuda and Sunda, 1982.)

C. APPLICABILITY OF THE FIAM — STUDIES WITH SEDIMENTS

All of the studies considered in the preceding section were carried out in the absence of sediments (water-only exposures); in this section we return to our area of primary concern, the sediment-water boundary, and consider several studies that attempt to relate metal bioavailability to the free-metal ion concentration under conditions of simultaneous water-sediment exposure. Two types of studies have been carried out: (1) bioassays on spiked sediments (Di Toro et al., 1990, 1992; Carlson et al., 1991; Hare et al., 1994) or on natural sediments collected from known contaminated sites (Ankley et al., 1991, 1993, 1994); and (2) field surveys of indigenous benthic organisms (Couillard et al., 1993; Tessier et al., 1993; also see reviews by Campbell and Tessier, 1989, 1991; Tessier and Campbell, 1990). Both approaches support the general idea that benthic organisms respond to the free-metal ion concentration in the ambient water* in or near the surficial sediments.

---

* The definition of "ambient" water will differ among benthic species and may include, in different cases, (a) anoxic waters for organisms in contact with deeper sediment, (b) oxic waters towards the sediment surface for organisms living within the sediment, but whose primary environmental contact is above the redox interface, and (c) oxic waters overlying the sediments in the case of suspension feeders (NRCC, 1988).

i. Bioassays

a. Laboratory Studies

Under the auspices of the U.S. Environmental Protection Agency (EPA), a major research effort is currently underway to define criteria for assessing contaminated sediments (Adams et al., 1992; Burton and Scott, 1992; Ankley et al., 1994). One of the approaches being explored is to estimate the pore-water concentrations of different contaminants on the basis of chemical equilibria, and to use these concentrations as a measure of the contaminant's chemical potential. The goal of this chemical equilibrium approach, as applied to metal-contaminated sediments, is to predict the (free-)metal concentration in the sediment pore water. As outlined earlier (Section II.B), the EPA-sponsored researchers have assumed that this concentration is controlled by precipitation reactions with amorphous sulfides (AVS). Since in the presence of excess AVS the concentrations of free metal should be vanishingly low (Di Toro et al., 1990, 1992), these workers have hypothesized that provided the molar ratio of {SEM}/{AVS} is less than unity, the sediment should not exhibit any toxicity (SEM, simultaneously extracted metal, i.e., metal extracted by the same cold acid reagent used to determine the amorphous acid volatile sulfide fraction). The corollary is that if the {SEM}/{AVS} ratio exceeds 1, the sediment could be toxic.

In a number of recent papers, this hypothesis has been tested in the laboratory with various benthic bioassay organisms. Of the eight studies summarized in Table 4, seven derive from this laboratory approach: four of these laboratory experiments were performed on spiked sediments (#1–4) and three on "naturally" contaminated sediments that had been collected along a contamination gradient (#5–7). The studies share a number of common features: the sediments were collected as a grab sample, normally with no attempt to distinguish between the oxic or suboxic strata, and were then homogenized and analyzed for {AVS} and for the simultaneously extracted metals, {SEM}. In experiments 1–4, the {SEM}/{AVS} ratio was varied by spiking the sediment with increasing amounts of the test metal. After a short "equilibration" period, the sediments were placed in small containers and allowed to settle (normally 24 h). The test organisms were then introduced and their survival monitored over the next 4–10 d. In most of these experiments the sediments were recovered at the end of the bioassay, rehomogenized, and reanalyzed for {SEM} and {AVS}. Virtually the same protocol was followed for the experiments with naturally contaminated sediments (#5–7), except that no additional metal was added to the sediments; {SEM}/{AVS} ratios were determined on the original sediment and again at the end of the bioassay, both times on homogenized samples.

The laboratory bioassays run with cadmium, nickel, lead, or zinc conformed to the AVS model (Table 4: experiments #1–5): both for spiked sediments and for the naturally contaminated sediments, no significant mortality occurred relative to controls if the molar concentration of simultaneously

Table 4   Tests of the Hypothesis that Acid Volatile Sulfide (AVS) Controls the Toxicity of Metals in Sediments

Metal	Organism	Response	Geochemical predictor	Sediment	Ref.
1. Cd	Marine amphipods *Ampelisca abdita; Rhepoxynius hudsonii*	Mortality (96 h)	Ratio {added Cd}/{AVS}	Spiked marine sediments	Di Toro et al., 1990
2. Cd	Freshwater oligochaete *Lumbriculus variegatus;* snail *Helisoma sp.*	Mortality (10 d) Bioaccumulation	Ratio {added Cd}/{AVS}	Spiked fresh-water sediments {SEM}/{AVS} varied from 0.1 to 10	Carlson et al., 1991
3. Cd	Marine amphipods *Ampelisca abdita; Rhepoxynius hudsonii;* freshwater oligochaete *Lumbriculus variegatus;* snail *Helisoma sp.*	Mortality	Ratio {added M}/{AVS}	Spiked freshwater or marine sediments	Di Toro et al., 1992
4. Cu Pb Zn	Marine polychaete *Capitella capitata*	Mortality (20 d)	Ratio {SEM}/{AVS}	Spiked anoxic marine sediments representing a gradient in AVS and organic carbon: {SEM}/{AVS} varied from 0.02 to 2.8 (Cu), 0.27 to 3.0 (Pb), 0.38 to 1.6 (Zn)	Casas and Crecelius, 1994
5. Cd Ni	Freshwater amphipod *Hyalella azteca;* oligochaete *Lumbriculus variegatus*	Mortality (10 d) Bioaccumulation	Ratio {SEM}/{AVS}	Estuarine sediments sampled along contamination gradient; {SEM}/{AVS} varied from 0.1 to 220	Ankley et al., 1991
6. Cu	Freshwater amphipod *Hyalella azteca*	Mortality (10 d)	Ratio {SEM}/{AVS}	Freshwater sediments sampled along contamination gradient; {SEM}/{AVS} varied from 0.2 to 68	Ankley et al., 1993
7. Cd Cu Ni Pb Zn	Freshwater oligochaete *Lumbriculus variegatus*	Bioaccumulation (30 d)	Ratio {SEM}/{AVS}	Freshwater sediments representing a contamination gradient; {SEM}/{AVS} varied from 0.44 to 0.62	Ankley et al., 1994
8. Cd	Benthic invertebrates (natural community)	Abundance of individual taxa; bioaccumulation (14 months colonization)	Ratio {added Cd}/{AVS}	Spiked freshwater sediments; {SEM}/{AVS} varied from 0.5 to 10	Hare et al., 1994

Note: All experiments but one (#8) were performed in the laboratory.

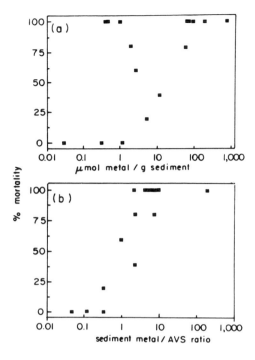

Figure 6 Sediment bioassays on contaminated sediments with the amphipod *Hyalella azteca*. Toxicity expressed as percent mortality relative to (a) total metal (Cd + Ni) concentrations in the samples, or (b) sediment metal concentrations normalized to acid-volatile sulfide (AVS). (Adapted from Ankley et al., 1991.)

extracted Cd, Ni, Pb, or Zn was less than the molar concentration of acid volatile sulfide (AVS) in the sediment. Mortality began to occur when the {SEM}/{AVS} ratio reached unity, and total mortality was observed when the ratio exceeded ~3 (Figure 6). The results for copper were, however, markedly different (Table 4: experiment #6). Normalization of sediment copper concentrations to AVS correctly predicted that sediments were nontoxic when the {Cu}/{AVS} ratio was less than one, but toxicity was frequently absent in samples with {Cu}/{AVS} ratios significantly greater than unity. The authors determined the pore-water copper concentration and showed that it was still an excellent predictor of amphipod mortality: toxicity curves for water-only exposure and for pore-water exposure were superimposed (Ankley et al., 1993). The authors concluded that "AVS is not the appropriate partitioning phase for predicting copper bioavailability in freshwater sediments." Very similar results were reported by Casas and Crecelius (1994) for bioassays run with anoxic marine sediments that had been spiked with copper (experiment #4). From a purely geochemical point of view, copper presumably does react with sulfide in suboxic sediments containing AVS. However, the results of experiments 4 and 6 indicate that AVS is not the only binding phase for Cu in the sediments to which the test organisms were exposed.

Relationships between metal bioaccumulation and {SEM}/{AVS} ratios have also been investigated in the laboratory, but the data base is more limited than for organism mortality (Table 4, column 3). In experiment 2 (*Lumbriculus variegatus*, *Helisoma* sp.), whole-body Cd concentrations were measured in worms and snails that survived to the end of the bioassay (i.e., for sublethal conditions, where the majority of the test organisms survived). Cadmium bioaccumulation tended to increase over the range of experimental {Cd}/{AVS} values from 0.05 to 3, with separate and distinct plots for each sediment. In addition, contrary to what might have been predicted, there was no distinct threshold effect when the ratio {Cd}/{AVS} exceeded 1.0 (Carlson et al., 1991). In contrast to these results, in experiment 5 (*L. variegatus*) negligible bioaccumulation of Cd or Ni occurred at {SEM}/{AVS} ratios less than one. However, for ratios greater than one, the magnitude by which {SEM}/{AVS} exceeded unity was an unreliable predictor of bioaccumulation. As pointed out by Ankley et al. (1991), this is not surprising, since the {SEM}/{AVS} ratio is dimensionless; high ratios can, in principle, occur in sediments with relatively low or relatively high metal contamination. When the {SEM}/{AVS} ratio exceeds one, a normal toxicological dose-response would be anticipated. However, when the authors sought to improve the prediction of bioaccumulation by taking into account the "excess" metal in the sediment, i.e., the difference {SEM} – {AVS}, they did not find any "definable relationship."

In a recent laboratory test of metal bioaccumulation in the presence of AVS (Table 4: experiment 7), Ankley et al. (1994) exposed the oligochete, *L. variegatus* to three freshwater sediments that contained elevated levels of Cd, Cu, Ni, Pb, and Zn, but for which the {SEM}/{AVS} ratios were less than one. Metals in the sediments were thus predicted to be of minimal biological availability and indeed, after 30 d exposure, metal concentrations in organisms that had been exposed to the test sediments were not significantly greater than those in control organisms that had been exposed to Lake Superior water only. It is not clear, however, that starved organisms held in oxic water without sediments constitute a valid control for those that were able to burrow in the sediments and feed on sedimentary organic matter.

b. Field Studies

All the bioassays described to this point have been performed in small laboratory microcosms. In an important extension of this approach to field conditions, Hare et al. (1994) have recently carried out a long-term study designed to test if Cd toxicity and accumulation in benthic invertebrates are related to the {Cd}/{AVS} molar ratio, as predicted by the AVS model. The experiment was designed as an *in situ* spiked sediment bioassay (Table 4: experiment 8). Sufficient Cd was added to a natural, uncontaminated lake sediment to give a range of nominal {Cd}/{AVS} ratios: ≈0.05 (control), 0.1, 0.5, 2, and 10. The spiked sediments were then returned to the lake bottom at the original collection site and the effects of the Cd on colonizing invertebrates were monitored after 14 months.

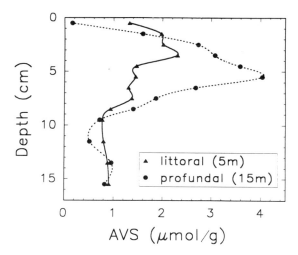

Figure 7   Depth distribution of acid-volatile sulfide (AVS) in lake sediment cores, showing subsurface maximum and surface depletion in AVS. Samples were col-
lected at two sites, one profundal and the other littoral, in Lake Tantaré, near
Québec City, Québec. (Adapted from Hare et al., 1994.)

Periodic sampling of water and sediments at the sediment-water boundary
confirmed the anticipated vertical geochemical gradients. Depth profiles of
AVS concentrations in the sediments (0, 5, 8 cm) were initially uniform, but
after placement of the sediments on the lake bottom {AVS} decreased near the
sediment-water interface, presumably as a result of sulfide oxidation; depth
profiles taken outside the enclosures also showed a surface minimum in AVS
(Figure 7). In contrast, depth profiles of acid-extractable Cd concentrations
showed little variation; as a result, ratios of {Cd}:{AVS} tended to decrease
with depth in the sediments (Figure 8). Above the sediment-water interface,
dissolved Cd concentrations were generally undetectable (<0.28 nM) — only
at the highest {Cd}/{AVS} ratio was dissolved Cd detected above the sediment. Below the sediment-water interface, dissolved Cd concentrations in the
control enclosure were uniformly low (< 1.5 nM); at three intermediate {Cd}/
{AVS} levels (0.1, 0.5, 2), aqueous Cd showed a peak around the 2- to 3-cm
level; at the highest {Cd}/{AVS} molar ratio, interstitial Cd concentrations
were either uniformly high at all sediment depths (May, October) or showed
a peak around 2 cm (August). This vertical heterogeneity in dissolved metal
concentrations and in {Cd}/{AVS} ratios illustrates the inherent difficulty in
defining a single representative {SEM}/{AVS} ratio for a particular sediment!
In the present study, the authors chose to average the {Cd}/{AVS} ratios over
a 6-cm depth.

To detect the biological effects of the contaminated sediment, Hare et al.
(1994) monitored the recolonization of the enclosures (abundance of various
insect larvae), measured the growth rates of two chironomid taxa, and determined levels of Cd bioaccumulation as a measure of larval exposure to Cd.
Cadmium bioaccumulation varied markedly among taxa and could be related

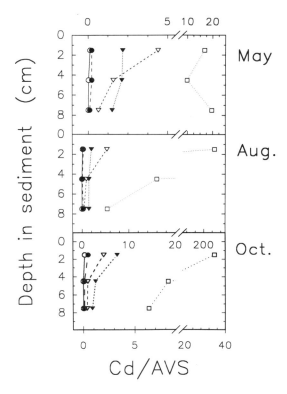

Figure 8 Long-term *in situ* spiked sediment bioassay. Seasonal fluctuations and sediment depth profiles of Cd/AVS molar ratios within the experimental enclosures. The Cd/AVS ratios are indicated for each of the nominal treatment levels: ○, 0.05; ●, 0.1; ▽, 0.5; ▼, 2; □, 10. (Adapted from Hare et al., 1994.)

to their behavior: for example, at all but the lowest {Cd}/{AVS} ratio, Cd levels tended to be higher in organisms that burrowed deeply into the sediment than in those that remained near the sediment surface or migrated daily into the overlying water column. At {Cd}/{AVS} molar ratios less than one, Cd concentrations in animals increased only slightly, whereas Cd bioaccumulation increased markedly at {Cd}/{AVS} ratios near or above one (Figure 9, right-hand panels). As pointed out by Hare et al. (1994), this overall trend is consistent with the AVS model, although according to this model practically no bioaccumulation should have occurred at {Cd}/{AVS} ratios below unity. Despite this bioaccumulation, however, the abundances of all but one of the insect taxa present were independent of the {Cd}/{AVS} molar ratios (Figure 9, left-hand panels). Furthermore, larval growth rates were unrelated to the {Cd}/{AVS} ratios, at least for the two taxa for which growth measurements were feasible.

These limited community and population responses to increases in {Cd}/{AVS} ratios above one differ from the laboratory bioassay results discussed earlier (Table 4, studies #1–7). Hare et al. (1994) discuss possible biological

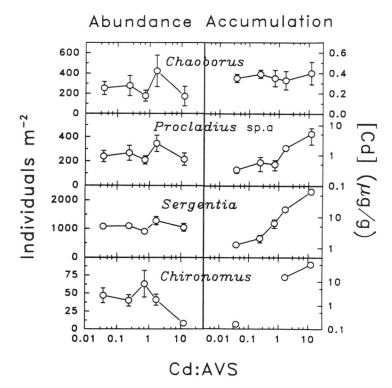

Figure 9 Long-term *in situ* spiked sediment bioassay. Mean abundances and Cd accumulation of six insect taxa at various {Cd}/{AVS} ratios in the manipulated
sediments. Variation about the means is indicated by standard errors ($n = 9$)
for larval abundance and standard deviations for bioaccumulation. (Adapted
from Hare et al., 1994.)

reasons for this discrepancy in detail (e.g., laboratory bioassay procedures may
stress the test organisms and render them more metal-sensitive; the common
laboratory test species may be more sensitive to Cd than are the species making
up the natural benthic community). Geochemical considerations may also be
involved, in that the {Cd}/{AVS} ratio is dimensionless and high molar ratios
can in principle occur in sediments with relatively low or relatively high metal
contamination (see Ankley et al., 1991) — total Cd concentrations were in fact
much lower in the field study than in the laboratory bioassays.

c. Summary

Three aspects of the spiked bioassay approach warrant attention: sediment
diagenesis, bioturbation/bioirrigation, and microhabitats. As pointed out in the
first paper in the AVS series (Di Toro et al., 1990), the normal method of
preparing sediments for bioassays is to produce a uniform mixture of sediment
and pore water by thorough mixing. For such systems, the AVS and the metals
will be uniformly distributed and the concentrations to be used for calculating

the {SEM}/{AVS} ratio are relatively unambiguous. However, as soon as the sediment is placed in the test container, even before the bioassay organisms are introduced, it begins to change! Under the influence of the indigenous sediment microorganisms, a redox cline is rapidly established across the sediment-water interface (Hargrave, 1975). Vertical gradients develop in the pore water, for such variables as dissolved oxygen, iron(II), manganese(II), sulfate, and sulfide, and in the solid phase for such key phases as Fe(III) and Mn(IV) oxyhydroxides and Fe(II) sulfides (the latter being major contributors to the AVS fraction — Di Toro et al., 1990). Thus, even in the absence of benthic invertebrates, the distribution of AVS in settled sediments would be expected to vary vertically and temporally.*

The introduction of test organisms into the experimental systems further complicates matters. In reworking the sediment, either to create their own microhabitat or in the course of their normal feeding/excretion behavior, and in irrigating their burrows, benthic animals will exert an influence on the diagenetic reactions described above. The influence of this "bioturbation/bioirrigation" on sediment chemistry was ignored in all but the most recent of the laboratory bioassay experiments summarized in Table 4 (i.e., the determinations of {SEM} and {AVS} were performed on sediments collected from parallel control containers without test organisms). Note, too, that benthic invertebrates can create their own microhabitat within the sediment (Aller, 1978; Aller and Yingst, 1978; Krantzberg, 1985). The activity of the animals (burrowing, feeding) creates microscale variations in sediment chemistry, which will be superimposed on the general vertical redox-cline (Davis, 1983; Abu-Hilal et al., 1988). For example, the physical exchange of water between the sediment pore-water compartment, the burrow, and the overlying water column will tend to favor the diffusion of oxygen into an otherwise anoxic environment.

It is clear from the preceding discussion that, under the combined influence of sediment microorganisms and benthic macrofauna, SEM and AVS concentrations in sediments may well vary temporally and vertically (Herlihy et al., 1988; Howard and Evans, 1993; Nriagu and Soon, 1985). Given this potential variability, it is difficult to decide which SEM and AVS concentrations should be used to evaluate the potential toxicity of the metals in sediments. Concentrations of SEM and AVS can be measured reproducibly on a homogeneous sediment suspension. But, how do these values relate to the true exposure conditions experienced by benthic organisms *in situ*?

It should be emphasized that the preceding critical analysis applies to the AVS approach, especially as it has been tested in the laboratory, but does not imply a rejection of the chemical equilibrium paradigm per se. On the contrary, we would argue that the chemical equilibrium approach (with its emphasis on

---

* These reactions occur sufficiently rapidly to affect sediment chemistry during the course of a 4- to 10-d bioassay. Indeed, marked changes in {SEM}/{AVS} ratios are often noted when samples collected at the beginning and end of the bioassay are compared (e.g., Di Toro et al., 1990; Ankley et al., 1993). Note that these differences cannot be attributed to a sampling artifact — the sediments are normally analyzed at the end of the bioassay after rehomogenization.

the prediction of the free-metal ion concentration, $[M]_i$, in the sediment-water interstitial see Table 1) constitutes a valid conceptual framework. What remains to be determined is the nature of the geochemical phases/reactions that are involved in controlling $[M]_i$. Normalization of metal concentrations with respect to AVS is, of course, invalid if {AVS} is zero, as would be the case in an undisturbed fully oxidized sediment. For sediments with trace amounts of AVS, Di Toro et al. (1990) suggest that it is likely that "other phases" (e.g., sedimentary organic matter, iron oxyhydroxides) would be important. Indeed, in recent laboratory experiments with copper, Ankley et al. (1993) and Casas and Crecelius (1994) conclude that strong binding phases other than AVS exist for copper in freshwater and marine sediments, respectively, and suggest that this additional binding capacity may be correlated with sedimentary organic matter.

All the studies discussed to this point have been carried out on "manipulated" sediments (i.e., transferred to the laboratory and/or spiked with added metals). Field studies on undisturbed *in situ* sediments and indigenous benthic organisms might be expected to yield a clearer indication as to which geochemical phases/reactions are involved in controlling the free-metal concentration in the ambient water near the sediment-water interface. Such studies are the subject of the following section.

ii. Field Studies on Natural Sediments

As an alternative to the bioassay approach described in the preceding section, one could in principle also study plant and animal populations living *in situ* in metal-contaminated sediments. The literature is replete, however, with unsuccessful examples of this approach, notably where researchers have sampled along a contamination gradient and have sought relationships between metal levels in the indigenous benthic organisms and total metal concentrations in the host sediments (see NRCC, 1988 for a review). More recently, as the link between metal speciation and metal bioavailablity has become better recognized, attempts have been made to refine the geochemical approach used to define the contamination gradient. In particular, several recent papers have used a geochemical modeling approach to relate sedimentary variables to the free-metal ion concentration in the ambient water near the sediment-water interface, and have then defined the contamination gradient in terms of $[M^{z+}]$. The present section focuses on such investigations.

All the studies considered in this section were performed in the field, at sites chosen to represent a metal contamination gradient (Table 5). Surficial oxic sediment samples were collected at each site and analyzed to define and quantify the contamination gradient from a geochemical point of view. Similarly, to evaluate the bioavailability of the sediment-associated metals, indigenous benthic organisms were collected and analyzed. The common goal of these studies was thus to relate the metal concentrations observed in the benthic organisms to the geochemistry of the host sediments. In each case, the geochemical gradient has been defined in terms of either the free-metal ion (as estimated

Table 5   Relationships Between Metal Bioavailability, as Sensed by Indigenous Benthic Organisms, and Geochemical Estimates of the Free-Metal Ion Concentration Present in the Oxic Sediment-Interstitial Water

Metal	Organism	Response	Geochemical predictor	Site[a]	Ref.
1. Cu Pb	Rooted freshwater aquatic plants, e.g., *Potomogeton richardsonii*	Metal bio-accumulation	{Fe–OM}/{Fe–ox}	Rouyn-Noranda ($N = 10$; $r^2 = 0.72$ for Cu; $r^2 = 0.62$ for Pb; $p < .01$)	Campbell and Tessier, 1991
2. Cd	Filter-feeding freshwater mollusc *Anodonta grandis*	Metal bio-accumulation	[Cd^{2+}], as estimated from oxic sediment-water equilibria	Rouyn-Noranda; Chibougamau; Eastern Townships, Quebec; Sudbury; Muskoka ($N = 19$; $r^2 = 0.82$; $p < .01$)	Tessier et al., 1993
3. Cd	Filter-feeding freshwater mollusc *Anodonta grandis*	Metallothionein induction	[Cd^{2+}], as estimated from oxic sediment-water equilibria	Rouyn-Noranda ($N = 11$; $r^2 = 0.56$; $p < .01$)	Couillard et al., 1993
4. Cu	Filter-feeding freshwater mollusc *Elliptio complanata*	Metal bio-accumulation	{Fe–OCu}/{Fe–ox}, both extracted with $NH_2OH \cdot HCl$	Rouyn-Noranda ($N = 8$; $r^2 = 0.95$; $p < .01$)	Tessier et al., 1984
5. Pb	Estuarine deposit feeder *Scrobicularia plana*	Metal bio-accumulation	{Fe–OPb}/{Fe–ox}, both extracted with HCl	U.K. estuaries ($N = 37$; $r^2 = 0.88$; $p < .01$)	Luoma and Bryan, 1978
6. Hg	Estuarine deposit feeder *Scrobicularia plana*	Metal bio-accumulation	{Hg}/{OM}, Hg extracted with HNO_3; organic matter determined by loss on ignition	U.K. estuaries ($N = 78$; $r^2 = 0.63$; $p < .01$)	Langston, 1982
7. As	Estuarine deposit feeder *Scrobicularia plana*	Metal bio-accumulation	{Fe–OAs}/{Fe–ox}, both extracted with HCl	U.K. estuaries ($N = 75$; $r^2 = 0.93$; $p < .01$)	Langston, 1980

[a] N = number of field sites.

from sediment-water equilibria, Section II.A) or the ratio of sorbed metal to sorbent. Provided that the ambient pH varies little along the contamination gradient, this latter ratio can be taken as a surrogate measure of the free-metal ion concentration (NRCC, 1988; Tessier and Campbell, 1990; Tessier et al., 1993; see Appendix 1).

Pathways for metal accumulation differ for the two types of benthic organisms represented in Table 5. Rooted aquatic plants live in intimate contact with the sediments but can only take up metals from the dissolved phase (sediment interstitial water and/or overlying water column). In the single study relating to this class of organism, regression analysis was used to examine the relationship between metal concentrations in the plant roots/rhizome ($[M]_r$)* and metal concentrations in the oxic sediment layer. The pH of the interstitial water close to the plant roots was assumed to be constant over the limited study region (a reasonable assumption given the known buffering capacity of lake sediments — Baker et al., 1985; Schindler et al., 1986); consequently the $[H^+]^n$ term was dropped from Equation 5 and relationships were sought between the metal levels in the plant (dependent variable) and the ratio {Fe–OM}/{Fe–ox} in the host sediment. The usefulness of this ratio as a predictor varied among metals

* For submerged rooted species in a field setting, the least ambiguous indicator of metal bioavailability in the sediments should be the metal concentrations in the underground parts of the plants (Schierup and Larsen, 1981; NRCC, 1988).

(Cu > Pb, Zn) and among plants (*Potamogeton richardsonii* > *Eleocharis smallii, Nuphar variegatum, Sparganium americanum* > *Glyceria borealis*) (see Campbell and Tessier, 1991). These preliminary results suggested that the pondweed *P. richardsonii* may prove to be useful as a biomonitor species for sediment-associated metals. However, to confirm the potential of this species and the usefulness of the expression $\{Fe\text{–}OM\}$ $[H^+]^n/\{Fe\text{–}ox\}$ as a surrogate for $[M^{z+}]$, it would be necessary to extend the study to a wider geological setting.

Benthic invertebrates are directly exposed to sediment-bound metals and may accumulate metals from the ambient water and/or from ingested food (Luoma, 1983; NRCC, 1988). If water is the more important exposure vector, $[M]_{benthos}$ should be correlated with the free-metal ion concentration in the ambient water ($[M^{z+}]$ or its surrogate).* This hypothesis has been tested for a variety of filter- and deposit-feeding invertebrates (Table 5). The general experimental approach has involved field studies at lacustrine or estuarine sites located along a spatial contamination gradient. Measured biological variables have included metal concentrations in the indigenous organisms and, in one case, metallothionein levels. Regression analysis was used to examine relationships between the biological response and either the free-metal ion (as estimated from sediment-water equilibria, Section II.A) or the ratio of sorbed metal to sorbent. Relationships between the biological response and these geochemical predictors were consistently highly significant ($p < .05$; Table 5). In contrast, correlations with total metal concentrations in the sediments were consistently weaker and often statistically insignificant (Figure 10).

The approach to metal bioavailability described in this section is based on the FIAM. It draws on surface complexation (SC) concepts (sorption) to predict the free-metal ion concentrations that should prevail in equilibrium with oxic sediments, and then assumes that the biological response should vary as a function of this concentration, $[M^{z+}]$. The success of the approach under field conditions, as demonstrated in Table 5, suggests that the underlying concepts may well be of general applicability. It should be pointed out, however, that all but one of the organisms represented in Table 5 live at the sediment-water boundary (as opposed to within the sediment, e.g., in burrows, where AVS might be expected to play a more important role). It remains to be seen if the FIAM-SC approach also applies to infaunal organisms whose behavior brings them into contact with suboxic sediments.

IV. INTERACTIONS BETWEEN PARTICULATE TRACE METALS AND AQUATIC ORGANISMS

Having dealt with dissolved metals (geochemical controls, bioavailability) in the previous two sections, we now focus on those metals that remain in

---

* Similar observations would be expected even if ingestion of food were the dominant exposure vector, provided metal concentrations in the ingested food were correlated with $[M^{z+}]$.

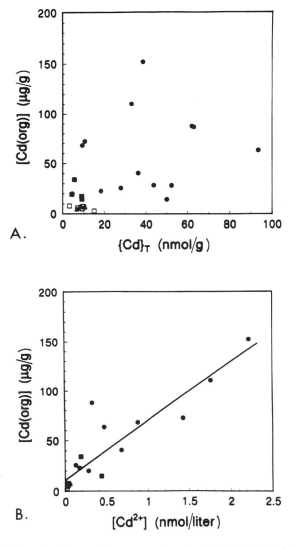

Figure 10 Relationship between Cd concentrations in the soft tissues of the freshwater
mollusc *Anodonta grandis* and the Cd concentrations in its habitat. The
molluscs were collected from 19 lakes chosen to represent a metal contamination gradient. The abscissa is expressed as (A) total cadmium in the
sediments (upper panel) or (B) the free dissolved Cd^{2+} concentration in the
overlying water (lower panel: $r^2 = 0.82$; [Cd(org)] = 59 ($\pm$7) [Cd^{2+}] + 11 ($\pm$18);
$n = 19$; $p < .01$). (Adapted from Tessier et al., 1993.)

particulate form, i.e., the direct route of exposure. The reasoning described
below applies to organisms for which particulate material constitutes an important vector for metal uptake. We describe possible links between digestive
processes in benthic organisms and the fate of ingested sediment-bound metals

in the intestinal tract, and review the results of laboratory experiments designed to determine the availability of metals present in different forms in the "food."

A. BACKGROUND INFORMATION

Benthic organisms that ingest particles tend to select and ingest the smaller and lighter particles in their environment. This nutritional strategy is "designed" to maximize the ingestion of organic carbon in the form of attached bacteria or surface-bound organic coatings, but it also results in the ingestion of fine-grained particles that tend to be enriched in metals.*

Assimilation of particle-bound metals will normally involve their conversion from particulate to dissolved form in the gut, followed by their facilitated diffusion across the intestinal membrane (Luoma, 1983). Digestive processes and the chemical conditions prevailing within the intestinal tract thus assume considerable importance (e.g., pH, digestion times, redox status). Digestive pH values tend to vary for different taxa and, within the gut, may vary longitudinally (Bärlocher and Porter, 1986; Dow, 1986). For lower trophic level invertebrates, such as those found in the benthic environment, circumneutral pH values are the norm: pH 5 is the lowest reported for the digestive tract of suspension feeders such as oysters (Owen, 1966), whereas the digestive pH of most deposit-feeding organisms is in the range 6 to 7 (Luoma, 1983; NRCC, 1988). More extreme pH conditions are observed in upper trophic level organisms. Intuitively, acidic conditions within the gut might be expected to favor metal desorption and dissolution, leading to higher concentrations of dissolved metals in the lumen. Counteracting this effect, however, will be the tendency of the hydrogen-ion, H^+, to compete with the dissolved metal for uptake sites on the intestinal membrane (Luoma, 1983; Campbell and Stokes, 1985). Indeed, the link between (low) digestive pH and (enhanced) metal uptake is weak — it would appear that while low pH may indeed extract more metal from the ingested particles, the membrane carriers in such an environment are less efficient in complexing the metal for transport (Luoma, 1983).

The residence time of the sediments in the digestive tract will also be of importance, since the desorption/dissolution processes involved in the digestive process may be slow. Longer digestive times would tend to favor more complete extraction of the metal. Extended residence times in the gut might also affect the redox status of the ingested sediment. Microorganisms associated with the ingested sediment will continue to consume oxygen — given sufficient time an originally oxic sediment might well find itself in an anaerobic environment, with obvious consequences for the amorphous Fe- and Mn-oxyhydroxides and their associated metals.

Endocytosis is a major digestive pathway in some benthic invertebrates, but the quantitative importance of this route in metal uptake is unknown. Endocytosis is the engulfment of a (metal-bearing) particle by the epithelial

---

* For example, Tessier et al. (1984) recovered the gut contents from two freshwater molluscs (*Elliptio complanata* and *Anodonta grandis;* discussed in Section III.C.ii), and showed that more than 90% of the particles were smaller than 80 μm.

membrane, which then pinches off and forms a membrane-limited vesicle within the cell. In a variation of this mechanism, specific amoebocytes and/or digestive vesicles may engulf the particles outside the cell membrane, e.g., in the lumen of the digestive tract or outside the gills, and then move back into the tissues carrying their particulate burden (George et al., 1978; Luoma, 1983). In many bivalves, particles are sorted in the stomach and then passed through a rapid "intestinal" digestion; a fraction of the food is further processed by a slower, intensive, largely intracellular "glandular" digestion (Decho and Luoma, 1994). Some metal forms are not assimilated during intestinal diges-tion but do become available to the organism during the glandular phase (Decho and Luoma, 1991). Note that although a metal may find itself "within" a tissue as a result of endocytosis, strictly speaking it has not been truly "assimilated" since it is still separated from the cytosol by the original vesicle membrane. Digestive processes within the vesicle might be expected to result in the desorption/solubilization of the metal and its subsequent movement from the vesicle into the cytosol, as discussed above for the intestinal tract. Although such a scenario seems plausible, it remains speculative because little is known about the chemical conditions within digestive vesicles or amoebocytes.

B. LABORATORY FEEDING EXPERIMENTS

Having described the conditions likely to prevail in the digestive tract of a typical benthic invertebrate, we can now ask the obvious question: do different metal forms behave differently in this internal environment? The same ques-tion might be formulated in a more general manner: is the bioavailability of sediment-bound metals affected by their partitioning among various solid phases?

To answer these questions one could, in principle, study metal bioaccumulation from different sediment phases, and indeed several such feeding experiments have been performed. The general approach in these experiments has been to (radio)label various "model" substrates, chosen to represent geochemical phases known to be present in natural sediments (e.g., organic detritus, amorphous Fe(III) or Mn(IV) oxyhydroxides, synthetic cal-cite, biogenic carbonate), and then to offer these substrates as "food" to a sediment-ingesting organism. Accumulation of the labeled metal within the test organism is then taken as evidence of the metal's "bioavailability" (Luoma and Jenne, 1976, 1977; Harvey and Luoma, 1985a,b; Decho and Luoma, 1994).

Though simple and direct, this approach is not without its drawbacks (Luoma, 1983, 1989). The most important of these is related to the tendency of the (radio)label to re-equilibrate with the solution phase. It is relatively straightforward to spike a model substrate with radiolabeled metal. However, as soon as the solid phase is recovered and placed in a "clean" aqueous system, the radiolabel will tend to redistribute between the model sediment and the aqueous phase. It follows that the test organism will be exposed not only to the particulate metal, but also to metal in solution. To evaluate the bioavailability of the particulate metal, it will be necessary to correct for any "accidental" uptake from solution. Similarly, metal redistribution among sediment phases may occur if one type of labeled phase is mixed with several types of unlabeled particles (Tessier and Campbell, 1988). Note too that the retention of undigested radiolabeled sediment in the digestive tract of the test organism may bias the results of whole-body metal analyses (Chapman, 1985; Hare et al., 1989).

Luoma and Jenne (1976, 1977) managed to minimize the influence of these factors in their studies of *Macoma balthica*, an estuarine deposit feeder that ingests both suspended particulates and surficial bottom sediments. In a series of elegant experiments, these authors exposed *M. balthica* to various model substrates, each of which had previously been radiolabeled with ^{110m}Ag, ^{60}Co, ^{65}Zn, or ^{109}Cd. For each experiment, a thin layer of labeled sediment of a single type was placed in an aquarium and covered with seawater. Eight clams were placed in the aquarium and allowed to feed on the sediments. To correct for uptake from solution, four clams were enclosed individually in dialysis bags and placed in the aquarium — the dialysis bag separated the enclosed animals from the sediment, but allowed exposure of both groups of animals to the same concentration of dissolved metal. The experiments were run for 14 d and at various intervals the organisms were removed from their microcosm, counted for radiolabel (nondestructively), and returned to their container. Nuclide concentrations in enclosed animals were subtracted from those in feeding animals to obtain an estimate of uptake from ingestion alone. Corrections were also applied to account for the presence of undigested radiolabeled substrate in the digestive tracts of the clams.

Significant bioaccumulation by ingestion alone was observed in the Jenne and Luoma experiments, but the efficiency of uptake varied greatly among different sediment types (Table 6). For example, uptake of silver varied over three orders of magnitude when normalized for the concentration of radionuclide in each substrate: uptake from calcite or Mn(IV) oxide was much greater than from the other model sediments (amorphous Fe(III) oxide, organic detritus). With the same suite of substrates, uptake of zinc varied over two orders of magnitude; in this case, uptake from biogenic carbonates and organic detritus was greater than from the Fe or Mn oxides. This difference between silver and zinc illustrates an important general conclusion from these feeding experiments, namely, that relative metal availability from a given sink varied from metal to metal.* The differences in metal availability were inversely related to the strength of metal binding to the particulates — sediments that exhibited the highest affinity for the metal (i.e., that released the least amount of metal back into solution) were also the substrates from which metal bioavailability was the lowest. Within the same sediment type (e.g., Fe(III)

* In other words, one cannot simply generalize that "metals" are more available from phase A (e.g., organic detritus) than from phase B (e.g., Mn(IV) oxyhydroxide); while the sequence A>B may be true for one metal, it will not necessarily hold for the next.

Table 6   Assimilation of Sediment-Bound Metals by the Estuarine Deposit-Feeder, *Macoma balthica* — Summary of Controlled Feeding Experiments with Radiolabeled (Model) Sediments

Metal	Relative Bioavailability Sequence	Reference
Cd	Uncoated Fe–ox $\gg$ coated Fe–ox, organic detritus	Luoma and Jenne, 1976
Ag	Calcite > Mn-ox $\gg$ biogenic $CaCO_3$ > Fe–ox > detritus	Luoma and Jenne, 1977
Zn	Biogenic $CaCO_3$ > detritus > calcite > Fe-ox, Mn-ox	
Co	Biogenic $CaCO_3$ > calcite ~ detritus > Fe-ox > Mn-ox	
Cd	(Exopolymer + Fe-ox) > (bacteria + Fe-ox) ~ uncoated Fe-ox	Harvey and Luoma,1985a
	Natural sediment $\gg$ alkaline extracted sediment	
	Extracted sediment + exopolymer ~ original sediment	
Zn	(Exopolymer + Fe-ox) slightly > uncoated Fe-ox	
	Natural sediment $\gg$ alkaline extracted sediment	
	Extracted sediment + exopolymer ~ original sediment	

oxide), metal bioavailability from ingested particles appeared to be sensitive to the degree of crystallinity of the model sediment. Uptake of radiolabel from freshly precipitated Fe_2O_3 (3 h) was much greater than from slightly aged oxyhydroxides (16 to 24 h).

A similar experimental design was employed by Harvey and Luoma (1985a,b) in their more recent studies on *Macoma balthica*. Instead of individual dialysis bags, a filter chamber device was used to separate the feeding from the nonfeeding clams. Their experiments were designed to quantify the effects of adherent bacteria and bacterial extracellular polymer ("exopolymer") on the uptake of particle-bound metals. Earlier experiments had shown that adherent bacteria were an important food source for *M. balthica*, but that the bacterial exopolymer (used in the process of bacterial adhesion) was not assimilated in the time frame of the feeding experiments. Various radiolabeled test sediments were used (Table 6): amorphous Fe(III) oxyhydroxides (Fe–ox); amorphous Fe(III) oxyhydroxides to which either *Pseudomonas atlantica*, an adherent bacterium, or its exopolymer had been added; unaltered San Francisco Bay sediments; sediments that had been extracted with an alkaline solution to remove the fulvic/humic acid component; extracted sediments that had been recoated with bacterial exopolymer.

As in the earlier experiments, amounts of metal taken up from a given model sediment varied greatly from one metal to another. For example, ingestion of radiolabeled but uncoated iron oxyhydroxides did not contribute appreciably to uptake of Ag or Cd, but accounted for 89 to 99% of Zn uptake. Similarly, amounts of metal taken up from ingested sediment were sensitive to the nature of the particle surface. For example, exopolymer adsorbed onto the Fe-ox surface caused an increase in the availability of particle-bound metals in the order Ag > Cd > Zn. Adherent bacteria had no effect on metal uptake, even though Cd and Zn were taken up from suspensions of bacteria that had been prelabeled with metal (Harvey and Luoma, 1985b).

In the trials with labeled sediments from San Francisco Bay, ingested particles contributed more than 50% of the Cd uptake. In marked contrast, uptake from alkaline-extracted sediments was negligible. Interestingly, addition of bacterial exopolymer to the extracted sediments restored the bioavailability of sediment-bound cadmium (note the parallel with the effect noted earlier for amorphous Fe(III) oxyhydroxides amended with exopolymer). A similar but less dramatic effect was noted for zinc.

Harvey and Luoma (1985a) were unable to explain how the alteration of inorganic particles with bacterial exopolymer enhanced the availability of particle-bound Ag, Cd, and Zn to *M. balthica*. They discounted the possibility that the exopolymer simply affected the affinity of the particles for the metal, or that feeding rates changed in the presence of coated particles. They suggested instead that the presence of exopolymer may have stimulated the secretion of enzymes capable of competing with Fe–ox for metals in the clam's digestive tract. The observation that adsorbed bacterial exopolymer and NaOH-extractable organics had a similar effect on the availability of Cd and Zn from ingested natural sediments suggests that the organic component of a natural sediment may have a marked influence on the biological availability of metals.

In a recent study, Decho and Luoma (1994) have extended this approach to look, not at organic polymers of bacterial origin, but rather at humic and fulvic acids extracted from marine sediments. The humic and fulvic acids were bound as organic coatings either to iron oxyhydroxides or to silica particles. Radiolabeled $^{109}Cd(II)$ and $^{51}Cr(III)$ were then adsorbed to the organic coatings, or directly to the original uncoated inorganic phases, and pulse-chase feeding experiments carried out with two estuarine bivalves: the suspension feeder *Potamocorbula amurensis* and the facultative deposit feeder *Macoma balthica*. Absorption efficiencies for Cr(III) were consistently low (<11%) for all types of nonliving particles. Humic organic coatings slightly enhanced Cr bioavailability to *P. amurensis* and *M. balthica*, but, given the generally low absorption efficiencies for all these forms of Cr, Decho and Luoma concluded that little of this element would be transferred from any of these particle types to detritus-feeding bivalves. The absorption efficiencies for Cd were generally higher than for Cr, ranging from 9 to 56%. Both bivalves absorbed Cd bound to uncoated iron oxyhydroxide or silica particles (i.e., with no humic acid or fulvic acid present); the presence of organic coatings on particles reduced Cd bioavailability compared to the uncoated particles.

Based on the studies described above, it is tempting to conclude that the physicochemical form of a sediment-bound metal will affect its availability in the digestive tract of a given benthic invertebrate. However, it should be emphasized that all but one of the studies were performed on the same deposit feeder, *M. balthica*, and as the authors themselves point out, it would be premature to attempt to generalize these results to all deposit feeders.

The results of the feeding experiments are admittedly qualitative and thus are not yet very useful from a quantitative or predictive point of view. Nevertheless, there is an interesting parallel between Luoma and Jenne's "affinity" concept, whereby they suggest that the availability of a particle-bound metal will be inversely related to the strength of the metal-particle association (Luoma and Jenne, 1976,1977) and the more recent suggestion by Di Toro et al. (1990) that the concentration of the metal in the interstitial water can be taken as a measure of its "chemical potential" in surficial sediments and thus its availability. This inference seems intuitively reasonable, provided that the chemical environment within the animal's digestive tract is similar to that in its immediate environment (i.e., provided the chemical potential of the metal does not change drastically on passing from the outside environment into the digestive system). In this context, better knowledge of the chemistry of invertebrate digestion would be very useful.*

V. ECOTOXICOLOGICAL CONSIDERATIONS — METALLOTHIONEIN AS A BIOCHEMICAL INDICATOR OF METAL-INDUCED STRESS

A. BACKGROUND INFORMATION

Even in contaminated lakes, the upper sediment layer is normally the habitat for a diverse and often abundant benthic community. The contamination of surficial sediments by metals of anthropogenic origin is thus of potential ecological importance. Two related questions have attracted the attention of researchers and regulators: (1) To what extent are the sediment-bound metals bioavailable? and (2) Are there demonstrable adverse effects on the benthic community, or on the predatory wildlife that depend on this community, that can be attributed to metal contamination? With regard to question 1, potentially toxic metals can attain very high levels in benthic organisms living in contaminated sediments (particularly in the invertebrate community). Considerable progress has been made towards the development of predictive models that relate metal body burdens in benthic invertebrates to the geochemical conditions prevailing in the host sediments (NRCC, 1988; see Section III.C.ii), and partial responses to question 1 are already available. However, as discussed below, benthic invertebrates possess a variety of metal detoxification mechanisms, and metal bioaccumulation per se is not necessarily an indication of adverse effects. The response to question 2 thus remains unclear.

Traditionally, attempts to define the impacts of contaminants on aquatic ecosystems have involved laboratory experiments under defined conditions (toxicity tests) and, to a lesser extent, field observations on impacted indigenous

---

* Reinfelder and Fisher (1991) studied the efficiency with which various ingested elements were assimilated by marine calanoid copepods (zooplankton). These authors fed their animals radiolabeled diatoms rather than sediments, but their results were nevertheless pertinent: assimilation efficiencies ranged from 0.9 to 97%, and were directly related to the cytoplasmic content of the diatoms. Reinfelder and Fisher speculate that these zooplankton, with their short gut residence times, have developed a gut lining and digestive strategy that only provide for assimilation of soluble material.

populations. Extrapolation of laboratory-derived toxicological data to the field is fraught with difficulties (Kimball and Levin, 1985; Cairns, 1986), as is the unambiguous interpretation of field observations (Ford, 1989). An alternative and complementary approach involves the use of biochemical indicators to monitor the response of individual organisms to toxic chemicals and to provide a measure of organism health (see Shugart, Chapter 5, this volume).

B. BIOCHEMICAL INDICATORS

The biochemical indicator concept is based on the principle that biological effects of toxic chemicals in the environment are initiated by the interaction of the toxic chemical with a biological receptor in a living organism (NRCC, 1985; Huggett et al., 1992). The assumption is made that effects at the ecosystem level are preceded by chemical reactions in individual organisms, and that concentrations of the contaminant needed to initiate these reactions are lower than those required to provoke a life-threatening situation for the target organism or perceptible degradation of the ecosystem. The detection and quantification of these chemical reactions could then be developed as a sensitive, specific indicator of environmental stress.

A successful biochemical indicator should satisfy a number of criteria (Stegeman et al., 1992; Haux and Förlin, 1989):

- The indicator should have an early warning capacity, i.e., the biochemical response should be predictive of effects at higher levels of biological organization and should precede them.
- The indicator should be specific to a particular contaminant or for a class of contaminants.
- The indicator should respond in a concentration-dependent manner to changes in ambient levels of the contaminant.
- The basic biology/physiology of the biomonitor organism should be known so that sources of uncontrolled variation (growth and development, reproduction, food sources) can be minimized.
- Levels of the indicator should be related to the health or "fitness" status of the organism.

For metals, much of the attention in the area of biochemical indicators has focused on metal-binding proteins, in particular on metallothionein (MT) and metallothionein-like compounds. Metallothionein, a low-molecular-weight, soluble, metal-binding protein or group of proteins, has been identified in mammals, invertebrates, plants, and prokaryotes (Roesijadi, 1981, 1992; NRCC, 1985; Engel and Roesijadi, 1987). Much of the early research on the structure and function of MT was performed on mammals, but over the last 10 to 15 years marine invertebrates have been studied intensively and the role of MT in their metal metabolism has been clarified (Engel and Roesijadi, 1987; Roesijadi, 1992). In distinct contrast, freshwater invertebrates have been studied much less intensively. Metallothioneins isolated from marine invertebrates were initially thought to possess properties that differed slightly from those of mammalian MT (e.g., lower cysteine content, presence of low levels of aromatic amino acids, propensity to form dimers). However, the low cysteine content and the presence of aromatic compounds have been shown, for invertebrate MTs that have been properly purified and characterized, to be artifacts (G. Roesijadi, University of Maryland, personal communication). In the present context, we shall use the term metallothionein (MT) in its generic sense.

In marine invertebrates, MT has been implicated in the storage, transport, and exchange of essential metals (Cu, Zn), and in the detoxification of these and nonessential metals (Ag, Cd, Hg). As a soluble protein present in the cytosol, with a demonstrably high *in vitro* affinity for such toxic metals as Ag, Cd, and Hg, MT is poised to sequester incoming metals and reduce their availability to critical biochemical sites within the cell (Roesijadi, 1981). In such a scheme, as applied to (marine) invertebrates, MT would be involved in the initial detoxification of the incoming metal, and would play a role in funnelling the metal to such complementary sinks as lysosomes and granules /concretions (Viarengo, 1985, 1989).

Support for the putative role of MT in metal detoxification is derived from the observation that exposure to elevated concentrations of Cd, Cu, Hg, or Zn often induces the synthesis of MT; in the case of Cd and Hg, the toxic metals have been reported to displace the essential metals normally associated with MT. Consistent with such a role is the observation of enhanced metal tolerance associated with the induction of MT. For example, prior exposure of the marine mussel *Mytilus edulis* in the laboratory to Cu or Cd, at concentrations sufficiently high to induce MT-like proteins, conferred increased tolerance to the toxicity of inorganic mercury (Roesijadi and Fellingham, 1987). Similar results have been reported for freshwater fish (Klaverkamp et al., 1984; Roch and McCarter, 1984a; Klaverkamp and Duncan, 1987).

In addition to its apparent role in metal detoxification, as suggested by the correlation between acclimation to metal toxicity and induction of MT synthesis, MT may also be involved in the homeostatic control of such essential metals as Cu and Zn. For example, MT has been reported as a constitutive (Cu, Zn) protein in organisms collected from pristine sites, and it has been shown to play a central role in the molting cycle of marine crustaceans (Engel and Brouwer, 1987). As Petering and Fowler (1986) point out, if the protein has normal functions to play in cells, then these might be disrupted when a toxic metal such as Cd binds to it — binding of Cd to basal MT, below levels of exposure that induce extra MT, might hinder the normal function of the protein in intracellular ligand-exchange chemistry. In other words, it is not clear that binding of toxic metals to MT is purely protective and is not itself a deleterious event.

C. METALLOTHIONEIN AS A MONITORING TOOL

Given its molecular properties and present knowledge of its role in metal uptake, transport, storage, and excretion, MT offers considerable potential as a contaminant-specific biochemical indicator of metal exposure and/or stress (Roesijadi, 1992; Stegeman et al., 1992). Possible approaches include:

- Direct measurement of [MT] as a simple indicator of prior exposure to toxic metals. In this case, it is assumed that constitutive levels of MT are low, and that any increase in concentration above these low basal levels is attributable to the induction of MT in response to an influx of toxic metals. Such an approach would, in principle, furnish a measure of the toxicologically significant intracellular fraction of metals (Olafson et al., 1979).
- Examination of the relative distribution of toxic metals in cytosolic ligand pools as a means of evaluating metal stress at the biochemical level. It has been suggested that excessive accumulation of metals beyond the binding capacity of available MT should result in their binding to other intracellular ligands (notably those of high molecular weight, HMW), a phenomenon termed "spillover" — metals bound to these other ligands are considered to be capable of exerting cellular toxicity (Brown and Parsons, 1978). In principle, this condition could be considered as symptomatic of metal stress and would be amenable to detection.

Field evidence in support of the first approach is slowly accumulating in the ecotoxicological literature. All studies summarized in Table 7 involved the sampling of indigenous populations of aquatic animals from sites chosen to represent a spatial (metal) contamination gradient. In all cases, the concentrations of metallothionein-like proteins did indeed prove to be higher at the more contaminated sites. Note, however, that in most cases the contamination gradient was defined only in a qualitative sense (e.g., upstream and downstream from a known point source of metals, or as a function of distance from a base metal smelter) and a true dose-response relation could not be derived. In only two cases (Roch et al., 1982; Couillard et al., 1993) were strict relations demonstrated between tissue MT concentrations and ambient metal levels.

In the studies of organisms living along a contamination gradient, tissue MT concentrations were often positively correlated with tissue Cd levels (Table 7, Figure 11). This correlation between [MT] and tissue levels of Cd is consistent with the reported potency of Cd in inducing MT biosynthesis in laboratory experiments (e.g., Roesijadi et al., 1988)* and suggests that MT levels will be particularly sensitive to increases in ambient Cd.

In addition to spatial trends in tissue MT concentrations, several workers have demonstrated that MT levels also respond temporally to changes in the degree of metal contamination (Table 8). For example, Roch and McCarter (1984b) kept hatchery-raised rainbow trout in pens at three contaminated sites in the Campbell River drainage system (May–June 1982: S. Buttle Lake, N. Buttle Lake, John Hart Lake): elevated levels of hepatic MT had developed after four weeks exposure *in situ*. In a companion study, the levels of hepatic

* Jones et al. (1988) showed that the relative ability of metals to induce metallothionein synthesis is inversely correlated with their softness parameter, ρ. A "soft" electron acceptor is characterized by a high polarizability of its outer electronic shell and a tendency to form stable bonds with "soft" ligands, e.g., those containing free thiol groups, RS^-. Lower values of ρ correspond to softer ions; ρ values for Cd^{2+}, Cu^{2+}, and Zn^{2+} are, respectively, 0.081, 0.104, and 0.115 (Ahrland, 1968).

Table 7   Field Validation of the Use of Metallothionein as a Biochemical
Indicator of Exposure to Bioavailable Metals — Spatial Studies
(N = Number of Sites)

Field site	Sentinel organism	Tissue	Metal gradient	Result	Ref.
Campbell River Watershed, British Columbia (N = 4)	Rainbow trout (*Oncorhynchus mykiss*)	Liver	Zn, Cu, (Cd) Defined in terms of dissolved metal, $[M]_d$	Hepatic metallothionein increased ~4-fold in the indigenous trout populations along the contamination gradient; [MT] correlated with hepatic Cu or Cd	Roch et al., 1982
Emån River, Sweden (N = 2)	Perch (*Perca fluviatilis*)	Liver	Cd Gradient undefined (up- and downstream from point source)	Hepatic metallothionein higher at downstream site; [MT] correlated with hepatic Cd in individual specimens (N = 20)	Olsson and Haux, 1986
Flin Flon Lakes, Manitoba (N = 6)	White sucker (*Catastomus commersoni*)	Liver, kidney	Cd, Cu, Zn Defined in terms of distance from smelter	Hepatic and renal metallothionein inversely correlated with distance from smelter	Klaverkamp et al., 1991
Rouyn-Noranda Lakes, Quebec (N = 11)	Freshwater mollusc (*Anodonta grandis*)	Gills; whole organism	Cd Defined in terms of $[Cd^{2+}]$ at sediment-water interface	Metallothionein increased 2.5- to 4-fold in the indigenous mollusc populations along the contamination gradient; [MT] correlated with tissue Cd	Couillard et al., 1993
Canadian Atlantic coast (N = 7)	Various seabirds	Kidney	Gradient undefined	[MT] and [Cd] correlated in the kidneys of Leach's storm-petrels, Atlantic puffin, and herring gull	Elliot et al., 1992

Table 8   Field Validation of the Use of Metallothionein as a Biochemical Indicator
of Exposure to Bioavailable Metals — Temporal Studies

Field site	Sentinel organism	Tissue	Metal gradient	Result	Ref.
Campbell River Watershed, British Columbia (Buttle Lake; John Hart Lake) (exposure period = 4 weeks)	Rainbow trout (*Oncorhynchus mykiss*)	Liver	Zn, Cu, (Cd) Defined in terms of $[M]_d$	Hepatic metallothionein increased ~3-fold in the trout held for 4 weeks in net pens at 3 locations contaminated by metals; after 4 weeks exposure *in situ,* hepatic [MT] was correlated with degree of contamination as measured by $[Zn]_d$	Roch and McCarter, 1984b
Campbell River Watershed, British Columbia (Buttle Lake) (exposure period = 4 years)	Rainbow trout (*Oncorhynchus mykiss*)	Liver	Zn, Cu, (Cd) Defined in terms of $[M]_d$	Hepatic metallothionein decreased ~4-fold in the indigenous trout populations from 1981 to 1985, presumably in response to the decrease in ambient dissolved metal levels	Deniseger et al., 1990
Rouyn-Noranda, Quebec (Lake Vaudray) (exposure period = 400 d)	Freshwater mollusc (*Anodonta grandis*)	Gills; mantle; whole organism	Cd Defined in terms of $[Cd^{2+}]$ at sediment-water interface	Metallothionein increased 2.5-fold over the first 400 d in molluscs transferred from control lake to highly contaminated Lake Vaudray; increase in tissue [MT] correlated with increase in tissue Cd	Couillard et al., 1995a,b

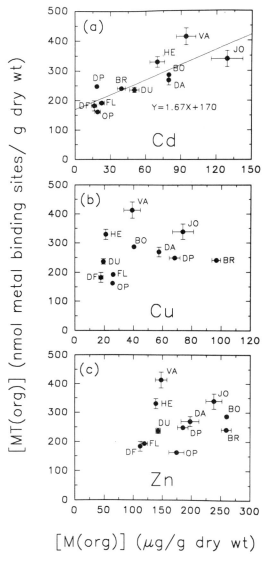

Figure 11 Relationships between metallothionein (MT) concentrations in the soft tissues of the freshwater mollusc *Anodonta grandis* and tissue concentrations
of Cd, Cu, and Zn. The molluscs were collected from 11 lakes chosen to
represent a metal contamination gradient. Pearson correlation coefficients
between whole organism MT concentrations and whole organism [Cd], [Cu],
and [Zn] are, respectively, 0.83 ($n = 11$; $p < .01$), 0.22 ($n = 11$; $p > .05$), and
0.21 ($n = 11$; $p > .05$). (Adapted from Couillard et al., 1993.)

MT in the indigenous Buttle Lake trout population were monitored over the period 1981 to 1985, i.e., as the ambient metal levels declined in response to changes in the tailings disposal methods. Hepatic MT concentrations decreased over the same period from a high of 269 (±23) nmol/g wet weight in August 1981 to a low of 64 (±22) nmol/g wet weight in June 1985, this latter value being comparable to levels found in fish from control lakes (Deniseger et al., 1990).

As for the second possible use of MT, i.e., as a means of evaluating metal stress at the biochemical level, the responses of aquatic organisms to excess metal have proven rather more diverse than originally postulated. In some cases the initial binding is not to MT but to the HMW fraction (Roesijadi, 1982; Langston and Zhou, 1987), in other cases both the HMW and the MT fractions increase in parallel (Roesijadi et al., 1982), and in still other examples the onset of toxic effects appears to be correlated not with the appearance of excess metals in the HMW fraction but rather with the detection and subsequent increase of metals in the very low molecular weight (VLMW) fraction (Sanders and Jenkins, 1984). The use of the spillover model in a diagnostic sense is thus premature; any such application must await specific information on the metal detoxification mechanisms that are operative in the proposed biomonitor species (Engel and Roesijadi, 1987; Roesijadi, 1992, 1994).

D. SUMMARY

Metallothioneins offer a number of advantages as biomarkers for exposure to toxic metals in the (aquatic) environment (Stegeman et al., 1992), e.g., presence of an extensive and increasing scientific information base, availability of sensitive analytical methods, concentration-dependent response to changes in ambient levels of (soft) metal cations. However, the biological functions of MT are not fully understood and it is not yet possible to link changes in MT status unambiguously to injury at the cellular or organismal level. In order to exploit MT as a predictive indicator of toxicity, it will be necessary to expand our knowledge base relative to its normal physiological function. The basic biology/physiology of the proposed biomonitor organism(s) must be known so that sources of variation other than metal exposure (e.g., growth and development, reproduction, food sources) can be taken into account.

ACKNOWLEDGMENTS

This chapter evolved from a report on possible means of evaluating the biological effects of subaqueous disposal of mine tailings, written for the Rawson Academy of Aquatic Science. The financial support of the Academy is gratefully acknowledged, as are the helpful comments of the liaison officer, Dr. Peter Sly. Research in our own group on the ecotoxicology of metals in the aquatic environment has been supported by grants and contracts from a variety of sources: the Natural Sciences and Engineering Research Council of Canada, Fisheries and Oceans Canada, Environment Canada, Environment Québec, and the Fonds pour la Formation de Chercheurs et l'Aide à la Recherche du Québec.

APPENDIX 1 — DERIVATION OF THE RELATIONSHIP BETWEEN DISSOLVED METAL CONCENTRATIONS AND METAL SORBED ON SURFICIAL SEDIMENTS

According to surface complexation theory, sorption of metal M on a solid substrate such as iron oxyhydroxide can be expressed in a simplified manner as:

$$\equiv FeOH_n + M^{z+} \overset{*K_i}{\longleftrightarrow} \equiv Fe - OM + nH^+ \tag{A1}$$

$$*K_i = \frac{\{\equiv Fe - OM\}[H^+]^n}{\{\equiv FeOH_n\}[M^{z+}]} \tag{A2}$$

where charges on the solid species are omitted for simplicity; $FeOH_n$ = sorbent (amorphous Fe–ox); $*K_i$ is an apparent overall equilibrium constant; "$\equiv$" refers to sorption sites, either free ($\equiv Fe-OH_n$) or occupied by M; and n = the apparent average number of protons released per metal ion sorbed.

At low sorption densities, i.e., when the concentration of occupied sorption sites is low relative to the free site concentration, the condition $\{\equiv Fe-OH_n\} \approx \{\equiv Fe-O-\}_T$ should apply, where $\{\equiv Fe-O-\}_T$ is the total concentration of sorption sites. This term can in turn be related to the analytical concentration of the sorbent,

$$\{\equiv Fe - O-\}_T = N_S\{Fe - ox\} \tag{A3}$$

where $\{Fe–ox\}$ = the analytical concentration of iron oxyhydroxide; N_S = the number of moles of sorption sites per mole Fe–ox. If it is further assumed that each sorbed metal ion occupies a single site, then

$$\{\equiv Fe - OM\} = \{Fe - OM\} \tag{A4}$$

Combining equations (A2), (A3), and (A4) yields

$$[M^{z+}] = \frac{\{Fe - OM\}}{\{Fe - ox\}} \cdot \frac{[H^+]^n}{N_{Fe} * K_i} \tag{A5}$$

which shows how one might calculate the free-ion concentration, $[M^{z+}]$, provided that the values of "$N_{Fe}*K_i$" and "n" are known. The symbols { } and [ ] refer to concentrations of solids and dissolved species, respectively. Note that if the ambient pH were relatively constant, as would normally be the case for the roots of an aquatic plant, the term $[H^+]^n$ would be constant in Equation A5 and the ratio {Fe–OM}/{Fe–ox} could be used directly to estimate $[M^{z+}]$.

To estimate $N_{Fe}*K_a$ and n, Equation A5 can be rearranged as follows:

$$K_{Fe-M} = \frac{N_{Fe} \cdot *K_i}{[H^+]^n} = \frac{\{Fe - OM\}}{\{Fe - ox\}[M^{z+}]} \tag{A6}$$

where K_{Fe-M} is an apparent equilibrium constant which is a function of pH. Values of {Fe–OM} and {Fe–ox}, needed to estimate K_{Fe-M}, can be obtained by selective extraction of the surficial sediments; values of $[M^{z+}]$ can be obtained by *in situ* dialysis (pore-water peepers) (see Tessier, 1992 and Tessier et al., 1993).

Equation A6 can then be linearized

$$\log K_{Fe-M} = n \ pH + \log(N_{Fe} \cdot *K_a) \tag{A7}$$

and a plot of log K_{Fe-M} vs. pH should yield a slope of n and a Y-intercept of log $(N_{Fe} \cdot *K_a)$. Values of "n" and "log $(N_{Fe} \cdot *K_a)$" obtained in this manner are summarized in Table 2.

Similar reasoning applies to other sorbents, such as sedimentary organic matter (Tessier et al., 1993) and manganese oxyhydroxides.

REFERENCES

Abu-Hilal, A., M. Badran and J. de Vaugelas, 1988. Distribution of trace elements in *Callichirus laurae* burrows and nearby sediments in the Gulf of Aqaba, Jordan (Red Sea). Mar. Environ. Res. 25, 233–248.

Adams, W.J., R.A. Kimerle and J.W. Barnett, 1992. Sediment quality and aquatic life assessment. Environ. Sci. Technol. 26, 1864–1875.

Ahrland, S., 1968. Thermodynamics of complex formation between hard and soft acceptors and donors. Struct. Bonding 5, 118–149.

Aller, R.C., 1978. Experimental studies of changes produced by deposit feeders on pore water, sediment and overlying water chemistry. Am. J. Sci. 278, 1185–1234.

Aller, R.C. and J.Y. Yingst, 1978. Biogeochemistry of tube dwellings: a study of the sedentary polychaete *Amphitrite ornata* (Leidy). J. Mar. Res. 36, 201–254.

Ankley, G.T., V.R. Mattson, E.N. Leonard, C.W. West and J.L. Bennett, 1993. Predicting the acute toxicity of copper in freshwater sediments: evaluation of the role of acid-volatile sulfide. Environ. Toxicol. Chem. 12, 315–320.

Ankley, G.T., G.L. Phipps, E.N. Leonard, D.A. Benoit, V.R. Mattson, P.A. Kosian, A.M. Cotter, J.R. Dierkes, D.J. Hansen and J.D. Mahony, 1991. Acid-volatile sulfide as a factor mediating cadmium and nickel bioavailability in contaminated sediments. Environ. Toxicol. Chem. 10, 1299–1307.

Ankley, G.T., N.A. Thomas, D.M. Di Toro, D.J. Hansen, J.D. Mahony, W.J. Berry, R.C. Swartz, R.A. Hoke, A.W. Garrison, H.E. Allen and C.S. Zarba, 1994. Assessing potential bioavailability of metals in sediments: a proposed approach. Environ. Manag. 18, 331–337.

Baker, L.A., P.L. Brezonik, E.S. Edgerton and R.W. Ogburn, 1985. Sediment acid neutralization in softwater lakes. Water Air Soil Pollut. 25, 215-230.

Bärlocher, F. and C.W. Porter, 1986. Digestive enzymes and feeding strategies of three stream invertebrates. J. North Am. Benthol. Soc. 5, 58–66.

Belzile, N., R.R. De Vitre and A. Tessier, 1989. In situ collection of diagenetic iron and manganese oxyhydroxides from natural sediments. Nature 340, 376–377.

Bernhard, M., F.E. Brinckman and K.J. Irgolic, 1986. Why "speciation"?. In: The Importance of Chemical "Speciation" in Environmental Processes, edited by M. Bernhard, F.E. Brinckman and K.J. Irgolic, Springer-Verlag, Berlin, pp. 7–14.

Brown, D.A. and T.R. Parsons, 1978. Relationship between cytoplasmic distribution of mercury and toxic effects to zooplankton and chum salmon (*Oncorhynchus keta*) exposed to mercury in a controlled ecosystem. J. Fish. Res. Board Can. 35, 880–884

Burton, G.A. and K.J. Scott, 1992. Sediment toxicity evaluations — their niche in ecological assessments. Environ. Sci. Technol. 26, 2068–2075.

Cairns, J., Jr., 1986. The myth of the most sensitive species. BioScience 36, 670–672.

Campbell, P.G.C., 1995. Interactions between trace metals and aquatic organisms: a critique of the Free-Ion Activity Model. In: Metal Speciation and Bioavailability in Aquatic Systems, edited by A. Tessier and D.R. Turner, John Wiley & Sons, New York, Chapter 2, pp. 45–102.

Campbell, P.G.C. and P.M. Stokes, 1985. Acidification and toxicity of metals to aquatic biota. Can. J. Fish. Aquat. Sci. 42, 2034–2049.

Campbell, P.G.C. and A. Tessier, 1989. Geochemistry and bioavailability of trace metals in sediments. In: Aquatic Ecotoxicology, edited by A. Boudou and F. Ribeyre, CRC Press, Boca Raton, FL, pp. 125–148.

Campbell, P.G.C. and A. Tessier, 1991. Biological availability of metals in sediments: analytical approaches. In: Heavy Metals in the Environment, edited by J.-P. Vernet, Elsevier, Amsterdam, pp. 161–173.

Carlson, A.R., G.L. Phipps, V.R. Mattson, P.A. Kosian and A.M. Cotter, 1991. The role of acid volatile sulfide in determining cadmium bioavailability and toxicity in freshwater sediments. Environ. Toxicol. Chem. 10, 1309–1319.

Casas, A.M. and E.A. Crecelius, 1994. Relationship between acid volatile sulfide and the toxicity of zinc, lead and copper in marine sediments. Environ. Toxicol. Chem. 13, 529–536.

Chapman, P.M., 1985. Effects of gut sediment contents on measurements of metal levels in benthic invertebrates — a cautionary note. Bull. Environ. Contam. Toxicol. 35, 345–347.

Couillard, Y., P.G.C. Campbell and A. Tessier, 1993. Response of metallothionein concentrations in a freshwater bivalve (*Anodonta grandis*) along an environmental cadmium gradient. Limnol. Oceanogr. 38, 299–313.

Couillard, Y., P.G.C. Campbell, A. Tessier, J.C. Auclair and J. Pellerin–Massicotte, 1995a. Field transplantation of a freshwater bivalve, *Pyganodon grandis*, across a metal contamination gradient. I. Temporal changes of metallothionein and metal (Cd, Cu, Zn) concentrations in soft tissues. Can. J. Fish. Aquat. Sci. 52(4), 690–702.

Couillard, Y., P.G.C. Campbell, J. Pellerin-Massicotte and J.C. Auclair, 1995b. Field transplantation of a freshwater bivalve, *Pyganodon grandis*, across a metal contamination gradient. II. Metallothionein response to metal exposure, evidence for cytotoxicity, and links to effects at higher levels of biological organization. Can. J. Fish. Aquat. Sci. 52(4), 703–715.

Daskalakis, K.D. and G.R. Helz, 1992. Solubility of CdS (greenockite) in sulfidic waters at 25°C. Environ. Sci. Technol. 26, 2462–2468.

Daskalakis, K.D. and G.R. Helz, 1993. The solubility of sphalerite (ZnS) in sulfidic solutions at 25°C and 1 atm pressure. Geochim. Cosmochim. Acta 57, 4923–4931.

Davis, W.R., 1983. Sediment-copper reservoir formation by the burrowing polychaete *Nephtys incisa*. In: Wastes in the Ocean, Vol. II: Dredged Material Disposal in the Ocean, edited by D.R. Kester, B.H. Ketchum, I.W. Duedall and P.K. Park, John Wiley & Sons, New York, pp. 173–184.

Davison, W., 1991. The solubility of iron sulphides in synthetic and natural waters at ambient temperature. Aquat. Sci. 53/54, 309–329.

Decho, A.W. and S.N. Luoma, 1991. Time-courses in the retention of food material in the bivalves *Potamocorbula amurensis* and *Macoma balthica*: significance to the absorption of carbon and chromium. Mar. Ecol. Prog. Ser. 78, 303–314.

Decho, A.W. and S.N. Luoma, 1994. Humic and fulvic acids: sink or source in the availability of metals to the marine bivalves *Macoma bathica* and *Potamocorbula amurensis*? Mar. Ecol. Prog. Ser. 108, 133–145.

Deniseger, J., L.J. Erickson, A. Austin, M. Roch and M.J.R. Clark, 1990. The effects of decreasing heavy metal concentrations on the biota of Buttle lake, Vancouver Island, British Columbia. Water Res. 24, 403–416.

Di Toro, D.M., J.D. Mahony, D.J. Hansen, K.J. Scott, A.R. Carlson and G.T. Ankley, 1992. Acid volatile sulfide predicts the acute toxicity of cadmium and nickel in sediments. Environ. Sci. Technol. 26, 96–101.

Di Toro, D.M., J.D. Mahony, D.J. Hansen, K.J. Scott, M.B. Hicks, S.M. Mayr and M.S. Redmond, 1990. Toxicity of cadmium in sediments: the role of acid volatile sulfide. Environ. Toxicol. Chem. 9, 1487–1502.

Dow, J.A.T., 1986. Insect midgut function. Adv. Insect Physiol. 19, 187–328.

Elliot, J.E., A.M. Scheuhammer, F.A. Leighton and P.A. Pearce, 1992. Heavy metal and metallothionein concentrations in Atlantic Canadian seabirds. Arch. Environ. Contam. Toxicol. 22, 63–73.

Engel, D.W. and M. Brouwer, 1987. Metal regulation and molting in the blue crab, *Callinectes sapidus*: metallothionein function in metal metabolism. Biol. Bull. 173, 239–251.

Engel, D.W. and G. Roesijadi, 1987. Metallothionein: a monitoring tool. In: Pollution Physiology of Estuarine Organisms, edited by W.B. Vernberg, A. Calabrese, F.P. Thurberg and F.J. Vernberg, University of South Carolina Press, Columbia, SC, pp. 421–438.

Ford, J., 1989. The effects of chemical stress on aquatic species composition and community structure. In: Ecotoxicology — Problems and Approaches, edited by S.A. Levin, M.A. Harwell, J.R. Kelly and K.D. Kimball, Springer-Verlag, New York, pp. 99–144.

George, S.C., B.J.S. Pirie, A.R. Cheyne, T.L. Coombs and P.T. Grant, 1978. Detoxication of metals by marine bivalves: an ultrastructural study of the compartmentation of copper and zinc in the oyster *Ostrea edulis*. Mar. Biol. 45, 147–156.

Hare, L., P.G.C. Campbell, A. Tessier and N. Belzile, 1989. Gut sediments in a burrowing mayfly (Ephemeroptera, *Hexagenia limbata*): their contribution to animal trace element burdens, their removal and the efficacy of a correction for their presence. Can J. Fish. Aquat. Sci. 46, 451–456.

Hare, L., R. Carignan and M.A. Huerta-Diaz, 1994. A field experimental study of metal toxicity and accumulation by benthic invertebrates: implications for the acid volatile sulfide (AVS) model. Limnol. Oceanogr. 39, 1653–1668.

Hargrave, B.T., 1975. Stability in structure and function of the mud-water interface. Verh. Intern. Verein. Limnol. 19, 1073–1079.

Harrison, G.I. and F.M.M. Morel, 1983. Antagonism between cadmium and iron in the marine diatom *Thalassiosira weissflogii*. J. Phycol. 19, 495–507.

Harvey, R.W. and S.N. Luoma, 1985a. Effect of adherent bacteria and bacterial extracellular polymers upon assimilation by *Macoma balthica* of sediment-bound Cd, Zn and Ag. Mar. Ecol. Prog. Ser. 22, 281–289.

Harvey, R.W. and S.N. Luoma, 1985b. Separation of solute and particulate vectors of heavy metal uptake in controlled suspension-feeding experiments with *Macoma balthica*. Hydrobiologia 121, 97–102.

Haux, C. and L. Förlin, 1989. Nine selected assays for health status in natural fish populations. In: Chemicals in the Aquatic Environment: Advanced Hazard Assessment, edited by L. Landner, Springer-Verlag, New York, pp. 197–215.

Herlihy, A.T., A.L. Mills and J.S. Herman, 1988. Distribution of reduced inorganic sulfur compounds in lake sediments receiving acid mine drainage. Appl. Geochem. 3, 333–244.

Howard, D.E. and R.D. Evans, 1993. Acid-volatile sulfide (AVS) in a seasonally anoxic mesotrophic lake: seasonal and spatial changes in sediment AVS. Environ. Toxicol. Chem. 12, 1051–1057.

Huggett, R.J., R.A. Kimerle, P.M. Mehrle and H.L. Bergman (Eds.), 1992. Biomarkers — Biochemical, Physiological and Histological Markers of Anthropogenic Stress, Lewis Publishers, Boca Raton, FL.

Jones, M.M., M.J. Meredith, M.L. Dodson, R.J. Topping and E. Baralt, 1988. Metallothionein synthesis and its induction mechanism: correlation with metal ion electronic configurations and softness parameters. Inorg. Chim. Acta 153, 87–92.

Kimball, K. D. and S.A. Levin, 1985. Limitations of laboratory bioassays: the need for ecosystem-level testing. BioScience 35, 165–171.

Klaverkamp, J.F. and D.A. Duncan, 1987. Acclimation to cadmium toxicity by white suckers: cadmium binding capacity and metal distribution in gill and liver cytosol. Environ. Toxicol. Chem. 6, 275–289.

Klaverkamp, J.F., M.D. Dutton, H.S. Majewski, R.V. Hunt and L.J. Wesson, 1991. Evaluating the effectiveness of metal pollution controls in a smelter by using metallothionein and other biochemical responses in fish. In: Metal Ecotoxicology: Concepts and Applications, edited by M.C. Newman and A.W. McIntosh, Lewis Publishers, Chelsea, MI, pp. 33–64.

Klaverkamp, J.F., W.A. Macdonald, D.A. Duncan and R. Wagemann, 1984. Metallothionein and acclimation to heavy metals in fish — a review. In: Contaminant Effects on Fisheries, edited by V.W. Cairns, P.V. Hodson and J.O. Nriagu, John Wiley & Sons, Toronto, pp. 99–113.

Krantzberg, G., 1985. The influence of bioturbation on physical, chemical and biological parameters in aquatic environments: a review. Environ. Pollut. A39, 99–122.

Langston, W.J., 1980. Arsenic in UK estuarine sediments and its availability to benthic organisms. J. Mar. Biol. Assoc. U.K. 60, 869–881.

Langston, W.J., 1982. The distribution of Hg in British estuarine sediments and its availability to deposit-feeding bivalves. J. Mar. Biol. Assoc. U.K. 62, 667–684.

Langston, W.J. and M. Zhou, 1987. Cadmium accumulation, distribution and elimination in the bivalve *Macoma balthica*: neither MT nor MT-like proteins are involved. Mar. Environ. Res. 21, 225–237.

Lion, L.W., R.S. Altmann and J.O. Leckie, 1982. Trace metal adsorption characteristics of estuarine particulate matter: evaluation of the contributions of Fe/Mn oxide and organic surface coatings. Environ. Sci. Technol. 16, 660–666.

Livett, E.A., 1988. Geochemical monitoring of atmospheric heavy metal pollution: theory and applications. Adv. Ecol. Res. 18, 65–177.

Luoma, S.N., 1983. Bioavailability of trace metals to aquatic organisms — a review. Sci. Total Environ. 28, 1–22.

Luoma, S.N., 1989. Can we determine the biological availability of sediment-bound trace elements? Hydrobiologia 176/177, 379–396.

Luoma, S.N. and G.W. Bryan, 1978. Factors controlling the availability of sediment-bound lead to the estuarine bivalve *Scrobicularia plana*. J. Mar. Biol. Assoc. U.K. 58, 793–802.

Luoma, S.N. and J.A. Davis, 1983. Requirements for modeling trace metal partitioning in oxidized estuarine sediments. Mar. Chem. 12, 159–181.

Luoma, S.N. and E.A. Jenne, 1976. Factors affecting the availability of sediment-bound cadmium to the estuarine, deposit–feeding clam, *Macoma balthica*. In: Radioecology and Energy Resources, edited by C.E. Cushing, Ecol. Soc. Am. Spec. Publ. No. 1, Dowden, Hutchinson and Ross, Stroudsbourg, PA, pp. 283–290.

Luoma, S.N. and E.A. Jenne, 1977. The availability of sediment-bound cobalt, silver and zinc to a deposit-feeding clam. In: Biological Implications of Metals in the Environment, edited by H. Drucker and R.E. Wildung, ERDA Symp. Ser. No. 42, Energy Research and Development Administration, Washington, DC, pp. 213–230.

Morel, F.M.M., 1983. Principles of Aquatic Chemistry. Wiley-Interscience, New York, NY, pp. 300–309.

Nirel, P.M.V. and F.M.M. Morel, 1990. Pitfalls of sequential extractions. Water Res. 24, 1055–1056.

NIST, 1993. Critical Stability Constants of Metal Complexes Database, NIST Standard Reference Database 46, U.S. Department of Commerce, National Institute of Standards and Technology, Gaithersburg, MD.

NRCC, 1988. Biologically Available Metals in Sediments. National Research Council Canada, Associate Committee on Scientific Criteria for Environmental Quality, NRCC Report No. 27694.

NRCC, 1985. The Role of Biochemical Indicators in the Assessment of Ecosystem Health — Their Development and Validation. National Research Council Canada, Associate Committee on Scientific Criteria for Environmental Quality, NRCC Report No. 24371.

Nriagu, J.O. and Y.K. Soon, 1985. Distribution and isotopic composition of sulfur in lake sediments of northern Ontario. Geochim. Cosmochim. Acta 49, 823–834.

Olafson, R.W., A. Kearns and R.G. Sim, 1979. Heavy metals induction of metallothionein synthesis in the hepatopancreas of the crab *Scylla serata*. Comp. Biochem. Physiol. 62B, 417–424.

Owen, G., 1966. Digestion. In: Physiology of Mollusca. Vol. II., edited by K.M. Wilbur and C.M. Young, Academic Press, New York, pp. 53–96.

Olsson, P.-E. and C. Haux, 1986. Increased hepatic metallothionein content correlates to cadmium accumulation in environmentally exposed perch (*Perca fluviatilis*). Aquat. Toxicol. 9, 231–242.

Pagenkopf, G.K., 1983. Gill surface interaction model for trace-metal toxicity to fishes: role of complexation, pH and water hardness. Environ. Sci. Technol. 17, 342–347.

Petering, D.H. and B.A. Fowler, 1986. Discussion summary. Roles of metallothionein and related proteins in metal metabolism and toxicity: problems and perspectives. Environ. Health Perspect. 65, 217–224.

Phinney, J.T. and K.W. Bruland, 1994. Uptake of lipophilic organic Cu, Cd and Pb complexes in the coastal diatom, *Thalassiosira weissflogii*. Environ. Sci. Technol. 28, 1781–1790.

Reinfelder, J.R. and N.S. Fisher, 1991. The assimilation of elements ingested by marine copepods. Science 251, 794–796.

Roch, M. and J.A. McCarter, 1984a. Hepatic metallothionein production and resistance to heavy metals by rainbow trout (*Salmo gairdneri*), I. Exposed to an artificial mixture of zinc, copper and cadmium. Comp. Biochem. Toxicol. 77C, 71–75.

Roch, M. and J.A. McCarter, 1984b. Hepatic metallothionein production and resistance to heavy metals by rainbow trout (*Salmo gairdneri*), II. Held in a series of contaminated lakes. Comp. Biochem. Toxicol. 77C, 77–82.

Roch, M., J.A. McCarter, A.T. Matheson, M.J.R. Clark and R.W. Olafson, 1982. Hepatic metallothionein in rainbow trout (*Salmo gairdneri*) as an indicator of metal pollution in the Campbell River system. Can. J. Fish. Aquat. Sci. 39, 1596–1601.

Roesijadi, G., 1981. The significance of low molecular weight, metallothionein-like proteins in marine invertebrates: current status. Mar. Environ. Res. 4, 167–179.

Roesijadi, G., 1982. Uptake and incorporation of mercury into mercury-binding proteins of gills of *Mytilus edulis* as a function of time. Mar. Biol. 66, 151–157.

Roesijadi, G., 1992. Metallothioneins in metal regulation and toxicity in aquatic animals. Aquat. Toxicol. 22, 81–114.

Roesijadi, G., 1994. Metallothionein and its role in toxic metal regulation. 15th Conf. European Soc. Comp. Physiol. Biochem., Biochem. Physiol. Effects Pollutants, Toxicol. Assess. Environ. Qual. (personal communication).

Roesijadi, G., A.S. Drum, J.T. Thomas and G.W. Fellingham, 1982. Enhanced mercury tolerance in marine mussels and relationship to low molecular weight, mercury-binding proteins. Mar. Pollut. Bull. 13, 250–253.

Roesijadi, G. and G.W. Fellingham, 1987. Influence of Cu, Cd and Zn pre-exposure on Hg toxicity in the mussel *Mytilus edulis*. Can. J. Fish. Aquat. Sci. 44, 680–684.

Roesijadi, G., M.E. Unger and J.E. Morris, 1988. Immunochemical quantification of metallothioneins of a marine mollusc. Can. J. Fish. Aquat. Sci. 45, 1257–1263.

Sanders, B.M. and K.D. Jenkins, 1984. Relationships between free cupric ion concentrations in sea water and copper metabolism and growth in crab larvae. Biol. Bull. 167, 704–712.

Schierup, H.-H. and V.J. Larsen, 1981. Macrophyte cycling of zinc, copper, lead and cadmium in the littoral zone of a polluted and a non-polluted lake. I. Availability, uptake and translocation of heavy metals in *Phragmites australia* (Cav.) Trin. Aquat. Bot. 11, 197–210.

Schindler, D.W., 1987. Detecting ecosystem responses to anthropogenic stress. Can. J. Fish. Aquat. Sci. 44 (Suppl. 1), 6–25.

Schindler, D.W., K.H. Mills, D.F. Malley, D.L. Findlay, J.A. Shearer, I.J. Davies, M.A. Turner, G.A. Linsey and D.R. Cruickshank, 1985. Long-term ecosystem stress: the effects of years of experimental acidification on a small lake. Science 228, 1395–1401.

Schindler, D.W., M.P. Turner, M.P. Stainton and G.A. Linsey, 1986. Natural sources of acid neutralizing capacity in low alkalinity lakes of the Precambrian Shield. Science 232, 844-847.

Schindler, P.W., 1967. Heterogeneous equilibria involving oxides, hydroxides, carbonates and hydroxide carbonates. In: Equilibrium Concepts in Natural Waters, edited by W. Stumm, American Chemical Society, ACS Adv. Chem. Ser. No. 67, Washington, DC, pp.196–221.

Shea, D. and G.R. Helz, 1989. Solubility product constants of covellite and a poorly crystalline copper sulfide precipitate at 298°K. Geochim. Cosmochim. Acta 53, 229–236.

St-Cyr, L., P.G.C. Campbell and K. Guertin, 1994. Evaluation of the role of submerged plant beds in the trace metal budget of a fluvial lake. Hydrobiologia 291, 141–156.

Steemann Nielsen, E., L. Kamp-Nielsen and S. Wium-Andersen, 1969. Effect of deleterious concentrations of copper on photosynthesis of *Chlorella pyrenoidosa*. Physiol. Plant. 22, 1121–1133.

Stegeman, J.J., M. Brouwer, R.T. Di Guiulo, L. Forlin, B.A. Fowler, B.M. Sanders and P.A. Van Veld, 1992. Molecular responses to environmental contamination — enzyme and protein systems as indicators of chemical exposure and effect. In: Biomarkers — Biochemical, Physiological and Histological Markers of Anthropogenic Stress, edited by R.J. Huggett, R.A. Kimerle, P.A. Mehrle and H.L. Bergman, Lewis Publishers, Chelsea, MI, pp. 235–335.

Sunda, W.G., 1991. Trace metal interactions with marine phytoplankton. Biol. Oceanogr. 6, 411–442.

Sunda, W.G. and S.A. Huntsman, 1983. Effect of competitive interactions between manganese and copper on cellular manganese and growth in estuarine and oceanic species of the diatom *Thalassiosira*. Limnol. Oceanogr. 28, 924–934.

Tessier, A., 1992. Sorption of trace elements on natural particles in oxic environments. In: Environmental Particles — Environmental, Analytical and Physical Chemistry Series, edited by J. Buffle and H.P. Van Leeuwen, Lewis Publishers, Boca Raton, FL, pp. 425–453.

Tessier, A. and P.G.C. Campbell, 1988. Comments on the testing of the accuracy of an extraction procedure for determining the partitioning of trace metals in sediment. Anal. Chem. 60, 1475–1476.

Tessier, A. and P.G.C. Campbell, 1990. Partitioning of trace metals in sediments and its relationship to their accumulation in benthic organisms. In: Metal Speciation in the Environment, edited by J.A.C Broekaert, S. Guçer and F.B. Adams, NATO ASI Series G — Ecological Sciences, Vol. 23, Springer–Verlag, Berlin, pp. 545–569.

Tessier, A. and P.G.C. Campbell, 1991. Comments on "Pitfalls of sequential extractions" by P.M.V. Nirel and F.M.M. Morel. Water Res. 25, 115–117.

Tessier, A., P.G.C. Campbell, J.C. Auclair and M. Bisson, 1984. Relationships between the partitioning of trace metals in sediments and their accumulation in the tissues of the freshwater mollusc *Elliptio complanata*. Can. J. Fish. Aquat. Sci. 41, 1463–1472.

Tessier, A., R. Carignan and N. Belzile, 1994. Reactions of trace elements near the sediment-water interface in lakes. In: Transport and Transformation of Contaminants near the Sediment-Water Interface, edited by J.V. De Pinto, W. Lick and J.F. Paul, Lewis Publishers, Boca Raton, FL, pp. 129–152.

Tessier, A., Y. Couillard, P.G.C. Campbell and J.C. Auclair, 1993. Modeling Cd partitioning in oxic lake sediments and Cd concentrations in the freshwater bivalve *Anodonta grandis* (Mollusca, Pelecypoda). Limnol. Oceanogr. 38, 1–17.

Viarengo, A., 1985. Biochemical effects of trace metals. Mar. Pollut. Bull. 16, 153–158.

Viarengo, A., 1989. Heavy metals in marine invertebrates: mechanisms of regulation and toxicity at the cellular level. Rev. Aquat. Sci. 1, 295–317.

Zamuda, C.D. and W.G. Sunda, 1982. Biovailability of dissolved Cu to the American oyster *Crassostrea virginica*. I. Importance of chemical speciation. Mar. Biol. 66, 77–82.

CHAPTER 3

Ecotoxicants at the Cell-Membrane Barrier

Ken Simkiss

I. BACKGROUND

In studying an ecotoxicological problem, the first question that arises is clearly whether the toxicants that are entering the system gain access to the cells of the organisms in that ecosystem. The answers to that apparently simple question may be complicated for several reasons. First, the molecules that poison the system (e.g., CH_3Hg) may not be the same as those that originally entered it (e.g., Hg) as in the Minamata disaster (Ottaway, 1980). Second, the delivery of these molecules to the cells may depend upon many local conditions. The use of the word "ecotoxicant" emphasizes these problems by indicating that the situation may not be the same as in a laboratory toxicity test. As an example of the difficulties, consider the chances of a particular annelid surviving in a polluted environment. In this situation, the toxicant may exist in a variety of forms, both free in solution and adsorbed onto particulate surfaces (Campbell and Tessier, 1995). The worm will be exposed to these materials by routes as diverse as the respiratory current and the digestive system. The respiratory current may change the form of the pollutant, by exposing it to a higher oxygen pressure and it may change further during its passage across the gill because carbon dioxide gradients may change the pH of the water. If the pollutant irritates the gill, the tissue may secrete mucus, covering the surface with an "unstirred layer" and the current may even be reversed, thereby changing all the uptake factors. Alternatively, because the particulate form of the toxicant may attract bacteria, it may be preferentially processed into the digestive system as a nutrient source. The pollutant may then be freed by digestive processes or the particles may be sorted by a variety of mechanisms so that some are phagocytosed into specific digestive cells. Whether the worm survives these processes will depend upon which biochemical lesions are

1-56670-1127-9/95/$0.00+$.50

induced, either in its own cells or in those of the other organisms in its ecosystem. These metabolic changes may in fact be species-specific so that it is the vulnerability of a particular biochemical pathway that eventually dictates the critical point in a particular ecosystem (Shugart, 1995).

The term "ecotoxicant" also indicates that the way that these poisons gain access to organisms relates directly to the respiratory, digestive, circulatory, and excretory physiology of the organisms in that environment. For many species, these physiological processes are themselves poorly understood so that the first opportunity for making any comparisons is by relating ecotoxicants to their effects at the cell membrane level. The cell membrane is the first barrier to the entry of natural and xenobiotic chemicals into the cell. As such, it protects the biochemical pathways from the environment. But it does this in subtle rather than absolute ways, and the purpose of this chapter is to provide some insights into those processes.

There are innumerable introductory texts that deal with the cell membrane from the viewpoints of basic biology, cell physiology, or biochemistry. This level of background knowledge is therefore assumed and may be built upon in the more advanced approaches that are available in Hille (1992), Rudy and Iverson (1992), Turner (1990), or Aloia et al. (1988). In this chapter, there is a brief historical introduction to the three main routes that materials can take to gain access to the cell. These are then discussed in relation to those factors that are likely to be of theoretical importance (e.g., lipid solubility) or practical relevance (e.g., dietary effects). It should be realized, however, that there is virtually no literature that relates ecotoxicological problems to causes at the cell membrane level. To that extent, this book is ahead of its time. There are, however, a considerable number of examples where it looks as if a better understanding of the processes that occur at the cell membrane could provide a valuable insight. The final section of this chapter is therefore concerned with demonstrating that a number of ecotoxicology problems may be caused by perturbations at this level.

II. ORIGIN OF THE CONCEPTS ON MEMBRANE PERMEABILITY

About 100 years ago, it became apparent that materials could enter cells by at least three routes. The first of these was revealed by the observations of Ernst Overton, who recognized that the ability of many molecules to enter cells was correlated with their lipid solubility. He observed that each time a methyl group was added to a compound it became less water soluble, more lipid soluble, and more cell permeable. Ecotoxicologists will instantly recognize in this the basis of a quantitative structure-activity relationship (QSAR), since it relates chemical structure to biological activity and, in fact, Overton was to concern himself particularly with the observation that such molecules were also narcotic. His work was to leave two conclusions: (1) "that the boundaries of each cell are impregnated with fatty oils or cholesterol and that lipid-soluble molecules cross the boundary by selective solubility," and (2) "that narcosis commences when any chemically indifferent substance has attained a certain molar concentration in the lipids of the cell" (Meyer, 1937). The first of these conclusions has been confirmed many times (Collander, 1954) while the second remains controversial (Franks and Lieb, 1994).

Contemporaneous with this work was the proposal by Julius Bernstein (1902) that excitable cells were surrounded by a membrane that was selectively permeable to potassium ions. He deduced that during excitation the permeability to other ions increased by some sort of "membrane breakdown." The idea was not universally accepted but it was a testable hypothesis and, as recording techniques improved, it became apparent that the membrane potential did not simply break down during nerve conduction, it reversed. This could only be explained if there was selective permeability of the cell membrane to certain ions. We may conclude, therefore, that (3) "the cell membrane is selectively permeable to some ions."

The final concept that we have to consider goes back to Ilya Metchnikoff, who at around the same time observed the particulate uptake of materials by cells. He related this activity to disease resistance and formulated a theory of phagocytosis for which he received the Nobel prize in 1908. The concept was a phenomenological one and the mechanisms have only recently been investigated. In terms of ecotoxicology, the system has been poorly studied, but we should note (4) "that materials may be surrounded by the membrane and carried into the cell."

In the past 50 years, major advances have been made in what will be referred to as the lipid, the aqueous, and the endocytotic routes for the entry of materials into the cell. These will, therefore, be outlined before considering the relatively neglected aspects of how pollutants may be involved in these systems.

III. ENTRY INTO THE CELL

A. THE LIPID ROUTE

The cell membrane is a bilayer of lipid molecules arranged with outwardly facing hydrophilic and inwardly directed hydrophobic groups. In this form, the lipids are able to move relatively freely in the plane of the membrane at a rate of about 10^{-8} cm^2/s (Storch and Kleinfeld, 1985) but it is energetically unlikely that they will be able to flip from one surface to the other. A number of factors influence the composition and hence fluidity of this membrane, but diet and temperature are the best understood. Dietary lipids are incorporated into cell membranes at different rates in different tissues, and they are mixed with lipids that are synthesized in the cell cytoplasm. These cytoplasmic lipids are selectively inserted into the cell membranes so that the inner surface has a different composition from the outer surface. These lipids only appear on the noncytoplasmic surface through the action of special "flip-flop" enzymes (Zackowski and Devaux, 1990). It is a key feature of the fluid/mosaic model of the membrane (Singer and Nicholson, 1972) that lipids are free to move in the plane of the membrane so that they can self-aggregate to form mosaic patches. In all cases, however, the polar lipids, which are predominantly anionic, remain with their charged groups protruding from the bilayer. In the Gouy-Chapman-Stern theory, this charge density gives rise to a surface potential and, in a saline solution, it attracts a diffuse electrical double layer for a distance of about 1 nm from the surface. This layer of ions screens the surface charge but it also affects the approach of charged molecules to the membrane and could influence access to any channels (Simkiss and Taylor, 1995).

It is generally considered that the flux of materials across the cell membrane is largely dependent upon their lipid solubility and this is conveniently modeled by the octanol/water partition coefficient. The more lipid soluble a molecule is, the greater will be its octanol/water partition coefficient and the greater will be its flux into the organism for any given concentration gradient. For this reason, many biocides are deliberately constructed so as to have a high lipid solubility and, for the same reason, many ecotoxicologists regard the octanol/water partition coefficient as a crucial test for the likely penetration of a pollutant into an organism (e.g., Esser, 1986). In many situations, this gives a good correlation between the predicted and actual diffusion rates across a bilayer (Orbach and Finkelstein, 1980), but it should be remembered that a bilayer is a much more structured system than a simple solution such as octanol. In a paper entitled "Can regular solution theory be applied to bilayer membranes," Simon et al. (1979) came out with a very clear answer, No! Despite this, it is clear that many substances enter the cell according to their solubility in the membrane lipids and that this is conveniently modeled by the octanol/water partition coefficient.

It should be realized, however, that some materials entering the environment in a hydrophilic form may be converted into a lipophilic form by bacterial action. The production of organometallic compounds by biomethylation has been the cause of a number of environmental disasters, but bacteria also produce specific complexing agents by which they gain access to other environmental metals. Bacteria produce and resorb a large number of ionophores. These act either by enclosing ions and shuttling them across the lipid barrier or by forming channels in the membrane itself. The selectivity of these systems depends upon their size and ion binding sites, and a number of naturally occurring ionophores, such as valinomycin which transports K^+, can act as antibiotics (Eisenman and Dani, 1987). Thus, lipid solubility is an important route for the penetration of many materials into the cell.

B. THE AQUEOUS ROUTE

The concept that there are aqueous pores through an otherwise lipid membrane has been around for a long time, but it was always difficult to explain the selectivity that these channels appeared to impose. And, as if that was not difficult enough, this selectivity varied from membrane to membrane (Table 1). The best understood channels are those that are selectively permeable to cations and an explanation for this came from an unexpected source.

Table 1 Series of Metal Uptake Systems

Series	Example
1. $Cs^+ > Rb^+ > K^+ > Na^+ > Li^+$	Black membrane with monoactin
2. $Rb^+ > Cs^+ > K^+ > Na^+ > Li^+$	Erythrocyte
3. $Rb^+ > K^+ > Cs^+ > Na^+ > Li^+$	Gallbladder
4. $K^+ > Rb^+ > Cs^+ > Na^+ > Li^+$	Malphighian tubules
5. $K^+ > Rb^+ > NA^+ > Cs^+ > Li^+$	Antibiotic nigericin
6. $K^+ > Na^+ > Rb^+ > Cs^+ > Li^+$	Blowfly salt receptor
7. $Na^+ > K^+ > Rb^+ > Cs^+ > Li^+$	Ionophore dianemycin
8. $Na^+ > K^+ > Rb^+ > Li^+ > Cs^+$	Squid axon
9. $Na^+ > K^+ > Li^+ > Rb^+ > Cs^+$	Squid action potential
10. $Na^+ > Li^+ > K^+ > Rb^+ > Cs^+$	Frog skin
11. $Li^+ > Na^+ > K^+ > Rb^+ > Cs^+$	Cornea

George Eisenman (1962) had been working on cation-selective glass electrodes with particular reference to Li^+, Na^+, K^+, Rb^+, and Cs^+. Of the 120 possible sequences for these ions, he found that only 11 occurred. He explained this on the basis of two forces, one between the ion and water, and the other between the ion and binding site in the glass. The energy of interaction was inversely proportional to the radii of the ion (r_c) and the binding site (r_{site}), i.e., $1/r_{c + site}$. This is, in effect, a measure of the reactivity of the ion to water in competition with an anion lining the route of permeation. It did not escape Eisenman's attention that the 11 series that occurred with glass electrodes were the same as those that occurred in biology (Table 1). The importance of this work is twofold. First, it demonstrated that selectivity was based upon an interaction between the hydration and dehydration of an ion, and upon the binding and release of the ion from surrounding ligands. Second, it provided a simple model that could be used to explain the properties of membrane channels.

A large number of conceptual models have been used to explain the mechanisms of transfer of hydrophilic molecules across the lipid bilayer. Channels, pores, and carrier proteins are frequent descriptions of these systems but, purely for the sake of simplicity, they will initially all be considered as membrane channels.

Membrane channels consist of distorted cylinders formed from proteins that loop in and out of the membrane. They have a hydrophobic surface facing the membrane lipids and a hydrophilic lining that forms an aqueous pore. It is clear from the dimensions of these proteins that they are very long relative to the size of ions and they protrude above the Stern layer of the membrane surface. Since ion channels are effectively aqueous pores through an otherwise insulating layer, they can enhance the rate of ion flow by a factor of up to 10^{39} and a single Na^+ channel can pass 10^7 ions/s (Hille, 1992).

A number of channel proteins are known that show almost no discrimination against the ions that pass through them. Such "porins" are found in the outer membranes of bacteria and mitochondria, where they form only a crude filter for the more selective channels on the inner membranes. Porins are also used by "killer cells" since, once the cell membrane becomes permeable to ions, the cell is unable to maintain its integrity and dies (Nikaido, 1992). It is clear, therefore, that selectivity is one of the crucial factors of channel function and, as Eisenman discovered, it is based upon an interaction between hydration and dehydration of the ion, and upon its binding and release from ligand groups on the channel wall.

If an ion such as Na^+ was to diffuse through a pore 0.6 nm wide and 1 nm long, it would emerge in about 0.4 nsec. But, if that process involved it in one dehydration and one rehydration cycle, an extra 1.0 nsec would be involved. For a full-sized channel about 10 nm long, the ion binding effect is likely to take 5 to 10 nsec, so that it would dominate the flux rate and ions that bind water tightly (e.g., Ni^{2+}, Mg^{2+}, Co^{2+}, and Mn^{2+}) would take from twenty to several thousand times longer than for normal diffusion. Thus, selectivity is a major factor retarding ion movement through channels and it is generally considered that most channels are funnel-shaped at their entrance and exit so as to reduce this effect to a small selectivity filter (Figure 1). From this, it will be apparent that channels with a large conductance will inevitably have a lower selectivity than those with small conductances.

When an ion channel acts as an aqueous pore through the lipid membrane, it is likely that ion movements within the channel will interact with water movements or even with other ions in the same channel. If the ions in a channel cannot pass each other, all the ions ahead of a new one entering the system will have to move before there is space available. This "long pore" effect has serious consequences since the movement of one ion is no longer independent of the other ions that are present. As a salt solution becomes more concentrated, the chances that a channel will be occupied increase, so that interference occurs and the channel shows saturation kinetics. Ions within a channel will also repel each other if they have the same charge and, if there is a procession in one direction, they may oppose the movement of ions in the opposite direction even if the channel is permeable to them. Thus, four of the major factors that will influence the movement of ions through a channel are the following: (1) shape of the pore, (2) features of the selectivity filter, (3) single ion or multiple ion occupancy, and (4) blocking by other ions.

In 1957, Skou (1988) discovered a membrane-bound enzyme that split adenosine triphosphate (ATP) but which was dependent upon Na^+ and K^+ ions for maximal activity. Subsequently, it was observed that there was a close correlation between cation fluxes across membranes and the amount of this ATPase. This formed the basis for the sodium pump and, with it, the concept that membrane channels could not only be selective but they could also use chemical energy to drive ions against their electrochemical pumps.

It also soon became apparent that many channels were gated, i.e., the aqueous passage was capable of being blocked by a gate that transiently opened according to some stimulus. Specific regions of the channel protein acted as sensors (Neeley et al., 1993) and the channel might respond to changes in electrical potential, specific chemical "transmitters," or even stretch effects by opening the channel. A number of naturally occurring molecules (e.g., tetratoxin) can also block channels, while artificial pesticides such as DDT activate them

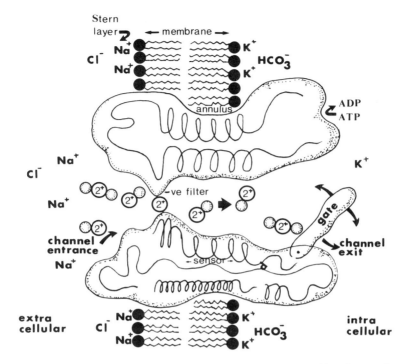

Figure 1 Diagram of a generalized ion channel in a cell membrane. Note the membrane is composed of a bimolecular layer of lipids through which the channel protein loops repeatedly to form an aqueous channel. The membrane lipids carry a negative surface charge that attracts a number of counterions to form the Stern layer. An annulus of specific lipid molecules may be attracted to the lipophilic outer surface of the channel protein. The channel that is shown is gated, i.e., may be closed or opened according to a number of specific stimuli, some of which may be detected by a sensor protein. The channel is also capable of transporting ions by using ATP (i.e., it can be an ion pump). The hydrophilic channel through which the ions move has a negatively charged selectivity filter. A divalent cation is shown passing through the channel and in the process it undergoes dehydration by the loss of water molecules (stippled) followed by rehydration once it passes the selectivity filter. Monovalent ions (Na^+) are repelled from the channel by these divalent cations. This diagram represents a composite picture incorporating concepts of pores, channels, and pumps and does not illustrate any real structure.

(Narahashi, 1978). There are, therefore, many opportunities for organic molecules to influence inorganic ion fluxes and for organic pollutants to affect inorganic processes. Many of these concepts are summarized in Figure 1.

C. THE ENDOCYTOTIC ROUTE

A number of materials enter the cell by being enclosed within a vesicle of membrane material that detaches from the cell surface and becomes part of an elaborate series of interactions with cytoplasmic organelles. The phenomenon is very widespread and includes molecularly specific, receptor-mediated events

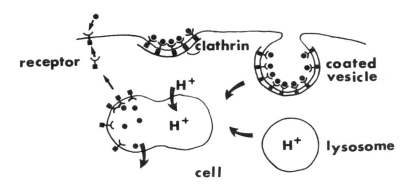

Figure 2   Endocytosis of material into the cell. Membrane receptors bind ligands and these accumulate at a capping region where they are enclosed in clathrin to form a coated vesicle. This enters the cytoplasm and may become associated with a number of organelles such as lysosomes. The influx of protons and a variety of enzymes may release the materials that have entered the cell by this route.

at one extreme and general food particle assimilation at the other. Ecotoxic materials may be absorbed in both these systems and this may pose an important problem, since pollutants are often concentrated to high levels on sediment particles that may subsequently be ingested (Bryan and Langston, 1992).

The uptake of iron is an example of a very specific type of endocytosis. In an aerobic environment at physiological pH values, the availability of iron is determined by the solubility product of $Fe(OH)_3$, which is roughly 10^{-38}. Iron is transported in association with transferrin, a protein that binds iron and HCO_3^- ions in a soluble form. Metabolizing cells bind the transferrin molecule to specific membrane receptor molecules which diffuse to specialized "coated pit" regions on the cell surface. The protruding tail region of these receptors may be involved in this process and in their attachment to clathrin molecules. These form a lattice structure that encloses a membrane vesicle that passes into the cytoplasm and fuses with another vesicle such as a lysosome (Figure 2). This organelle can transport protons into the vesicle which destroy the HCO_3^- on the transferrin, releasing the iron and facilitating the recycling of receptor and clathrin molecules to the cell surface. Cells with a rapid metabolism may have up to 800,000 transferrin receptors per cell, of which 60% are normally in the process of being recycled through the cell (Dautry-Versat et al., 1983).

The system sounds elaborate but it is a fundamental cell process that forms the basis of intracellular digestion, by linking the endocytosis of food particles to lysosomal digestion. A large number of materials are probably taken into the cell by this means and exposed to low pH and lysosomal enzymes. Materials that are not broken down by these agents may be egested, retained in a compartmentalized region (e.g., tertiary lysosomes), or released into the cytoplasm. The consequences of these possibilities are a major challenge to ecotoxicologists. The uptake of lipophilic molecules is understood and can be estimated by octanol/water partitioning; the role of aqueous channels is the subject of extensive research with good model systems; but the role of endocytosis and how to assess its importance is largely unexplored and poorly assessed.

IV. COMPLEX INTERACTIONS

The information that is discussed in Section III represents an attempt to describe a general model of a cell membrane. Each component of the model is itself an abstraction of a large number of very specific observations made under controlled conditions. To apply these concepts to ecotoxicology requires a major suspension of one's scientific training and the following is a "best guess" approach justified only on the basis that real problems need to be tackled.

A. LIPID EFFECTS

i. Membrane Fluidity

The fluid mosaic model envisages the lipids as being mobile in the plane of the membrane. As such, they are able to interact with each other and with other components of the membrane to form a mosaic pattern. Clearly, the permeability of these mosaic regions vary but more importantly it has been proposed that specific lipids will surround the channel proteins to form an annulus (Figure 1). The components of the channel proteins are mobile either because they show conformational changes associated with phosphorylation (e.g., pumping) or because of movements during gating. Thus, changes in the fluidity of the lipid components could influence membrane function by restricting channel movements.

It is well known that changes in the diet influence membrane composition and that cooling poikilothermic organisms leads to a decrease in membrane fluidity. In this latter case, there is an increase in microsomal desaturase activity and an increase in the desaturated (more fluid) lipids in the membrane (Hazel, 1984). These changes in membrane fluidity have been shown to influence rat hepatocyte Na^+/K^+ ATPase and to modify the enzyme activity in erythrocytes (Storch and Schachter, 1984; McMurchie, 1988). In an interesting experiment on a mutant cell line with defective cholesterol synthesis, it was shown that adding cholesterol could modify ATPase activities 10-fold (Sinensky et al., 1979). Clearly lipid changes in the cell membrane could have important ecotoxicological effects.

ii. Narcosis

In the initial experiments that led Overton to suggest that the cell membrane was a lipid layer, he used tadpole narcosis as an indicator of biological action (Meyer, 1937). The concept subsequently developed that anesthetic potency was due to solution into, and disruption of, the lipid membrane. Thus, it was proposed that the degree of narcosis was simply a function of the molar concentration of a compound in the lipid phase. This is, of course, what is predicted in the octanol/water partition coefficient that forms the basis of so many QSARs and the problem has been discussed in detail by Donkin (1994). There is, however, an alternative hypothesis described forcefully by Franks and Lieb (1994), who argue that anesthetics exert their action by binding to proteins and that their principal targets are ligand-gated ion channels. The site of action has not been clearly identified, but it is thought that they must be hydrophobic pockets exposed to water on the surface of the channel proteins. This is clearly an important problem, since narcotic effects are common in ecotoxicology (Donkin, 1994).

iii. Parabolic Responses

The concept that the cell membrane is a biomolecular layer of lipids provides a simple basis for considering that C, the concentration of a drug that will produce a biological response, would be linearly related to the octanol/water partition, P.* In fact, plots of log $1/C$ against log P frequently produce parabolic rather than linear relationships (Hansch and Clayton, 1973). A number of attempts have been made to explain this phenomenon. (See Chapter 4 of this volume for additional discussion.) It is intuitively apparent that a pollutant that is extremely lipid soluble will never leave the membrane, whereas a totally hydrophilic molecule will never enter it. Thus, it has been argued that it is only the intermediate values of P that produce a biological effect. An alternative explanation for the parabolic effect is that the biological response only occurs after the pollutant has reached a specific site of action and this may involve a continual series of partitioning exercises between hydrophobic and hydrophilic compartments in the organism. A crucial component of these explanations is the suggestion that the biological system never equilibrates with components in the environment. This implies that predictions of biological responses in these situations will be extremely difficult to quantify since there is no simple theoretical approach that can be applied to nonequilibrum states.

B. CHANNEL ROUTES

i. The Permeant Form

It is generally agreed that it is the ionic species of most elements that is the physiologically or toxicologically active form (e.g., Chapter 2). In the case of the bulk electrolytes (Na^+, K^+, Ca^{2+}, Mg^{2+}), the ionic is the predominant form

---

* Elsewhere, including several chapters in this volume, K_{ow} may be used instead of P to designate the octanol/water partition coefficient.

and the electrochemical equations for transmembrane potentials and fluxes correlate with this. For the trace elements that are often more covalent in their reactions, the metals often exist in solution largely as complexes. Despite this, there are good correlations between the ion activities in solution and physiological responses. This has been clearly shown for the effects of copper on crab larvae (Sanders et al., 1983). It should be realized, however, that the theory of channel permeation is based upon relative ligand binding and the ion speciation data that is frequently presented is only an approximation to this.

ii. One Ion, One Channel?

There are over 20 essential elements that might reasonably be expected to enter the cell via a channel. There are, in addition, a large number of organic metabolites that probably also enter the cells via channels. The impression is frequently given that each of these channels is a discrete structure. Thus, although Na^+ ions outnumber Ca^{2+} ions by 100-fold in the extracellular fluid, there is a discrimination of over 1000-fold against Na^+ for ions that enter by the calcium channel. Furthermore, the system is so precise that a mutation that changes a single amino acid can convert a Na^+ channel into a Ca^{2+} channel (Heinemann et al., 1992). It has, therefore, been generally assumed that there are specific channels for each of the ions of the essential elements. This is a reassuring concept since it implies that nonessential and pollutant ions should be excluded. But they are not.

iii. Channel Selectivity

Channel selectivity involves binding ions to ligands in the channel wall, thereby retarding flux rates. Clearly, the selectivity filter is a major influence and sets a maximum size for any permeant ion that attempts to pass through it. Below that level, however, there is a lot of variation. Potassium channels can pass at least 4 other ions, sodium channels 14 others, calcium at least 8 others, and the muscle end plate channels more than 50 others. In addition to the selectivity filter, ions that bind strongly to the walls of the channel can block them. Since movement through a multi-ion channel requires sites to be vacated, a strongly bound ion can produce such a high occupancy of a site that it reduces or blocks the passage for other ions. Thus, calcium channels can be blocked by Mn^{2+}, $Co^{2+,}$ or Ni^{2+} ions, while potassium channels can be blocked by Cs^+, Na^+, or Ba^{2+}. The blockage can have two effects. One is to act as a physical obstacle, while the other is electrically to repel other ions from approaching the pore channel. Thus, sodium ions may pass rapidly through the calcium channel as long as there are no divalent cations present. In the presence of trace amounts (1 µmol/l) of calcium, however, the flow abruptly stops. A calcium ion can bind to the channel wall, stopping the flux of sodium within the channel and repelling other cations from entering it. For this reason, some channels appear to be much more selective than they actually are. Thus, calcium channels are actually freely permeable to sodium ions if they are the only ions that are present. But, as soon as calcium ions are added, the channel becomes highly discriminatory since the calcium repels all of the univalent ions and only calcium will pass through the pore. It is now apparent that such selectivity is an operational rather than a design feature. If the toxicologist was similarly to consider the sodium channel in the end plate as a typical membrane pore, the results would actually show that 6 monovalent cations, 9 divalent ions (Mg^{2+}, Ca^{2+}, Sr^{2+}, Ba^{2+}, Mn^{2+}, Co^{2+}, Ni^{2+}, Zn^{2+}, Cd^{2+}), and 41 organic cations were all at least 10% as permeable as sodium (Adams et al., 1980; Dwyer et al., 1980). Thus, although under normal conditions channels are clearly both effective and selective, they are not the reliable sentinels that might be expected.

C. VESICULAR UPTAKE

i. Transferrin Specificity

The transferrin molecule provides a route for the uptake of iron but clearly any other metal that parasitizes this binding site will also be transported. There is evidence that zinc may enter cells along this route (Evans, 1976) and, more importantly in ecotoxicological terms, that Al^{3+} may substitute for iron (MacDonald and Martin, 1988).

ii. Particle Assimilation

Endocytosis or phagocytosis appears to be an important route for digesting food particles in many organisms. The system is particularly important in filter-feeding organisms and particles are often selected, not for their inherent nutrient value, but because they include surface-bound bacteria (Harvey and Luoma, 1985). Unfortunately sediments are frequently major concentrators of pollutants and may contain levels that are thousands of times higher than those of the overlying water column (Bryan and Langston, 1992). The biological availability of these sediment-bound pollutants and the routes of uptake are controversial (Tessier et al., 1984), but direct endocytosis is involved in a number of situations. The difficulty, in terms of ecotoxicology, is that there is no accepted model for predicting which pollutants may enter along this route or how they may be released. The relevant factors include (1) the presence or absence of bacterial exopolymers on the surface (Decho and Luoma, 1991), (2) the nature of the particle surface/pollutant interaction, (3) the specificity of cell absorption, (4) the effect of lowering the pH in the lysosome to a value of about 5, (5) the nature of lysosomal membrane channels (Winchester, 1992), and (6) the subsequent fate of the lysosomal contents. Outstanding among the requirements for understanding this route is the need to understand the particle surface reactivity and the effect of changes of pH on this binding (Simkiss, 1995).

V. MEMBRANE ECOTOXICOLOGY

It is now possible to try and relate some of these features of membrane function to the problems of ecotoxicology.

A. THE LIPID MEMBRANE

Lipid solubility is perhaps the single most important factor that is used in assessing the biological availability of environmental compounds because it reflects the ability of an organic compound to traverse the membrane lipids.

i. LC50 of Chlorophenols

In this example, Konemann (1981) showed that exposing 2- to 3-month-old guppies (*Poecilia reticulata*) to 72 industrial pollutants gave an LC50 with a high correlation ($r = 0.98$) to log P. This work is representative of a large number of studies that simply use lipid solubility but they can become much more sophisticated when terms are introduced into the QSAR equations by including a measure of the biochemical lesion that is involved (Karcher and Devillars, 1990). This study is chosen, however, because Konemann and Musch (1981) subsequently showed with chlorophenols that the LC50 also depended strongly on pH. The toxicity of these compounds increases with decreasing pH. The explanation for this appears to be that as the acidity of the water increases, the ionization, and hence hydrophilia of the chlorophenols is depressed. The more hydrophobic the compound is the more easily it penetrates the lipid membrane and this can clearly be influenced by simple environmental variables such as pH (Figure 3).

ii. Insecticide Effects

Lipid solubility is frequently incorporated into the structure of many pesticides. The work of Umeda et al. (1988) shows a log dose/response plot for a pyrethroid applied to the cockroach, *Blattella germanica*. The response is the knock down time (KT50) in minutes (Figure 4).

Pyrethroids, like many insecticides, penetrate the nerve membrane and bind to the gated sodium channel holding it open and facilitating the influx of ions. As a result, the nerve becomes depolarized and incapable of conducting impulses. Many insecticides are based upon this membrane depolarizing effect and can influence a range of nontarget organisms.

iii. Oil Spills

A less clearly understood effect is observed in many marine organisms exposed to oil spills. A large number of physiological and cellular responses

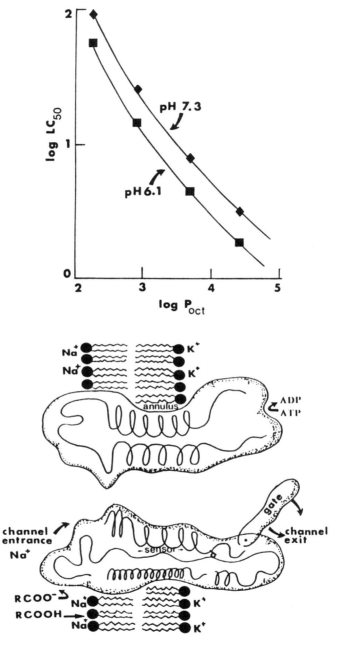

Figure 3   The effect of a fall in pH on the concentration necessary to kill 50% (LC50) of the guppies exposed to these chlorophenols (top). Note that the increased acidity increases the lipid solubility. An interpretation of this data for a generalized acid is given in the membrane model (bottom).

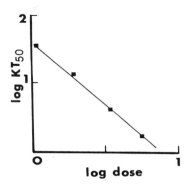

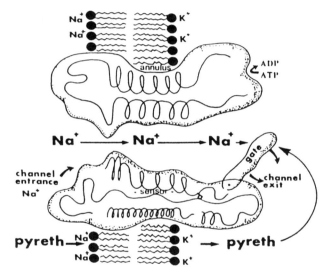

Figure 4   The knock down time (KT50) in minutes in relation to the dose (μg) of the pyrethroid allethrin applied to *B. germanica* (top) (data from Umeda et al., 1988). The pyrethroid affects the gated sodium channel of the nerve membrane resulting in an influx of sodium ions (bottom).

are found in such animals; Figure 5 is a plot derived from the data of Bayne et al. (1982). The "scope for growth" of *Mytilus edulis* exposed to 30 μg/l hydrocarbon in seawater is compared with the fragility of the lysosomal membrane. The lysosome appears to be particularly vulnerable to pollutants that influence its membrane fluidity. Thus, phenanthrene, like many other petroleum compounds, damages this organelle and leads to leakage, an increase in catabolism, and eventual death (Moore and Viarengo, 1987). This response is of particular interest because lysosomes may be involved in some endocytotic systems (Figure 2) where sediment particles with adsorbed pollutants may be incorporated into digestive cells.

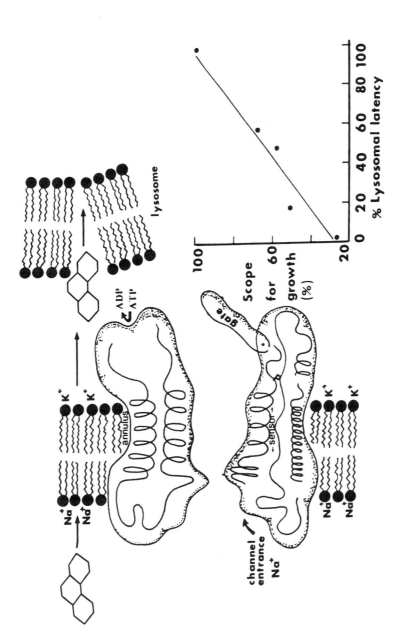

Figure 5   Effect of hydrocarbons on scope for growth and lysosomal latency (right) (data from Bayne et al., 1982). Petrochemicals of this type can influence the stability of the lysosome causing extensive cellular damage and an increase in catabolism (left).

Two other potential influences on the lipid component of the membrane are also worth noting. Lead is a well-known pollutant with effects on the nervous system and, at the membrane level, it has been shown to decrease the fluidity of the lipid bilayer (Cook et al., 1987). This is unlikely to have a direct effect on the cell but the additional secondary effects may be considerable and present a challenge for future studies. Another lipid effect has been noted by Green et al. (1980), who found that phospholipids could be involved in the transport of ions across membranes. Certainly phosphatidic acid can function as a calcium ionophore (Chauhan and Brockerhoff, 1984) and appears to act in this capacity during hormonal stimulation of the parotid gland (Serhan et al., 1981). The extent to which lipids in natural membranes can act as ionophores is not clear but Bulman and Griffin (1980) pointed out that thioacetamide, which increases phospholipid levels, resulted in an increase in Pu uptake when given to rats.

B. CHANNEL INTERFERENCE

The problems with interpreting ecotoxicological data in terms of membrane channels are that 1) few experiments have been formulated in these terms, 2) most of our knowledge relates to gated channels in excitable tissues whereas transport systems in epithelial cells are probably much more relevant, and 3) little is known of the true diversity of ion channels. The following are data that can be speculatively interpreted in terms of membrane events.

i. Cd^{2+} Uptake into O. gammarellus

In this experiment, Rainbow et al. (1993) exposed the amphipod *Orchestia* to artificial seawater at different salinities and measured the uptake of ^{109}Cd in relation to computed ion activities. The results are shown in Figure 6. The authors rightly consider that the experiment demonstrates that uptake is proportional to cadmium ion activity up to about 30 µg/l but then plateaus due to a "physiological response in apparent water permeability".

A more direct mechanistic interpretation would be that (1) Cd^{2+} enters the organism by parasitizing an epithelial ion channel (probably a Ca^{2+} channel, since it is unlikely that an organism would possess a channel for a nonessential toxicant ion, and Cd^{2+} [0.97 Å] and Ca^{2+} [0.99 Å] have similar ionic radii), and (2) the membrane channel is relatively long and ion movements are therefore not independent. The channel shows saturation kinetics of the Michaelis-Menton type at a $[Cd^{2+}]$ of 30 µg/l.

ii. Cu^{2+} Toxicity to Salmo clarki

In work on the cut-throat trout, Borgmann (1983) showed that decreasing the pH of the environmental water decreased the toxicity of copper (Figure 7). The expectation was that increasing the acidity would increase the free $[Cu^{2+}]$

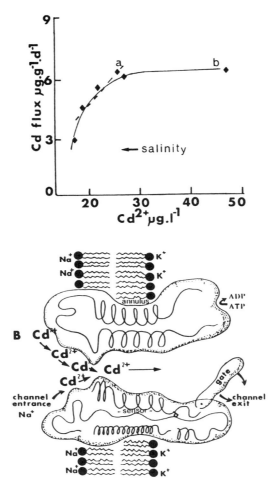

Figure 6   Effect of increasing Cd2+ ion concentration on Cd²⁺ influx into the amphipod, *Orchestia* (top). Note the response up to 30 µg/l is linear (a) but then plateaus (b). (After Rainbow et al., 1993.) An interpretation of this in terms of channel saturation is shown on the bottom.

and, therefore, increase the toxicity. The fact that the opposite effect was observed was interpreted as showing that protons competed with copper for ligand binding sites.

The simplest conceptual way to model this effect is to imagine the highly mobile proton as penetrating the channel and binding to the anionic selectivity filter. In doing this, it obstructs the narrowest part of the pore so that, although there are now more copper ions, less of them can penetrate the membrane and toxicity is reduced.

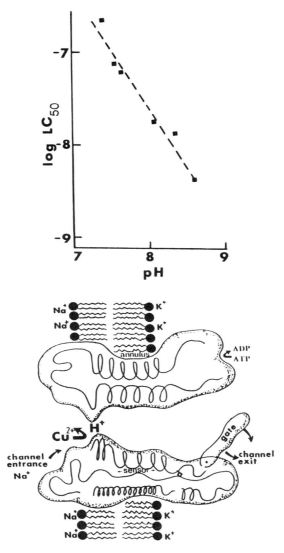

Figure 7 The effect of pH on the lethal concentration of copper necessary to kill 50% (LC50) of cut-throat trout (top). Note that making the water more acid protects against the effects of the copper ion. (Data recalculated to free copper levels by Borgmann, 1983.) The illustration on the bottom provides a possible explanation of this effect based upon channel blockage by protons.

iii. Zn²⁺ Effects on Fish Gill Ion Regulation

The effects of metal ions on ion regulation have been studied extensively. There is initially a shock phase (at about 2 d), with major disturbances to both electrolyte balance and gill morphology, followed by recovery and increased tolerance (McDonald and Wood, 1993). Part of the initial response is an increase in mucus secretion that effectively increases the unstirred layer effect over the cell membrane. There is evidence for a disruption of cell-cell contacts increasing intercellular fluxes and a direct effect on ion pumps (Simkiss, 1984). The metals Cu^{2+}, Al^{3+}, and Cr^{3+} appear mainly to affect Na^+ and Cl^- regulation, while Cd^{2+}, Zn^{2+}, and Mn^{2+} influence Ca^{2+} regulation (Shepherd and Simkiss, 1978). Recovery involves repair and often proliferation of the chloride cells of the epithelium.

A study on the effects of 2.3 µmol/l Zn on rainbow trout by Hogstrand et al. (1994) shows the importance of extending the models that have so far been invoked. Calcium is thought to enter fish gills along an electrochemical gradient using channels in the apical membrane of the chloride cells. Once within the cell, it is bound to a number of regulatory ligands and then transported across the basolateral membrane by a high-affinity Ca^{2+} ATPase channel. Note that, in order to explain this example of epithelial transport, it is necessary to invoke two calcium channels with different properties on spatially separated sites.

Zinc ions inhibit calcium transport across this epithelium, leading to hypocalcemia. A kinetic analysis of this phenomenon is shown in Figure 8. Changes in K_m indicate competitive inhibition, while changes in J_{max} suggest noncompetitive interactions. The results of this study suggest that initially Zn^{2+} enters the gill through the Ca^{2+} pathway and that there is competitive interaction within the Ca^{2+} channel (Figure 8). Subsequently, there may be a change in the apical permeability to calcium and the K_m of the channel may be modified so as to reduce Zn^{2+} binding without greatly affecting Ca^{2+} influx. In an in vitro study, Shephard and Simkiss (1978) suggested that competition with Cu^{2+} led to an increase in the number of Ca^{2+} ATPase sites. Thus, in both examples, it is necessary to invoke cellular regulation whereby the characteristics of the ion transporting membrane of the gill are actively modified so as to accommodate the effects of pollutant ions. The implication is that the cell membrane is not just a site of pollutant interaction but also a complex adaptive site.

These examples show that it is possible to apply some of the concepts of membrane function to ecotoxicological data, although it should be emphasized that alternative interpretations are often possible.

VI. MEMBRANES IN THE ECOTOXICOLOGY HIERARCHY

In 1960, strong accusations were made to the British government that Peregrine falcons (Falco peregrinus) were seriously predating homing pigeons. The ensuing investigation showed the exact opposite, for the Peregrine appeared to be in serious decline and likely to be exterminated (Ratcliffe, 1993), as were other avian predators such as the Sparrowhawk (Accipiter nisus) both in Britain and around the world (Newton, 1986). An appreciation of the

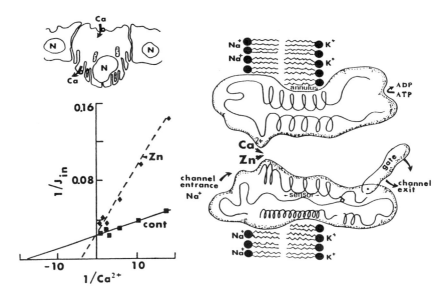

Figure 8 A chloride cell (left, top) showing an apical and a basolateral calcium pump. The relationship between flux rate and calcium concentration is shown in a Lineweaver-Burk plot (left, bottom) for control (cont.) and Zn-treated fish (redrawn from Hogstrand et al., 1994). The results demonstrate competitive interaction between the two ions and this is shown in the membrane model (right).

density-dependent and density-independent factors that affected these populations led to the diagnosis that there were localized declines in habitat quality leading to a reduction in reproductive success. The cause was shown to be a decrease in brood size (Figure 9a) associated with eggshell thinning. It subsequently became clear that this coincided with the use of organochlorine pesticides, of which DDT (and its persistent metabolite, DDE) was an important component (Figure 9b). For certain species of birds such as Northern Gannets (*Sula bassanus*), this decline in reproductive success was sufficient to seriously reduce the population (Elliot et al., 1988). Thus, the problem took the form of habitat pollution → DDE accumulated in prey species → DDE in predators → decline in brood size → potential extermination. The same phenomenon can, however, be written in a different form. Lipid soluble toxicant → bioaccumulation in organisms with poor detoxification systems (birds metabolize DDE very poorly when compared with mammals) → vulnerable target organs (i.e., the shell gland has a high Ca flux) → inhibition of membrane-bound ATPases at crucial periods (Miller et al., 1976; Lundholm, 1987) → potential extermination. Ecologists would claim a decline in population recruitment, biochemists an inhibition of membrane enzymes.

Perhaps the important factor in appreciating these different interpretations of the hierarchical approach is the critical pathway. At all levels of organization, there are likely to be interactions that favor the movement of materials along particular routes. Thus, at the food chain level, a pathway exists for the

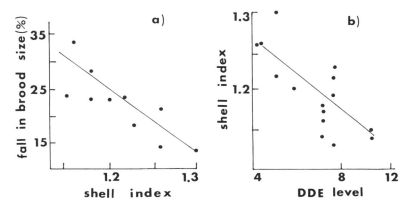

Figure 9 (a) Effect of a decrease in shell thickness on brood size, and (b) the way that DDE dose reduces this shell index (from Newton I., 1986. The Sparrowhawk. Academic Press, London. With permission.)

accumulation of lipid-soluble molecules into top predators. When this pathway coincides with a vulnerable process, such as inhibition by a lipophilic pollutant of a membrane-bound enzyme in a system with a low safety margin, then a similar critical pathway emerges. The result can have severe environmental consequences, which would not be predictable without simultaneous analyses at several different levels in the hierarchy.

REFERENCES

Adams, D. J., T. M. Dwyer and B. Hille, 1980. The permeability of end plate channels to monovalent and divalent metal cations. J. Gen. Physiol. 75, 493-510.

Aloia, R.C., C.C. Curtain and L.M. Gorden (Eds.), 1988. Physiological regulation of membrane fluidity. Alan Liss, New York.

Bayne, B. L., J. Widdows, M.N. Moore, P. Salkeld, C. M. Worrall and P. Donkin, 1982. Some ecological consequences of the physiological and biochemical effects of petroleum compounds on marine molluscs. Philos. Trans. R. Soc. London Ser. B. 297, 219-239.

Bernstein, J., 1902. Untersuchungen zur thermodynamik der bioelektrischen Strome Erster Theil. Pflugers Arch. 92, 521-562.

Borgmann, V., 1983. Metal speciation and toxicity of free metal ions to aquatic biota. Adv. Environ. Sci. Technol. 13, 47-72.

Bryan, G. N. and W. J. Langston, 1992. Bioavailability, accumulation and effects of heavy metals in sediments with special reference to United Kingdom estuaries: a review. Environ. Pollut. 76, 89-131.

Bulman, R. A. and R. J. Griffin, 1980. Actinide transport across cell membranes. J. Inorg. Biochem. 12, 89-92.

Campbell, P. and A. Tessier, 1995. Ecotoxicology of metals in the aquatic environment — Geochemical aspects, Chapter 2, this volume.

Chauhan, V. P. S. and H. Brockerhoff, 1984. Ca (phosphatidate)$_2$ can traverse liposomal bilayers. Life Sci. 35, 1395-1399.

Collander, R., 1954. The permeability of *Nitella* cells to non-electrolytes. Physiol. Plant. 7, 420–445.

Cook, L. R., S. J. Stohs and C. R. Angle, 1987. Erythrocyte membrane microviscosity and phospholipid composition in lead workers. Br. J. Ind. Med. 44, 841–844.

Crosa, J. H., 1989. Genetics and molecular biology of siderophore-mediated iron transport in bacteria. Microbiol. Rev. 53, 517–530.

Dautry-Versat, A., A. Ciechanover and H. F. Lodish, 1983. pH and the recycling of transferrin during receptor mediated endocytosis. Proc. Natl. Acad. Sci. USA 80, 2228–2262.

Decho, A. W. and S. N. Luoma, 1991. Time-courses in the retention of food material in the bivalves *Patamocorbula amurensis* and *Macoma balthica*: significance to the absorption of carbon and chromium. Mar. Ecol. Prog. Ser. 78, 303–314.

Donkin, P., 1994. Quantitative structure-activity relationships. In: Handbook of Ecotoxicology, Vol. 2, edited by P. Calow, Blackwell Scientific, Oxford, pp. 321–347.

Dwyer, T. M., D. J. Adams and B. Hille, 1980. The permeability of the end plate channel to organic cations in frog muscle. J. Gen. Physiol. 75, 469–492.

Eisenman, G., 1962. Cation selective glass electrodes and their mode of operation. Biophys. J. 2 (Suppl. 2), 259–323.

Eisenman, G. and J. A. Dani, 1987. An introduction to molecular architecture and permeability of ion channels. Annu. Rev. Biophys. Chem. 16, 205–226.

Elliot, J. E., R. J. Norstrom and J. A. Keith, 1988. Organochlorines and eggshell thinning in Northern Gannets (*Sula bassanus*) from Eastern Canada 1968–1984. Environ. Pollut. 52, 81–102.

Esser, H. O., 1986. A review of the correlation between physicochemical properties and bioaccumulation. Pestic. Sci. 17, 265–276.

Evans, G. W., 1976. Transferrin function in zinc absorption and transport. Proc. Soc. Exp. Biol. Med. 151, 775–778.

Foster, T. J., 1987. The genetics and biochemistry of mercury resistance. Crit. Rev. Microbiol. 15, 117–140.

Franks, N. P. and N. R. Lieb, 1994. Molecular and cellular mechanisms of general anaesthesia. Nature 367, 607–614.

Green, D. E., M. Fry and G. A. Blondin, 1980. Phospholipids as the molecular instruments of ion and solute transport in biological membranes. Proc. Natl. Acad. Sci. USA 77, 257–261.

Hansch, C. and J. M. Clayton, 1973. Lipophilic character and biological activity of drugs. II. The parabolic case. J. Pharm. Sci. 62, 1–21.

Harvey, R. W. and S. N. Luoma, 1985. Effect of adherent bacteria and bacterial extracellular polymers upon assimilation by *Macoma balthica* of sediment bound Cd, Zn and Ag. Mar. Ecol. Prog. Ser. 22, 281–289.

Hazel, J. R., 1984. Homeoviscous adaptation in animal cell membranes. In: Physiological Regulation of Membrane Fluidity, Vol. 3, edited by R. C. Aloia and L. M. Gordon, Alan Liss, New York, 149–188.

Heinemann, S. H., H. Terlau, W. Stuhmer, K. Imoto and S. Numa, 1992. Calcium channel characteristics conferred on the sodium channel by single mutations. Nature 356, 441–443.

Hille, B., 1992. Ionic Channels of Excitable Membranes. 2nd ed., Sinauer, Sunderland, MA, 607 pp.

Hogstrand, C., R. W. Wilson, D. Polgar and C. M. Wood, 1994. Effects of zinc on the kinetics of branchial calcium uptake in freshwater rainbow trout during adaptation to waterborne zinc. J. Exp. Biol. 186, 55–73.

Karcher, W. and Devillars, J. (editors) 1990. Practical applications of quantitative structure-activity relationships (QSAR) in environmental chemistry and toxicology. In: Euro Courses. Chemical and Environmental Sciences series. Vol. 1. Khuwer Academic, Dordrecht, Netherlands.

Konemann, H., 1981. Quantitative structure-activity relationships in fish toxicity studies. 1. Relationship for 50 industrial pollutants. Toxicology 19, 209–221.

Konemann, H. and A. Musch, 1981. Quantitative structure-activity relationships in fish toxicity studies. 2. Influence of pH on the QSAR of chlorophenols. Toxicology 19, 223–228.

Lundholm, E., 1987. Thinning of eggshells by DDE: mode of action on the eggshell gland. Comp. Biochem. Physiol. 88C, 1–22.

MacDonald, T. L. and R. B. Martin, 1988. Aluminium ion in biological systems. Trends Biochem. Sci. 13, 15–19.

McDonald, D. G. and C. M. Wood, 1993. Branchial mechanisms of acclimation to metals in freshwater fish. In: Fish Ecophysiology, edited by J. C. Rankin and F. B. Jensen, Chapman & Hall, London, pp. 297–321.

McMurchie, E. J., 1988. Dietary lipids and the regulation of membrane fluidity and function. In: Physiological Regulation of Membrane Fluidity, Vol. 3, edited by R. C. Aloia, C. C. Curtain and L. M. Gordon, Alan Liss, New York, pp. 189–237.

Meyer, K. H., 1937. Contributions to the theory of narcosis. Trans. Faraday Soc. 33, 1062–1064.

Miller, D. S., W. B. Kinter and D. B. Peakall, 1976. Enzymatic basis for the DDE induced eggshell thinning in a sensitive bird. Nature 259, 122–124.

Moore, M. N. and A. Viarengo, 1987. Lysosomal membrane fragility and catabolism of cytosolic proteins; evidence for a direct relationship. Experientia 43, 320–323.

Narahashi, T., 1978. Neurotoxicology of Insecticides and Phenomones. Plenum Press, New York, 308 pp.

Neeley, A., X. Wei, R. Olcese, L. Birnbaumer and E. Stefani, 1993. Potentiation by the ß subunit of the ratio of the ionic current to the charge movement in the cardiac calcium channel. Science 262, 575–578.

Newton, I., 1986. The Sparrowhawk. Academic Press, London, 396 pp.

Nikaido, H., 1992. Porins and specific channels of bacterial outer membranes. Mol. Microbiol. 6, 435–442.

Orbach, E. and A. Finkelstein, 1980. The non-electrolyte permeability of planer bilayer membranes. J. Gen. Physiol. 75, 427–436.

Ottaway, J.H., 1980. The Biochemistry of Pollution. Edward Arnold, London, 60 pp.

Rainbow, P. S., I. Malik and P. O'Brien, 1993. Physicochemical and physiological effects on the uptake of dissolved zinc and cadmium by the amphipod crustacean *Orchestia gammarellus*. Aquat. Toxicol. 25, 15–30.

Ratcliffe, D. 1993. The Peregrine Falcon. T. and A. D. Poyser, Carlton, 454 pp.

Rudy, B. and Iverson, L.E., 1992. Ion channels. Methods Enzymol. 207, 917 pp.

Sanders, B., K. D. Jenkins, W. G. Sunda and J. D. Costlow, 1983. Free cupric ion activity in seawater: effects on metallothionein and growth in crab larvae. Science 222, 53–55.

Serhan, C., P. Anderson, E. Goodman, P. Dunham and G. Weissmann, 1981. Phosphatidate and oxidised fatty acids are calcium ionophores. J. Biol. Chem. 256, 2736–2741.

Shephard, K. L. and K. Simkiss, 1978. The effects of heavy metals on Ca^{2+} ATPase extracted from fish gills. Comp. Biochem. Physiol. 61B, 69–72.

Shugart, L.R., 1995. Molecular markers to toxic agents, chapter 5, this volume.

Simkiss, K., 1967. Calcium in Reproductive Physiology. Chapman & Hall, London. 264 pp.

Simkiss, K. 1984. Effects of metal ions on respiratoty structures. In; Toxins, Drugs and Pollutants in Marine Animals, edited by L. Bolis, J. Zadunaisky and R. Gilles. Springer Verlag, Berlin, 137–146.

Simkiss, K., 1995. The application of controlled release and QSAR technology to sediment toxicity. Mar. Pollut. Bull. (in press).

Simkiss, K. and M. G. Taylor, 1995. Trace metals and organisms. In: Metal Speciation and Bioavailability, edited by A. Tessier and D. Turner, Int. Union. Pure Appl. Chem., John Wiley & Sons, New York, (in press).

Simon, S. A., W. L. Stone and P. B. Bennett, 1979. Can regular solution theory be applied to bilayer membranes. Biochim. Biophys. Acta 550, 38–47.

Sinensky, M., F. Pinkerton, E. Sutherland and F. R. Simon, 1979. Rate limitation of Na/ K ATPase by membrane acyl chain ordering. Proc. Natl. Acad. Sci. USA 76, 4893–4897.

Singer, S. J. and G. L. Nicholson, 1972. The fluid mosaic model of the structure of cell membranes. Science 175, 720–731.

Skou, J. C., 1988. Overview: the Na, K pump. Methods Enzymol. 156, 1–25.

Storch, J. and A. M. Kleinfeld, 1985. The lipid structure of biological membranes. Trends Biochem. Sci. 10, 418–421.

Storch, J. and D. Schachter, 1984. Dietary induction of acyl chain desaturates alters the lipid composition and fluidity of rat hepatocyte plasma membranes. Biochemistry 23, 1165–1170.

Tessier, A., P. G. C. Campbell, J. C. Auclair and M. Bisson, 1984. Relationships between the partitioning of trace metals in sediments and their accumulation in the tissues of the freshwater mollusc Ellyptio complanata in a mining area. Can. J. Fish Aquat. Sci. 41, 1463–1472.

Turner, A.J. (editor) 1990. Molecular and Cell Biology of Membrane Proteins: Glycolipid Anchors of Cell Surface Proteins. Ellis Horwood, New York, 220 pp.

Umeda, K., T. Yano and M. Hirano, 1988. Pyrethroid-resistance mechanisms in the German cockroach Blattella germanica (Orthoptera: Blattellidae). Appl. Entomol. Zool. 23, 373–380.

Winchester, B., 1992. The lysosome; structure, organisation and function. In: Fundamentals of Medical Cell Biology, Vol 4. edited by E. E. Bittar, JAL Press, Greenwich, CT.

Zachowski, A. and P. F. Devaux, 1990. Transmembrane movement of lipids. Experientia 46, 644–656.

Evaluation of Organic Contaminant Exposure in Aquatic Organisms: The Significance of Bioconcentration and Bioaccumulation*

Peter F. Landrum, Gail A. Harkey, and Jussi Kukkonen

I. INTRODUCTION

Bioaccumulation of compounds by an organism reflects its exposure to contaminants from various sources over time and represents the balance between the flux into the organism and the loss from the organism through processes such as biotransformation and elimination. Bioaccumulation is therefore an important direct link between the external contaminant concentrations in the sources and the potential effect of contaminants at various levels of biological structure and function. Bioaccumulated contaminants that attain sufficient concentrations at a receptor site for sufficient duration exert pharmacological and/or toxicological effects on the organism. Thus, the extent of bioaccumulation can be employed as a surrogate for the concentration at the receptor. This chapter will review our understanding of the accumulation of organic contaminants by aquatic organisms from water, sediment, and food, the factors that influence the accumulation processes, models for predicting accumulation, and the utility of estimating bioaccumulation for hazard assessment.

When aquatic organisms are only exposed to nonpolar contaminants via water, there is a strong relationship between the dose at the receptor and the concentration in the water. Thus, the paradigm developed for aquatic toxicology exposures is that the dose at the receptor is proportional to the dose in the organism which is, in turn, proportional to the concentration in the water. This paradigm allowed hazard evaluation based on concentrations in the external

---

* GLERL Contribution No. 911.

1-56670-1127-9/95/$0.00+$.50

media and the development of relationships between compound physical/chemical properties and toxicity, e.g., the relationship between LC50 and the octanol:water partition coefficient (K_{ow}) (Ikemoto et al., 1992). The use of external environmental concentrations for risk assessment has been generally accepted in environmental and aquatic toxicology (Suter, 1993) and has even been applied to exposure in sediments (Long and Morgan, 1990; Di Toro et al., 1991; Neff et al., 1988). The use of external concentrations relies on one dominant source for exposure and fails to recognize the relative importance of multiple routes of exposure and the rates at which transfers between sources and biota occur.

Bioaccumulation of contaminants is temporal (kinetic) in nature and depends on the conditions under which the accumulation takes place (Landrum et al., 1992a; 1994a). The steady state condition (e.g., the balance between the contaminant fluxes of infusion and loss) represents the maximal accumulation that can be attained for a given set of exposure conditions. However, conditions can change rapidly enough that steady state may not be attained except under controlled situations. Early work in aqueous media used exposures that were sufficiently long with constant conditions of external factors, such as temperature, to establish steady state. The methodology was difficult, however, and a kinetic approach for estimating steady state was developed for aqueous exposures (Branson et al., 1975; Neely, 1979). Short-term kinetic measures permitted determination of steady state through calculations which yielded similar results to the longer exposures and established the utility of a kinetic approach. This approach also allowed evaluation of the net bioaccumulation at other than steady-state conditions. The continued fostering of this approach has allowed understanding of relationships among the contaminant concentration, processes in the external environment, and accumulation of contaminants by biota (Landrum et al., 1992a).

In the following sections, we will discuss factors that affect bioavailability of organic contaminants to biota in the aquatic environment (Table 1). Special emphasis will be given to the measurement and prediction of bioavailable contaminants, along with the problems that hinder the accurate prediction of bioavailabile contaminants in real-world aquatic environments.

II. AQUEOUS EXPOSURES

A. BIOCONCENTRATION OF CONTAMINANTS

Bioconcentration is the accumulation of freely dissolved contaminant in water by aquatic organisms through nondietary routes. Many of the factors affecting this process have been reviewed (Barron, 1990). In water-only exposures, the primary route of uptake in fish is across the gill epithelium, but depending on the compound and body size, 25 to 40% of the total body burden may penetrate across the skin (Saarikoski et al., 1986). This dermal absorption is a particularly important route of exposure for non-polar contaminants into organisms containing a chitinous skeleton (Landrum and Stubblefield, 1991).

Table 1 Factors Affecting Bioavailability of Organic Contaminants in Aquatic Systems

Factors affecting bioavailability in aqueous exposures	Factors affecting bioavailability in sediment exposures
Dissolved Organic Material (DOM): Concentration of DOM Molecular structure of DOM pH of exposure media pK_a of contaminant Co-contaminants Temperature Organism elimination processes	Total organic carbon (TOC): Concentration of TOC TOC composition Sediment particle-size distribution Organism lipid content Sorption/desorption from particles Contaminant physical/chemical characteristics: Octanol/water solubility Contaminant concentration DOM and colloid concentrations in interstitial water Organism feeding behavior and life history

In many cases, the toxicokinetic behavior of the contaminant in aquatic organisms can be described by a first-order, one-compartment model (Spacie and Hamelink, 1985). The degree of bioconcentration at steady state, represented by the bioconcentration factor (BCF), depends both on the rate of absorption and the rate of elimination:

$$BCF = \frac{k_1}{k_2} = \frac{C_a}{C_w} \qquad (1)$$

where k_1 is the uptake clearance (e.g., ml/g organism/h), k_2 is the elimination rate constant for the compound (e.g., 1/h), C_a is the concentration in the organism at steady state, and C_w is the concentration in the water at steady state. Using the freely dissolved water concentration and assuming no biotransformation, this BCF represents the relative solubility of the compound in water versus the organism's tissues. (The BCF can be reduced by biotransformation processes and active elimination.) The BCF as used in hazard assessment is an estimate of the maximum potential for contaminant accumulation in aquatic organisms and is estimated from the log K_{ow} for nonmetabolized, nonionic compounds in fish (Neely et al., 1974; Veith et al., 1979, 1980; Mackay, 1982; Oliver and Niimi, 1983), invertebrates (Connell, 1988), and macrophytes (Gobas et al., 1991). This model assumes that accumulation is a partitioning process between the water and the organism, with no physiological barriers to affect the accumulation process. Differences in the regression lines representing the relationship between log BCF and log K_{ow} among organisms may well be related to differences in lipid content and composition (see below). Further, the variability in the relationships of log BCF and log K_{ow} can be reduced by working with separate relations that are limited by compound class (Schüürmann and Klein, 1988).

Linear relationships between log BCF and log K_{ow} with slopes approaching unity (Veith et al., 1979, 1980) suggested that thermodynamics dominated the steady-state condition and that molecules were essentially at chemical equilibrium. Closer study of log BCF and log K_{ow} regressions show that over a broad range of log K_{ow} (1 to 6), the BCF is described rather well, but outside this range or when the study compounds only cover a narrow range of log K_{ow}, the BCF may not be estimated accurately. This occurs because the majority of these log K_{ow}–log BCF regression models have been developed for specific classes of contaminants, e.g., halogenated hydrocarbons or polycyclic aromatic hydrocarbons (Veith et al., 1980; McElroy et al., 1989). The linear relation between log BCF and log K_{ow} breaks down for strongly hydrophobic compounds (log K_{ow} > 6; Sagiura et al., 1978; Bruggeman et al., 1984; Muir et al., 1985; Opperhuizen et al., 1985; Gobas et al., 1989a). The maximal observed value for log BCF seems to result with compounds having a log K_{ow} between 5 and 6. This effect is thought to be due to difficulty of the molecules to penetrate membranes because of diffusion and blood flow rate limitations (McKim et al., 1985; Gobas et al., 1986; Hayton and Barron, 1990), and because Log K_{ow} is a poor model for the partitioning between water and fish lipids (Opperhuizen et al., 1988; Ewald and Larsson, 1994).

As the data base for the log BCF–log K_{ow} model grows, more complex models are required to explain the bioaccumulation potential from water. Slow cavity formation in lipid membranes has been suggested to slow or limit accumulation of hydrophobic compounds and to cause variation between the octanol-water and membrane-water partition coefficients (Gobas et al., 1988b; Schüürmann and Klein, 1988). Therefore, adding a factor for the compound solubility in octanol improves ($S_{octanol}$) the log BCF–log K_{ow} relationship and accounts for some of the nonlinearity at the higher log K_{ow} values for compounds that are strongly nonideal in both octanol and lipid, e.g.,

$$\log BCF = -1.13 + 1.02 \log K_{ow} + 0.84\, S_{octanol} + 0.0004\,(mp = 25)$$

$$(n = 36, \quad r = 0.95) \tag{2}$$

where mp is the melting point of a solid in centigrade, and all liquid compounds are assumed to have a melting point of 25°C (Banerjee and Baughman, 1991). This equation allows additional prediction of bioconcentration for large hydrophobic molecules, such as dyes, and for polar compounds. However, compounds such as octachloronaphthalene are still not well predicted, presumably because of limited gill penetration. All these models contain the same assumptions of negligible metabolism and ignore the importance of organism physiology.

Molecular size and shape of a hydrophobic compound can affect or even inhibit its accumulation. The determining factor appears to be molecular size or molecular volume rather than molecular weight. The inclusion of steric factors to describe the bioaccumulation of polychlorinated biphenyls in fish in the traditional log BCF–log K_{ow} relationship improved predictions of PCB accumulation (Shaw and Connell, 1984). Further, the structure of the phospholipid bilayer of the gill epithelium can restrict uptake of hydrophobic molecules of long chain length or large cross-sectional area (>9.5 Å) by imposing a physical barrier to diffusion (Bruggeman et al., 1984; Opperhuizen et al., 1985). The toxicity of polydimethylsiloxane (PDMS), a large hydrophobic organic molecule, to different organisms is low (Hobbs et al., 1975; Aubert et al., 1985) because it does not significantly accumulate in fish or in benthic organisms, either through aqueous or dietary exposures (Opperhuizen et al., 1987; Kukkonen and Landrum, 1994a). The study of Saito et al. (1990) showed that, due to their large molecular size, the absorption of highly lipophilic macromolecules (log K_{ow} 14; size 2,000 to 50,000 daltons) was limited by low diffusion into gill membranes and resulted in no bioconcentration.

The role of stagnant water layers next to membranes during the uptake of various contaminants across fish gills has been studied (Gobas et al., 1986; Saarikoski et al., 1986). Two permeation processes, one membrane-controlled and one diffusion layer-controlled, are apparently responsible for the kinetics of accumulation processes (Gobas et al., 1986). The uptake rate of compounds under membrane control is proportional to the log K_{ow} of the contaminant, while for a diffusion layer-limited process, the uptake is independent of log K_{ow}. Such a two-level process control can explain the positive correlation between log K_{ow} and accumulation rates through fish gill for contaminants with log K_{ow} ranging from 1 to 4, but cannot explain the absence of correlation or even decreasing accumulation rate for contaminants with log K_{ow} higher than 4 (Pärt, 1989). Further, studies of the influence of log K_{ow} on the transport across the gill membrane suggest that the mechanisms controlling transport are nonspecific with respect to chemical structure. Additional factors, e.g., molecular volumes, molecular weight, and self-association, affect the transport of highly lipophilic compounds (McKim et al., 1985).

Despite these limitations, log K_{ow} remains an important parameter for estimating the BCF of compounds in aquatic organisms (Lyman et al., 1990). This relationship is dominated by the solubility of nonpolar compounds in the organism lipids; thus, organisms became viewed as bags of lipids floating in water, and BCF as a chemical equilibrium between the lipid and water. Further, the relationship between log BCF and log K_{ow} has led to the use of thermodynamic equilibrium models for describing the potential accumulation of nonpolar organic contaminants by aquatic organisms and has dominated the risk assessment of aquatic systems. This concept has dominated even when multiple sources and kinetic limitations are apparent.

B. FACTORS THAT INFLUENCE BIOAVAILABILITY IN AQUEOUS EXPOSURES

Only the bioavailable fraction of a compound in water can be accumulated by aquatic organisms (Hamelink and Spacie, 1977). Factors that may influence contaminant bioavailability include the exposure concentration and contaminant binding to dissolved organic matter or particles in water. The pH can also affect bioavailability of polar compounds through dissociation/association processes for acids and bases.

Dissolved organic material (DOM) affects the chemistry and ecology of aquatic habitats and, more importantly, the fate of pollutants. DOM can bind several types of hydrophobic organic contaminants in aquatic environments (Ogner and Schnitzer, 1970; Hassett and Anderson, 1979; Gjessing and Bergling, 1981; Carlberg and Martinsen, 1982; Carter and Suffet, 1982; Landrum et al., 1984; Hassett and Milicic, 1985; Sithole and Guy, 1985; Chiou et al., 1986; Lara and Ernst, 1989; Lee and Farmer, 1989; Kango and Quinn, 1992). Some studies suggest that the binding occurs rapidly (McCarthy and Jimenez, 1985a; Schlautman and Morgan, 1993). In surface waters from different sources, the affinity of DOM for a given compound appears to differ (Carter and Suffet, 1982; Landrum et al., 1985; Morehead et al., 1986; Kukkonen and Oikari, 1991). The causes of these differences in binding affinity are not fully described. However, several phenomena have been postulated, and two or more binding interactions may occur simultaneously, depending on the chemical characteristics of the compound and the DOM (Choudhry, 1983). The interactions most likely involved are van der Waals forces, hydrophobic bonding, hydrogen bonding, charge transfer, ion exchange, and ligand exchange (Choudhry, 1983). The exact mechanism for a particular binding site will depend on both the chemical characteristics of the contaminant and the DOM.

The magnitude of contaminant-DOM binding is generally expressed as a partition coefficient, K_{oc} (e.g., low K_{oc} values represent less contaminant-DOM binding than high K_{oc} values), and for nonpolar organic contaminants is generally linearly related to K_{ow} as an expression of contaminant hydrophobicity (McCarthy and Jimenez, 1985a; Chiou et al., 1986; Lara and Ernst, 1989). The partition coefficients determined for various hydrophobic organics vary by factors of 5 to 10 among various soil and aquatic humic sources of dissolved/colloidal-bound forms in water (Carter and Suffet, 1982; Chiou et al., 1986). A reason for this variation in partitioning may be that contaminant binding is affected by the structure of the DOM. For example, pyrene binding to DOM in natural waters correlated with DOM aromaticity (Gauthier et al., 1987); an increase in aromaticity was shown to increase strength of PAH binding. Further, the binding of benzo(a)pyrene to DOM correlated with aromaticity, molecular size, and hydrophobic acid content of DOM (McCarthy et al., 1989; Kukkonen and Oikari, 1991). The structural features of DOM that influence binding vary among compound classes. For example, 2,2',4,4',5,5'-hexachlorobiphenyl will only bind strongly to the hydrophobic neutral fraction of DOM, while BaP will bind to both the neutral and the hydrophobic acid fractions. These examples demonstrate the complex interaction of compounds with DOM (Kukkonen et al., 1991).

The bioavailability of organic compounds in natural waters is decreased by DOM (Leversee et al., 1983; Carlberg et al., 1986; Kukkonen et al., 1989; Servos et al., 1989; Servos and Muir, 1989). The magnitude of the decrease is related to the magnitude of the contaminant-DOM partition coefficient, generally calculated as an organic carbon normalized partition coefficient (K_{oc}) (Landrum et al., 1985, 1987; McCarthy and Jimenez, 1985b; McCarthy et al., 1985; Black and McCarthy, 1968). An exception is naphthalene accumulation from natural waters with a low DOM concentration (<4 mg carbon/l), which yielded two to three times higher BCF values than in DOM-free control water (Kukkonen and Oikari, 1991). A similar but less pronounced effect of DOM on naphthalene accumulation was detected by diluting a natural water sample to 5 mg carbon/l with DOM-free control water (Kukkonen et al., 1990). Likewise, the bioavailability of methylcholanthrene to *Daphnia magna* increased in water containing Aldrich humic acid (Leversee et al., 1983), but this could not be confirmed in another laboratory (McCarthy et al., 1985).

The observed bioaccumulation of model compounds in waters containing DOM can be compared to the predicted BCF values. The prediction is based on the assumption that contaminant bound to DOM is unavailable for uptake, mainly because the contaminant-DOM complex is too large to penetrate biomembranes and the dissociation rate is too slow to allow significant competition for uptake with the freely dissolved compound. Accordingly, bioaccumulation in water containing DOM is assumed to be proportional to the fraction of the contaminant that is freely dissolved. The present data on the effects of natural DOM on bioaccumulation suggest that correlation of freely dissolved contaminant with bioavailability is valid for short-term exposures, where little or none of the contaminant associated with DOM is available to aquatic animals, like *D. magna* (McCarthy et al., 1985; Kukkonen et al., 1989, 1990; Kukkonen and Oikari, 1991), *Diporeia* spp. (Landrum et al., 1985, 1987), *Crangonyx laurentianus* (Amphipoda) (Servos and Muir, 1989), and rainbow trout (Black and McCarthy, 1988). In addition to dissolved organic matter, the BCF of contaminants is reduced in systems where there is increased primary productivity. The ability of both dissolved and particulate organic matter to bind contaminants apparently reduces the amount that is bioavailable (Larsson et al., 1992; McCarthy and Black, 1988). The relative role of organic matter will depend on its binding capability (Eadie et al., 1990, 1992).

For ionizable compounds, the pH of the medium and the pK_a of the contaminant dictate the presence of nonionized and ionized forms. Nonionized forms are assumed to be the bioavailable forms according to the pH-partition hypothesis originally formulated for compound absorption in the intestine. Ionized forms are assumed not to penetrate the gut at significant rates (Klaassen, 1986). Consequently, for species exposed in an aqueous environment, the pH of the water strongly affects the toxicity and accumulation of organic weak acids, such as chlorophenols and dehydroabietic acid (McLeay et al., 1979a, 1979b; Saarikoski and Viluksela, 1981; Spehar et al., 1985; Fisher and Wadleigh, 1986; Saarikoski et al.,1986; Stehly and Hayton, 1990; Fisher, 1990, 1991). A similar potential is recognized for organic weak bases (Kalsch et al., 1991), although these compounds are not as well studied as weak acids.

The accumulation of pentachlorophenol (PCP) and dehydroabietic acid by *D. magna* and *Heptagenia fuscogrisea* is clearly pH dependent, both in artificial freshwater and in natural water containing aquatic DOM (Kukkonen, 1991). DOM reduced the uptake of PCP into *H. fuscogrisea* significantly when pH ranged from 4.5 to 7.5. However, at pH 3.5, the difference between humic and control waters was not statistically significant, although a 15% lower BCF value was noted in humic water than in control water. At pH 8.5, there were no differences between humic and control treatments. Thus, the nonionized form of PCP is more readily accumulated by animals. This is also in line with other studies showing that the nonionized forms of compounds are more available and toxic than the ionized forms at lower pH (Saarikoski and Viluksela, 1981; 1982; Spehar et al., 1985; Fisher and Wadleigh, 1986; Fisher, 1990; Howe et al., 1994a).

The accumulation of weak organic acids cannot be fully described by the pH-partition hypothesis (Saarikoski et al., 1986; Pärt, 1989). Accumulation data indicate that the ionized compounds contribute significantly to the uptake rates and can affect toxicity at higher pH values (Stehly and Hayton, 1983). This effect happens when the exposure pH is well above the pK_a value (1 to 2 pH units) of the compound in question. There are three possible explanations for this phenomenon. The first considers the role of stagnant water layers at the gill surface. The diffusion rates in water of both nonionized and ionized forms are similar, and the diffusion resistance in the stagnant layers is independent of pH. For the nonionized forms with the highest K_{ow}, the diffusion through the stagnant layer will be rate-limiting. The uptake will proceed at the rate proportional to that of nonionized forms. When the pH increases above the pK_a, the relative importance of the stagnant layer gradually decreases, and the resistance in the membrane becomes rate-limiting (Saarikoski et al., 1986). The second possibility is that the gill membranes are permeable to the ionized form of contaminants. The third possibility is that the pH at the gill surface is different from that of the bulk water such that a portion of the ionized compound becomes nonionized.

The pH also affects the toxicity and accumulation of neutral compounds, although this effect is not as obvious as in the case of acids (Fisher, 1985; Fisher and Lohner, 1986). Accumulation of the nonpolar compounds, benzo(a)pyrene and lindane, by *H. fuscogrisea* and *D. magna* are also pH dependent; the highest BCF values are obtained at pH 6.5 and the lowest at 8.5 (Kukkonen, 1991). The pH-dependent accumulation pattern for lindane in these organisms is similar to that observed for the midge, *Chironomus riparius* (Fisher, 1985). The phenomenon is attributed to a lower permeation of lindane into the animals at pH 4 and an enhanced degradation of lindane at pH 8, compared to that at pH 6 (Fisher, 1985). Alternatively, the pH effect on the uptake rates of nonpolar compounds within the range from 5 to 10 has not always been observed (Pärt, 1989).

Other compounds in the water can influence the relative accumulation of a particular contaminant. When trace amounts of co-contaminants were added with particular contaminants to water, the bioavailability declined. For BaP, the more aromatic and hydrophobic the co-contaminants, (e.g., toluene) the lower the co-contaminant concentration that is required to reduce the uptake clearance (Landrum, 1983). However, water-soluble co-contaminants such as acetone do not exert any effect until they are in the 10% concentration range (Landrum, 1983). These effects are hypothesized to occur because of some interaction between the BaP and solvent that produces a complex that is too large to penetrate the membrane or is slow to penetrate because of its size. In addition, solvent may disrupt membrane fluidity that may change membrane permeability to increase or decrease diffusion. However, the relative influence of co-contaminants has not always resulted in altered kinetics (McCarty et al., 1992).

In addition to organic matter and pH, temperature is an important environmental parameter that influences bioconcentration. Most exposures are conducted at constant temperature and the possible effect of temperature change is not known. That temperature has a profound effect both on observed toxicity and bioconcentration is known (Brown et al., 1967; Boryslawskyj et al., 1987; Landrum, 1988; Lohner and Fisher, 1990; Lydy et al., 1990; Lewis and Horning, 1991; Howe et al., 1994a). In general, there is a two- to fourfold (in some cases even tenfold) increase in toxicity with each 10°C rise in temperature (Fisher, 1991). This toxicity change corresponds with the concept that ectothermic metabolism increases approximately twofold for every 10°C increase. As the overall metabolism increases, the rate of uptake increases and the time required to reach a lethal body burden decreases (Gerould et al., 1983; Landrum, 1988). In some cases, the overall steady-state condition and the resulting effect remain constant, but the kinetics increase the time required to reach steady state at the lower temperature (Howe et al., 1994b). However, when biotransformation is important for determining contaminant disposition within an organism, increasing temperature may well increase the biotransformation rate, increase elimination, and decrease toxicity (Howe et al., 1994a; Niimi and Palazzo, 1985). Thus, the results of increasing temperature can enhance the rates of both the uptake processes and the biotransformation processes (Landrum, 1988; Lydy et al., 1990; Dabrowska and Fisher, 1993). There are, however, biological limits to such increases; at sufficiently elevated temperatures, organism tolerance will generally decline because its temperature tolerance is exceeded, e.g., anthracene biotransformation peaks at 25°C and declines at 30°C in chironomids (Gerould et al., 1983). Therefore, the influence of temperature on the overall uptake of contaminants can produce a confounding effect on the expected accumulation unless the temperature dependences on biotransformation, accumulation, and elimination processes are known.

In addition to uptake, elimination processes strongly affect contaminant concentration in the organism. Elimination occurs by two processes: (1) diffusion of the parent compound through the surface membranes of the organism (gills, skin), and (2) biotransformation and elimination of metabolites. In simple kinetic models, as described above, the mechanisms dictating the elimination are not always defined. Elimination is often measured as loss of radioactivity, which is loss of parent and nonparent metabolites. In rate constant terminology, this can be described as

$$k_e = k_p + k_m \qquad (3)$$

where k_e is the overall elimination rate constant (1/h), k_p is the rate constant for physicochemical loss (elimination of parent compound, 1/h), and k_m is the biotransformation rate constant (1/h) (de Wolf et al., 1992). The relative contribution of the k_m to the k_e will depend on the k_p. In instances when a high log K_{ow} and a low k_p are observed, even a limited change in the ability of the organism to metabolize the contaminant can have a substantial effect on the overall elimination rate (k_e). This phenomenon was demonstrated with 2,8-dichlorodibenzo-p-dioxin (DCBP) kinetics in fish (Sijm and Opperhuizen, 1988). Two groups of goldfish were exposed to DCBP; one was pretreated with piperonyl butoxide, an inhibitor of the cytochrome P-450-dependent monooxygenases. The elimination rate constant of DBCP in the pretreated fish was significantly lower than that in the untreated fish. The calculated rate constant for metabolism, k_m, was significantly larger than the elimination rate constant, k_e, in the treated fish. However, in instances when a compound of low K_{ow} is observed along with a high k_p relative to the accumulation rate, the compound will have to be biotransformed at very high rates before the effect on the overall elimination rate can be noticed (de Wolf et al., 1992). Additionally, biotransformation can reduce the rate of total elimination in invertebrates (Landrum and Crosby, 1981). The idea of enhanced elimination, i.e., loss of parent compound plus metabolites, with biotransformation presumes a kidney-like function, where polar materials are filtered and cannot be reabsorbed. In invertebrates, however, this function does not exist, and biotransformation to more polar compounds may trap the metabolites in the organism unless other processes can actively transport the compound out of the organism. Such processes may include the formation of peritrophic membranes and fecal elimination.

C. NORMALIZATION OF CONTAMINANTS ACCUMULATED FROM AQUEOUS EXPOSURES

BCF is traditionally defined as contaminant concentration in the organism divided by the concentration in the water. However, recent work suggests normalizing contaminant concentration in the organism to the fraction of organism lipid. This will reduce the variance among organisms of differing lipid concentrations if the source is the same (Barron, 1990). However, variation in lipid composition or food chain position may still result in differences among species. When organisms collected from Lake Baikal were examined, the magnitude of the lipid-normalized BCF values were similar for two fish species, although the relationship between log $BCF_{lipid\ normalized}$ and log K_{ow} had significantly different slopes (Kucklick et al., 1994). However, when significantly different trophic levels were compared, seals had concentrations that were often ten times higher than fish, even after lipid normalization (Kucklick et al., 1994).

The BCFs for lake trout and white fish from Siskiwit Lake Isle Royal exhibit significant variability. Even with lipid normalization, the correlation of log K_{ow} with log BCF was weak for pesticides, while the correlation for PCBs was even more variable (Swackhamer and Hites, 1988). More recently, the partitioning of 2,2',4,4'-tetrachlorobiphenyl (TCB) between water and fish lipid appeared to be influenced by lipid composition (Ewald and Larsson, 1994). Fish lipid with high phospholipid content had essentially the same accumulation of TCB as fish of a low phospholipid content, but differences in lipid normalized BCFs were shown, suggesting species-specific accumulation. With the trend toward lipid normalization of organism data, care must be taken to ensure which method is being employed, since differences in the method can result in different lipid measures and therefore different normalized BCF values (Randall et al., 1991).

III. SEDIMENT EXPOSURES

Accurate prediction and evaluation of contaminant exposure and accumulation from sediments remains difficult because of the complex interactions between the contaminant, the sediment, and the organism. Factors that contribute to these interactions include:

1. Chemical characteristics and concentration of the contaminant
2. Physical and chemical characteristics of sediments
3. The presence of complex mixtures that can produce confounding effects when related to sediment constituents and biota
4. Organism behavior and physiology influenced by such environmental factors as temperature, nutrient availability, and habitat that can modify the exposure between species and temporally within a species
5. The length of sediment/contaminant contact time, which can change bioavailability

One approach to normalizing exposure of neutral organics against principal controlling factors in sediments uses an equilibrium partitioning bioaccumulation model (Lake et al., 1987; U.S. EPA, 1989). The basic hypothesis of this model is thermodynamic equilibrium. Thus, contaminant chemical activity is the same in each phase of a sediment matrix, including sediment particles, sediment carbon, interstitial water, and biota. For neutral organic compounds, the amount of organic carbon in the system generally determines the extent of contaminant partitioning between the sediment particles, interstitial water, and dissolved organic carbon. Thus, the exact route of contaminant exposure is not needed to determine the exposure or biological effect produced in the equilibrated system, because all phases will be at equilibrium and equal chemical potential (Di Toro et al., 1991). If a system is at equilibrium, equilibrium partitioning theory predicts that bioavailability should be directly proportional to contaminant chemical activity in a particular phase (i.e., interstitial water) and inversely proportional to the organic carbon content of the sediment, since organic carbon, to a large extent, controls sorption to sediment particles (U.S. EPA, 1989).

The equilibrium partitioning model can be directly applied for estimating bioaccumulation potential of sediment-associated neutral organic contaminants in benthic macroinvertebrates (McFarland, 1984; Lake et al., 1987). Assuming that organic carbon is the only sink for neutral organic contaminants, and that lipids are the only sink in organisms, the following model can be constructed (Lee, 1992):

$$\frac{Ctss}{L} = AF\left(\frac{Cs}{TOC}\right) \qquad (4)$$

where Ctss = tissue concentration at steady state ($\mu g/g$ organism), L = fraction of lipids in organism (g lipid/g organism), AF = accumulation factor (g carbon/g lipid), Cs = concentration of contaminant in the sediment ($\mu g/g$ dry weight), and TOC = mass fraction of organic carbon in sediment (g carbon/g sediment). According to this model, the derived AF should not vary among neutral organic contaminants, because partitioning is not a function of lipid or carbon composition (Gobas et al., 1989b). Using empirical relationships between log K_{ow} and log BCF, and log K_{ow} and log K_{oc}, the AF was estimated as 1.7 (McFarland and Clarke, 1989). Accumulation factors less than this value would indicate less partitioning into lipids than predicted, while values greater than 1.7 would indicate a greater uptake of contaminant than can be explained by the model. In addition, if the systems are at equilibrium, AFs should be constant among species and among sediments, regardless of the amount of lipid contained in the species or the amount of organic carbon content contained in the sediment.

The validity of this approach has been tested primarily through the use of toxicity tests. A number of toxicity studies have accurately predicted exposure via equilibrium partitioning theory (Adams et al., 1984; Ziegenfuss et al., 1986; Nebeker et al., 1989; Swartz et al., 1990). Based on data derived from such studies, Swartz et al. (1990) and Di Toro et al. (1991) found correlations among organism survival, interstitial water concentrations, and organic carbon-normalized sediment concentrations for fluoranthene and kepone (Adams et al., 1984). Likewise, sediment toxicity conducted with *Hyalella azteca* in sediments dosed with DDT decreased with increasing carbon concentration (Nebeker et al., 1989). Equilibrium partitioning accurately predicted DDT toxicity which accounted for the reduced bioavailability in proportion to the amount of organic carbon. However, *H. azteca* exposed to endrin-spiked sediments under the same test conditions demonstrated an increase in toxicity with sediment organic carbon (Nebeker et al., 1989). Under the equilibrium partitioning approach, the extent of toxicity should be explained by the normalization to TOC in the sediment, and should be valid for contaminants with both high and low K_{ow} values. While this held true for DDT, it did not for endrin. Equilibrium partitioning should have predicted toxicity more accurately for endrin (log K_{ow} 3.23) than DDT (log K_{ow} = 6.6), since sorption capacity of sediment should have been much less for endrin than DDT (Nebeker et al., 1989).

Although equilibrium partitioning theory assumes that, at equilibrium, all phases in a sediment matrix have the same potential for contributing contaminant to an organism, this assumption has been found to be an oversimplification. Sorption was a strong function of organic carbon content when toxicity was determined for a mixture of chlorinated ethers in sediment (Meyer et al., 1993). The toxicity to *H. azteca*, *C. tentans*, and *D. magna* depended on sediment organic carbon, strengthening the validity of the equilibrium partitioning approach. However, the interstitial water concentrations did not correlate with observed effects for the two sediment-dwelling species, *C. tentans* and *H. azteca*. Further, toxicity did not correlate with interstitial water concentrations, and LC50 values varied by up to a factor of six among sediments containing various amounts of organic carbon. For equilibrium partitioning, the exposure and, in turn, the bioavailability of a contaminant should be similar for all test species after normalizing for contaminant concentration differences in interstitial water and organic carbon content. However, the bioavailability of fluoranthene to three species (*H. azteca*, *D. magna,* and *C. tentans*) was not similar in three laboratory-dosed sediments possessing the similar physical and chemical characteristics (i. e., similar organic carbon, particle size fractions). Significant differences in species response were observed (Suedel et al., 1993), and three- to five-fold differences in EC50 values were seen among the three sediments.

The aforementioned equilibrium partition studies have used toxicity as a surrogate for exposure in contaminated sediments. In order for toxicity to serve as a surrogate, the contaminant must be toxic at concentrations within its aqueous solubility limit, assuming that the organism is at the same chemical activity as the aqueous phase (i.e., interstitial water). A more direct approach to validate equilibrium partitioning theory is to measure bioaccumulation of sediment-associated contaminants. Bioaccumulation assays are perhaps more appropriate for testing equilibrium partitioning, since, depending on exposure length, a steady state body burden can be obtained. Equilibrium partitioning assumptions were tested in two infaunal species, *Macoma nasuta* and *Nereis virens*, by examining accumulation of PCB congeners and fluoranthene from sediments that varied in composition and organic carbon content (Brannon et al., 1993). Although calculated AFs bracketed the theoretical value of 1.7, the range over which AFs varied among the sediments and among the individual species was approximately 600%. Variation in bioavailability for three sediment-ingesting species was sevenfold, based on accumulation values normalized to contaminant concentration and organic carbon (Harkey, 1993). A wide variation in calculated AFs was found for infaunal organisms exposed to a variety of neutral hydrophobic contaminants over a range of hydrophobicities (Lee, 1992). Mean AFs ranged from 0.1 to 10.9 and were inconsistent both among species and contaminants. For example, mean AFs calculated for total PCBs were as low as 0.4 and as high as 5.9, using the same indicator species, *Macoma nasuta* (Rubenstein et al., 1987; Ferraro et al., 1990a). For compounds of similar hydrophobicity, mean AFs ranged over an order of magnitude, as with benzo(a)pyrene (mean AF = 0.3, Ferraro et al., 1990a) and chlordane (mean AF = 4.7, Lake et al., 1987). Overall mean AFs calculated for PCBs, DDD, and chlordane exceeded the 1.7 value by a range of 25 to 250% (Lee, 1992). These reports suggest that factors other than carbon and lipid content are responsible for the wide range of contaminant accumulation among sediment-dwelling species.

Where equilibrium partitioning theory fails to accurately predict exposure, one or more assumptions of the model have been violated. For significant differences between predicted and observed bioavailability, thermodynamic equilibrium may not have been achieved. Such would be the case where sediment is manipulated prior to sediment assays (i.e., using laboratory-dosed sediments for bioaccumulation studies). If equilibrium partitioning fails, then organisms exposed to environmentally resident contaminants, that we can assume to have reached equilibrium, should provide validation. Such field-validation studies are scarce. However, deviations from constant AF (Lake et al., 1990) and toxic response relative to interstitial water concentrations (Hoke et al., 1994) exhibit deviations with environmentally resident contaminants (e.g., historically contaminated sediments). The equilibrium partitioning approach tends to reduce the variance in toxicity response and bioavailability over the use of whole sediment concentrations, but is insufficient to predict bioavailability of hydrophobic organic contaminants closer than about a factor of ten. The method, however, does provide a point of departure for comparison with other approaches and an initial estimate of expected results.

Bioaccumulation varies with the characteristics of the contaminant. For oligochaete worms, the bioaccumulation factor for accumulation of chlorinated hydrocarbons from sediments was low and relatively stable between log K_{ow} of 3 and 4, increased at values between 4.5 and 6, and declined at log K_{ow} of greater than 6 (Oliver, 1987; Landrum et al., 1989; Figure 1). The variation in bioaccumulation with log K_{ow} does not appear to be complicated by exposures to mixtures. Contaminants in the environmental mixtures seem to behave independently when concentrations are at typical environmental levels (Oliver, 1987; Landrum et al., 1989; 1991). When the concentrations become extreme, then the chemistry will change and the bioavailability will change. One example is the influence of cosolvents on partitioning to suspended matter. Sorption coefficients decreased exponentially with increasing fractions of methanol-water and acetone-water cosolvents when soil solutions containing hydrophobic contaminants were evaluated (Nkedi-Kizza et al., 1985).

The predominant environmental factor affecting contaminant bioavailability is adsorption to particles and, for neutral hydrophobic organic compounds, the particle-associated organic matter. Yet it appears that normalizing accumulation data to the fraction of organic carbon in sediment is inadequate to account for all of the variability in the data. Partitioning studies indicate that contaminant sorption and bioavailability may be affected by different forms of organic carbon as well as by amount (Word et al., 1987; Suedel et al., 1993). For example, organic carbon composed of mineral forms such as coal may sorb

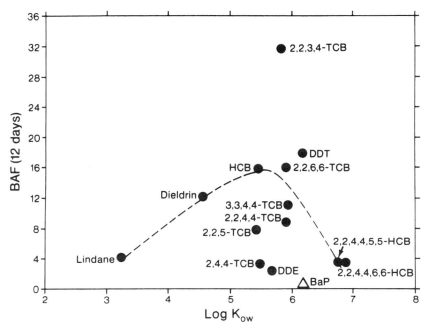

Figure 1   Scatter plot of the bioaccumulation factor (BAF), concentration in *Diporeia* spp. divided by the concentration in sediment, against log K_{ow}. The open triangle represents benzo[*a*]pyrene, the only compound that is not a chlorinated hydrocarbon. (Reprinted from Landrum et al., 1989. In: Aquatic Toxicology and Hazard Assessment XII, ASTM STP 1027, edited by U.M. Cowgill and L.R. Williams. American Society for Testing and Materials, Philadelphia, PA, pp. 315–329.)

hydrophobic contaminants more tightly or less tightly than dissolved or particulate forms of organic carbon (Suedel et al., 1993). The partitioning to soils was found to vary with soil organic matter composition. The organic carbon-normalized partition coefficient decreases for a particular compound with increases in the soil organic matter polarity expressed as the (O + N)/C ratio (Rutherford et al., 1992). The molecular structure as well as the amount of surface area comprising the various forms of organic carbon may largely control the bioavailability of neutral organics (Word et al., 1987). The differential bioavailability seen among sediments possessing the same organic carbon content but obtained from different sources (Word et al., 1987; DeWitt et al., 1992; Suedel et al., 1993) points heavily to the influence of sediment composition on exposure.

Not only does the partitioning apparently vary with the organic carbon content and composition, but compounds of different chemical classes appear to distribute differentially on various size classes of particulate material. Recently, the distribution of several nonpolar organic compounds among sediment particles was examined with respect to the organic matter content. In these studies, the distribution of pyrene, benzo(*a*)pyrene, and hexachlorobiphenyl on sediment particles differed from the organic carbon distribution on sediment particles that were <63 µm in diameter (Harkey et al., 1994a; Kukkonen and Landrum, 1994b). The distributions of contaminants among small particles also differed significantly even between compound classes (Harkey et al., 1994a) and when the contaminant concentration changed (Kukkonen and Landrum, 1994b; Landrum et al., 1994b). These relative distributions were directly attributed to bioavailability differences of the sediment-associated contaminants observed in *Diporeia* and *Lumbriculus variegatus* (Harkey et al., 1994a; Kukkonen and Landrum, 1994a,b; Landrum et al., 1994b).

As in aqueous exposures, the amount of dissolved organic matter (DOM) and colloids in interstitial water will also affect contaminant sorption and thus reduce contaminant bioavailability. If organic carbon concentration from all sources (e.g., sediment, DOM, colloids) is not considered, a large error may result when predicting bioavailability.

Determinations of contaminant binding to DOM depend on the method used, which can vary significantly among methodologies (Harkey et al., 1994b). McCarthy and Black (1988) estimated 50 to 90% errors in steady state predictions when DOM was not included as a factor regulating bioavailability. Likewise, the method for determining total organic carbon content in sediment, frequently determined by combustion and measuring CO_2 or gravimetric methods, will not yield similar values; the role of organic matter may appear to vary if the differences in methods are not considered.

Partitioning between porewater and sediment particles has been described in multikinetic processes that appear as two differentially bioavailable pools: one in a reversible pool and another in a resistant pool (Landrum and Robbins, 1990). The fraction of contaminant that resides in each of these pools changes, depending upon the sorption duration, until equilibrium is reached. This may take months to years to achieve (Karikhoff, 1980; Di Toro et al., 1982; Coats and Elzerman, 1986; Witkowski et al., 1988; Fu et al., 1994). During this equilibration phase, contaminant bioavailability can also be expected to change with time, which can be especially important when examining data obtained from bioassays that employ spiked sediment. The duration of the mixing process affected toxicity in *Daphnia* sp., where toxicity decreased as the mixing time increased (Stemmer et al., 1990). Similarly, the bioavailability of phenanthrene and pyrene was found to change upon spiked sediment aging from 3 to 180 d; however, BaP exhibited no significant bioavailability changes, as determined by uptake rate coefficients (Landrum et al., 1992c; Harkey et al., 1994c). Differential bioavailability with sediment aging has also been observed for field-collected sediments. When *Macoma nasuta* were exposed to surface and deeper sediments (4 to 8 or 8 to 12 cm) of a sediment core, significant increases in calculated AFs were seen in surface sediments for 20 of the 29 exposures reported (Lee, 1991; Ferraro et al., 1990b). The higher AFs from the surface sediments were for contaminants that had spent less time in the sediment and were more bioavailable to the clams (Lee, 1991). However, the composition of the subsurface organic matter is most likely different from that at the surface. Subsurface material is likely more reduced due to diagenetic

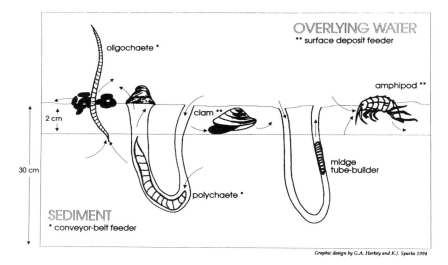

Figure 2 Representation of typical feeding behaviors of test species used in bioaccumulation assays. Surface deposit feeders ingest sediment from the sediment/water interface to the upper few centimeters, while conveyor-belt feeders and tube builders generally ingest sediment from deeper in the sediment core. Amphipods may occupy the sediment/water interface or burrow into the upper few centimeters of sediment, depending on species. Both oligochaete and polychaete worms ingest sediment from a variety of depths and deposit gut contents on the sediment surface so that uptake and elimination generally occur in surficial sediment. Arrows depict areas of contaminant uptake, elimination, and deposition by benthos via sediment particles and interstitial water. Biota are not drawn to scale.

(reconstructive) processes. As discussed above, changes in composition can influence contaminant bioavailability and may have contributed to the bioavailability differences observed with depth.

Epifaunal and infaunal sediment dwellers represent a myriad of feeding behaviors and life histories. Each of these behaviors can affect the relative contaminant exposure via manipulation of the environment surrounding the organisms (Figure 2). For example, infaunal oligochaetes burrow through sediment and obtain food from ingested sediment particles. These organisms, appropriately named "conveyor belt deposit-feeders," ingest sediment over a range of depths while they deposit gut contents on the sediment surface from posterior ends that protrude at the sediment-water interface (Karickhoff and Morris, 1985; Robbins, 1986). Bioturbation produced by these organisms disrupts any equilibrium established among sediment-associated contaminants, affecting bioavailability not only to the oligochaetes, but to all biota in the reworking zone. This behavior can also redistribute contaminants from buried deposits back into the feeding zone for shallower feeding organisms (Kielty et al., 1988a,b).

Selective feeding behavior is one of a number of biological characteristics affecting contaminant availability. In *Diporeia* spp., this selectivity combined with differential partitioning among sediment particles is suggested as a major reason for the differential accumulation of chlorinated hydrocarbons as opposed to polycyclic aromatic hydrocarbons from sediments (Landrum, 1989; Landrum and Faust, 1991; Harkey et al., 1994a). When the relative distribution among sediment particles is large, such selectivity is readily assumed (Harkey et al., 1994a). However, when the chemically measured differences in distribution are small, the picture is much more uncertain (Kukkonen and Landrum, 1995). The magnitude of contaminant association to particles may be equally as important as selective feeding.

Intraspecies feeding and physiological variation is responsible for the lower AFs generally seen in filter-feeding organisms, compared to deposit feeders. Surface deposit feeders, such as the clam *Macoma nasuta,* filter food particles from the upper few millimeters of sediment. These organisms would not be exposed to contaminants found at depth (Lee, 1991). Thus, sediment-ingesting species have been recommended for use in bioaccumulation studies, since they most likely represent a worst-case scenario (e.g., more apt to accumulate contaminants than other species) for sediment exposures (U.S. EPA, 1989).

Physiological adjustments made among infaunal species, such as the production of tubes or burrows, can also alter the bioavailability of contaminants contained in interstitial water. These structures effectively encase the organism and create a differential permeability that tends to change the chemical composition of the water inside the tube compared to the surrounding interstitial water, so that direct contact with interstitial water does not occur (Lee, 1991; Aller, 1983; Aller et al., 1983). The resulting change in water flux through the burrow has been estimated to be between 2 and 2000 ml/h (Lee and Swartz, 1980; Figure 3).

In instances where contamination can be detected by organisms, sediment avoidance can occur, either by the animal burrowing around or under the zone of contamination (Landrum and Robbins, 1990) or by emerging to settle on the surface (Keilty et al., 1988c; Kukkonen and Landrum, 1994b). Of a more subtle nature, the exposure and accumulation of contaminants can result in contaminant-related changes in physiology. These changes in the physiology of the organism, which are manifest as changes in toxicokinetics, represent changes in accumulation and, in turn, bioavailability (Landrum et al., 1991, 1994b; Kukkonen and Landrum, 1994b). Both enhancement and reduction in bioaccumulation that occur with dose are thought to be strongly related to feeding rate for sediment-associated contaminants, because enhanced feeding rate can result in increased accumulation (Harkey et al., 1994a).

Once ingested by an organism, uptake of sorbed contaminants can be modified by gut processes that cause contaminant fugacity (apparent chemical activity of a chemical substance from a phase) and concentration to increase as the volume of food decreases and lipids are hydrolyzed (Lee, 1991; Gobas et al., 1988a, 1993). These digestive processes alter the thermodynamic gradient in such a way that there is a net flux of contaminant from gut contents into the gut wall that ultimately causes tissue contaminant concentrations to rise above what is predicted by equilibrium partitioning. Tissue concentrations of highly

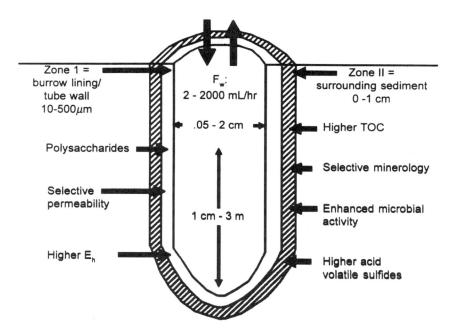

Figure 3 Properties of infaunal tubes and burrows. A generalized schematic to illustrate the structure and the range in the dimensions and flux of water (F_w or "irrigation") of tubes and burrows. The actual dimensions and shapes vary greatly among species. (Reprinted with permission from Lee, 1991. In: Organic Substances and Sediments in Water, Vol. III, Biological, edited by R. A. Baker, Lewis Publishers, Ann Arbor, MI, pp. 73–93.)

hydrophobic compounds have often been shown to exceed the concentrations predicted from thermodynamic partitioning (i.e., McFarland and Clarke's (1989) AF value of 1.7; Connolly and Pederson, 1988).

Other biological factors that may affect contaminant accumulation from sediments are organism age, sex, lipid content, and seasonality. Reproductive state and seasonal feeding behavior of species such as *Diporeia* affect accumulation and elimination of contaminants (Landrum, 1988). *Diporeia* spp. are particle-selective feeders that lack a continuous feeding pattern throughout the year (Quigley, 1988). As a result, their uptake and elimination kinetics are expected to differ from species that do not demonstrate a strong seasonal feeding behavior, e.g., oligochaetes.

IV. FOOD WEB TRANSFER

From various modeling efforts (i.e., the equilibrium model of organic chemical accumulation in aquatic food webs with sediment interaction; Thomann et al., 1992), it is apparent that identification of food web transfer and determination of assimilation efficiencies are major processes that require better description. The relative role of food chain transfer and biomagnification in the accumulation of contaminants was the subject of significant debate (Biddinger and Gloss, 1984). Whether food chain transfer significantly contributes to the total accumulation depends on (1) the extent to which contaminants are metabolized and subsequently eliminated by the prey (Neff, 1979), and (2) the differences between the concentrations in the water and food sources (Opperhuizen, 1991).

Laboratory studies have described conditions where food chain accumulation was significant. The accumulation of DDT from food ranged from 30 to 100% for fish (Macek and Korn, 1970; Macek et al., 1970; Rhead and Perkins, 1984); kepone accumulation was additive from water and dietary sources (Fisher and Clark, 1990). Even field data indicate the influence of food web transfer on contaminant accumulation. The BCFs for all species in Lake Ontario were linear with log K_{ow}, but the chlorine content of PCBs found at the top of the food chain was greater than at the lower end of the food chain (Oliver and Niimi, 1988). Further, the food web structure was found to influence accumulation of chlorinated dioxins and furans in a Canadian river system (Owens et al., 1994) and of chlorinated hydrocarbons in the Great Lakes ecosystem (Bierman, 1990).

Accumulation through the food chain route depends on feeding rate and assimilation efficiency. These two factors vary with contaminant and food characteristics and also with contaminant concentration. As the feeding rate increases, assimilation efficiency declines, because the residence time in the gut declines (Klump et al., 1987; Opperhuizen and Schrap, 1988; Weston, 1990). Likewise, as contaminant concentration in the food particles increases, assimilation efficiency declines (Opperhuizen and Schrap, 1988). Further, assimilation efficiencies from ingested food change with organism age (Sijm et al., 1992).

Assimilation efficiencies for chlorinated organics in fish exhibit a nonlinear relationship with K_{ow}; for compounds with log K_{ow}s up to approximately 7, uptake efficiency is constant. However, for compounds with log K_{ow}s exceeding 7, efficiencies decline as the K_{ow} increases such that $1/E_0 = 5.3\ (\pm 1.5) \times 10^{-8}$ $K_{ow} + 2.3\ (\pm 0.3)$, where E_0 is the assimilation efficiency (Gobas et al., 1988a). These efficiencies for fish were measured with high quality (e.g., highly nutritional) food. The quality of the food can strongly affect the observed assimilation efficiency. With zebra mussels feeding on suspended sediment or algae, assimilation efficiencies for the suspended sediment particles were approximately 30% and, for algae, nearly 90% for hexachlorobiphenyl (Bruner, 1993). The mechanism for accumulation from the intestinal tract is apparently increased fugacity resulting from the assimilation of food materials (Gobas et al., 1993). This mechanism would account for the effects of feeding rate and food composition on assimilation. The increased feeding rate would reduce transit time as well as the fraction of food assimilated. Thus, the fugacity of the contaminant would not be increased as much as when a more complete assimilation of food occurs with slower gut passage. The difference in assimilation efficiencies can be explained by this same mechanism, where one food type is better assimilated than another, e.g., algae versus sediment detritus. Thus, particle-selective feeding, coupled with differential contaminant partitioning to particles of differing composition, makes measurement of assimilation efficiencies for selective-feeding benthic organisms extremely difficult.

V. ESTIMATION METHODS

Because of the utility of log K_{ow} for predicting bioconcentration and partitioning to organic matter for nonpolar organic contaminants, methods that estimate log K_{ow} permit estimation of BCF. Beside direct measurement, empirical methods for estimating log K_{ow} include regression with aqueous solubility, partition coefficients with other solvents, or use of estimated activity coefficients (Lyman et al., 1990). Estimation of log K_{ow} from structure is best established through fragment analysis (Leo et al., 1971), and this method remains frequently used for log K_{ow} estimation. Other structural approaches have included the use of molecular volume (McGowan and Mellors, 1986), molecular connectivity (Kier and Hall, 1976; 1986), and more recently, linear solvation energy relationships (LSER, Kamlet et al., 1988). Among the methods that estimate log K_{ow} from structural properties of the molecules, the molecular volume method provides good estimates within a class of compounds. This is because the energy required for cavity formation within a solvent depends on molecular size, and the electronic character contributed by functional groups will remain relatively constant within a compound class. The molecular connectivity and LSER methods work well among a diverse mix of compounds, because both methods incorporate the change in molecular characteristics with changes in number and type of functional groups on the molecule.

The LSER model estimates a given property of a compound through multiple linear regressions of that property against several molecular parameters. The parameters are $V_I/100$ (intrinsic molecular volume divided by 100 to scale it to the other parameters), $\pi*$ (molecular polarizability), β (the ability to donate a proton to a hydrogen bond), and α (the ability to accept a proton in a hydrogen bond) (Kamlet et al., 1986). The LSER model works best when a wide variety of molecular classes are included in the data set. This occurs because of the inclusion of dipole and hydrogen bonding characteristics of molecules in the model. Critical to the use of LSER is the determination of the parameters for each molecule. This leads to the following equation for estimating log K_{ow} (Kamlet et al., 1988):

$$\log K_{OW} = 0.35 + 5.35\left(\frac{V_I}{100}\right) - 1.044(\pi* - 0.35\delta) - 0.35\beta + 0.10\alpha \quad (5)$$

($n = 245$, $r = 0.996$, sd = 0.131).

where δ is a modifier of the dipolarity (π^*) to account for the influence of the aromatic nature of molecules and the influence of halogen atoms. A set of rules for parameter estimation has been published (Hickey and Passino-Reader, 1991).

Molecular connectivity is also very versatile. An additional feature is that the three-dimensional character of the molecule is implicitly encoded in some of the higher-level connectivity indices. The simple molecular connectivity indices are generated from the δ values, which represent the number of bonds to non-hydrogen atoms or the valence-corrected molecular connectivity that accounts for the number of valence electrons not participating in a bond to hydrogen. The level of connectivity can represent the environment of single atoms to multilevel connections of various path lengths, rings, or clusters within a molecule. The formal description of single-bond simple molecular connectivity, ^1X, would be calculated as follows for the single-bond paths in a molecule:

$$^1X = \Sigma(\delta_i\delta_j)^{-0.5} \tag{6}$$

where δ_i and δ_j are the number of non-hydrogen bonds for each atom (i and j) of a single bond path.

As the complexity of the calculation increases at higher levels of connection within a molecule, the use of computer programs to perform the calculations is suggested (L.H. Hall, Hall Associates Consulting, Quincy, MA). Following this approach, K_{ow} has been calculated for a wide range of nonionic compounds (Murry et al., 1978):

(Hydrocarbons) $\log K_{OW} = 0.884\ (\pm 0.03)^1X + 0.41\ (\pm 0.09)$

$$(r = 0.975, n = 45) \tag{7}$$

(All other compounds) $\log K_{OW} = 0.95\ (\pm 0.01)^1X + 0.48\ (\pm 0.04)$

$$(r = 0.986, n = 138) \tag{8}$$

Use of either LSER or molecular connectivity methods makes little difference for the purposes of determining $\log K_{ow}$. In either case, to perform an estimate from a structure activity model, the parameters for the particular model need to be evaluated and the regression determined from a training set (a subset of the data representative of model parameters). In most cases, the greater the similarity of molecular structure in the training set to the compound of interest, the more likely the estimate will be accurate. Further, it is important to ensure that the conventions used for parameter estimation in the training set are consistent with those used for the compound of interest. There may well be different assumptions employed by different researchers (i.e., whether to use or not to use a vector sum or sum with a component group of hierarchy of importance) to estimate the parameters, particularly for the LSER model (Hickey and Passino-Reader, 1991).

In addition to estimating $\log K_{ow}$, both LSER (Park and Lee, 1993) and MC (Sabljic and Protic, 1982; Grovers et al., 1984; Connell and Schüürmann, 1988) can estimate BCF directly. For example:

Linear solvation energy relationships

$$\log BCF_{fish} = -0.95\ (\pm0.23) + 4.74\ (\pm0.25)\left(\frac{V_1}{100}\right) - 4.39\ (\pm0.62)\beta + 0.88\ (\pm0.38)\alpha$$

$$(r = 0.947, n = 51;\ \text{Park and Lee, 1993}) \tag{9}$$

Molecular connectivity

$$\log BCF_{fish} = -0.168\ (\pm0.013)(^2X^V)^2 + 2.22\ (\pm0.15)^2X^V - 2.32\ (\pm0.38)$$

$$(r = 0.947,\ SD = 0.3,\ n = 20;\ \text{Sabljic and Protic, 1982}) \tag{10}$$

$$\log BCF_{fish} = 0.58\,^1X^V + 0.54$$

$$(r = 0.88,\ n = 49;\ \text{Connell and Schüürmann, 1988}) \tag{11}$$

where $^1X^v$ is the valence-corrected single-bond connectivity index.

These direct approaches for estimating BCF have the advantage that the log K_{ow} for the compound need not be known. Further, there is not the added variance of first estimating a log K_{ow} and subsequently a BCF. The BCF is estimated directly from compound characteristics and no measurements need be performed. When such estimates are made, the closer the compound characteristics are to the training set used to generate the structure-activity relationship, the more likely the estimate will be accurate. As with any such estimation, the variance of the correlation used for estimation must be considered when accepting the predicted value. Direct estimation of BCF from molecular characteristics is hindered because only a few equations are available. Such direct equations are currently available only for fish and for a narrow range of compounds, generally only polycyclic aromatic hydrocarbons and polychlorinated biphenyls. Because most of these regressions come from log relationships, the BCF estimates represent the medians and not the mean values when converted from log values to arithmetic values. If the mean BCFs are desired, then a correction for bias must be incorporated (Newman, 1993). Direct estimation that uses either LSER or molecular connectivity methods also helps overcome some of the complexities of nonlinearity in the log K_{ow} – log BCF relationship (Sabljic and Protic, 1988). The advantage of these methods lies in their incorporation of electronic and spatial characteristics of the molecule in the relationship to estimate the BCF from the molecular structure. Disadvantages of these methods, however, ignore biological parameters, such as animal physiology and the role of biotransformation in determining contaminant accumulation.

VI. STEADY-STATE MODELS

The premise behind the use of equilibrium models for nonpolar organic contaminants is that accumulation of compounds is dominated by their relative solubility in the various phases, e.g., water, organism, and organic matter in other compartments (e.g., colloids, microparticulates). Equilibrium models, therefore, rely on the assumptions that (1) the compounds are not actively biotransformed or degraded, (2) there are no active (energy-requiring) processes dominating the distribution, (3) the conditions are stable enough that a quasi-equilibrium will occur, (4) environmental factors such as temperature do not change significantly to alter the equilibrium condition, and (5) organism and/or organic matter composition is not sufficiently variable to alter the distribution. For nonpolar organic contaminants, the solubility in organic phases as well as proportionality to K_{ow} has been widely recognized. This then led to development of relationships among various media for estimating bioaccumulation (Kenega and Goring, 1980) and has even been suggested as an approach for establishing sediment quality criteria (Di Toro et al., 1991).

The relative solubility of compounds among organic phases of the aquatic environment and in the lipids of organisms has led to development of equilibrium calculations between sediment and organisms. Because sediments represent the temporally integrated load to the aquatic system, the exposure of organisms should be better defined by relating bioaccumulation to the sediment concentration. These biota-sediment accumulation factors (BSAFs) are defined as the lipid-normalized concentration in organisms divided by the organic carbon-normalized concentrations in sediment (Ankley et al., 1992). These have been referred to in the literature as accumulation factors (AFs) while the inverse have been called preference factors (PF) (Landrum et al., 1992a; Lee, 1992). These BSAFs provide an estimate of accumulation from the sedimentary environment similar to the BCF from water. Such estimates were expected to reduce the variance compared to bioaccumulation factors (BAF, concentration in the organism divided by the concentration in the sediment or other reference compartment) which were typically reported on a wet weight of organism and dry weight of sediment basis. The first report that attempted to make this dual normalization suggested that the BSAF for nonpolar organic compounds would be a constant (McFarland, 1984). The utility of these concentration factors is limited due to the number of variables that affect accumulation from sediment (see above); the expected range for a given compound can exceed 100 (Lee, 1992). However, the relationship between sediment and bioaccumulation continues to be studied to describe the potential for ecological partitioning of compounds (Connor, 1984; Rowan and Rasmussen, 1992).

The use of thermodynamic equilibrium concepts to better estimate distribution among environmental compartments, and particularly to estimate accumulation in biota, was made simple and elegant through the use of the fugacity concept (Mackay, 1991). The fugacity approach even permits disequilibrium among compartments for modeling purposes so long as equilibrium within a compartment exists. When compounds can flow between compartments, the compounds will move down chemical activity gradients until the chemical activities (fugacities) are equal. This approach permits estimation of contaminants among environmental compartments either with simple equilibrium assumptions (level I) or nonsteady-state calculations (level III; Mackay, 1991). With level I calculations, the concentration in fish is estimated from the log K_{ow} – log BCF relationship; the concentration will depend on equilibrium distribution among the various phases. While the same relationship is employed at higher-level fugacity calculations, the concentration in the water that will dictate the BCF will be modified by various environmental processes prior to estimating the accumulation in fish (Mackay, 1991). Even processes such as biomagnification that allow the fugacity of fish to exceed that of water can be explained in terms of the enhanced fugacity in the intestinal tract of the fish feeding on various prey (Gobas et al., 1988a; Clark and Mackay, 1991). At the organism level, the fugacity approach can also accommodate more complex exposures and temporal changes to accommodate the complexities of the real world (Clark et al., 1990; Clark and Mackay, 1991). In all cases, the objective is to obtain a steady state estimate.

When multiple sources become significant for bioaccumulation, deviations from the expected equilibrium condition have been observed. The classic condition is food chain transfer and its relative importance for nonpolar persistent organic compounds. Even if the main source to the food web is water, the deviation from simple equilibrium becomes significant (Thomann, 1981, 1989; Thomann and Connolly, 1984; Connolly and Pedersen, 1988). Models that incorporate multiple contaminant sources must use some method of evaluating food chain transfer as well as accumulation from water. All these models demonstrate that the higher trophic level organisms have organism:water fugacity ratios greater than one. The elevated fugacity ratios demonstrate the incidence of biomagnification, i.e., the concentration in the organism relative to that of the water at steady state increases as the position in the food chain increases. Such biomagnification is apparently driven by the increased fugacity in the intestinal tract of organisms, as mentioned previously (Gobas et al., 1988a; Clark and Mackay, 1991). One absolute requirement for such biomagnification to take place is the inability of the organism to rapidly biotransform or eliminate the compound. The extent of biomagnification is also modified by the growth rate and alteration of the organism's lipid content. For example, increases in total lipid content decreased the elimination rate of polycyclic aromatic hydrocarbons (PAH) in *Diporeia* and *Hexagenia* (Landrum, 1988; Landrum and Poore, 1988).

Further, the recognized relationships between log K_{ow} and log BCF or partitioning to organic matter in sediments has allowed the development of equilibrium partitioning models that, combined with food chain models, estimate the importance of sediments as contaminant contributors to the food chain (Connolly, 1991; Thomann et al., 1992). Equilibrium partitioning between sediment organic matter, interstitial water, and sediment-associated benthos is the basis of these models (Di Toro et al., 1991). These models also rely on relationships among bioaccumulation, lipid content, and the relative hydrophobicity of the contaminants. The various assumptions made in the equilibrium partitioning and food chain models permit the model predictions to bracket observed data. The fact that reasonable model results are achieved may be, in part, fortuitous. Issues such as growth dilution, which causes disequilibrium in species from fish to phytoplankton, accumulation, and uptake processes that are not directly coupled to respiration from water, are not considered. Alternatively, these pathways and processes may have little impact on the overall model. The models do indicate two areas that require significant development: (1) improved identification of the food chain links, and (2) improved data on assimilation efficiencies for ingested materials. Identification of food chain links can be improved through analysis of the stable isotope composition of food web members. As the trophic level increases, the stable isotope ratio increases, thus providing a quantitative approach to describing food web links. By better defining the food web in this way, models of food chain transfer can be made more reliable (Broman et al., 1992).

VII. USES AND LIMITS OF TOXICOKINETICS

Although equilibrium models give good approximations of contaminant distribution, kinetic models are needed to predict nonsteady-state, nonequilibrium accumulation from temporally and spatially varying exposures when the simplifying assumptions of the equilibrium partitioning models are inappropriate. This can be seen in the above examples, where either bioenergetics approaches (Connolly, 1991) or some form of food chain transfer was assumed that depended on estimated feeding rates and assimilation efficiencies (Thomann et al., 1992). The question becomes, can temporal variations be observed and is their magnitude sufficiently significant that kinetic models are warranted? For kinetics to be important in the bioaccumulation of contaminants, rates of the exogenous kinetic processes must be of similar magnitude as the biological processes, e.g., the rate of desorption or load is similar to respiration rate. The rates of importance for particular species will differ, because biological rates vary with organism size (Landrum et al., 1994a). For the amphipod, *Diporeia* spp., seasonal variation in the accumulation of PAH was observed and ranged from approximately a factor of 4 to 10 (Landrum et al., 1992b). This seasonal variation was best described by a kinetic model that incorporated both environmental and physiological variation in the kinetics (Landrum et al., 1992b). Likewise, the accumulation of PCBs by algae was seasonally dependent, where PCB uptake by algae was slow relative to growth (Swackhamer and Skoglund,

1991, 1993). Subsequently, thermodynamic equilibrium was not achieved. These nonequilibrium conditions limit the accuracy of the thermodynamic models described above, which often rely on equilibrium conditions, particularly at the lower end of the food web.

Kinetic models are particularly applicable for evaluating exposure from multiple sources and for examining changes in the exposure that may take place with either changes in concentration, chemical speciation, physiology, or environmental conditions. Kinetic models are also extremely useful for evaluating the important processes involved in contaminant bioaccumulation. Kinetic models can incorporate pathways that describe the mechanisms affecting the accumulation and loss of contaminants in organisms, and permit prediction of adverse biological responses using the tissue residue approach (Landrum et al., 1992a, 1994a). The disadvantage of such models is their need for relatively large amounts of data. Also, the kinetics for each species and the factors that affect the kinetics need to be known to parameterize the models. It then becomes a question of the need for time-dependent estimates versus the amount of effort required to achieve accurate models.

Kinetic models and their role in hazard assessment were recently reviewed (Landrum et al., 1992a). The formalisms that are available for compartment-based models include rate coefficient models, clearance volume models, and fugacity-based models. In the simplest cases, these models are mathematically equivalent. For the case of water-only exposure, the simplest rate coefficient model would conform to the following equation:

$$C_a = \frac{k_u C_w}{k_e}(1 - e^{-k_e t}) \qquad (12)$$

where C_a is the concentration in the organism, k_u is the uptake clearance from water (ml/g organism/h), k_e is the elimination rate constant (1/h), C_w is the concentration in the water, and t is time (h). The uptake clearance in this and in other models is defined as the amount of source compartment scavenged of contaminant per mass of organism per unit of time. For this model, the concentration in the water is constant and there is no biotransformation. This model represents the simple uptake and loss of the parent compound such that at steady state, BCF = C_a/C_w = k_u/k_e.

In the clearance volume model, the BCF is equal to the volume of distribution (V_d). The V_d is the volume equivalent of a reference compartment, in this case water, that contains the same amount of compound as found in the organism. In this formulation, p is equivalent to k_u. This leads to the following integrated equation for accumulation:

$$C_a = V_d C_w \left(1 - e^{-\left(\frac{p}{V_d}\right)t}\right) \qquad (13)$$

The argument for using this formalism suggests that k_e is an artificial value describing the fractional loss in concentration from an organism with time, which will depend on both physiological and environmental factors, while V_d is an equivalent referenced volume. In the case of k_e, changes in magnitude may result from both changes in the size of the storage compartment as well as changes in physiological function. Specific studies may well be required to determine which mechanism is applicable. With the above formulation of the clearance volume model, changes in V_d can also result from either of the mechanisms stated above. Additional data or a different model formulation may permit separation of the mechanism of importance.

When converting the above models to the fugacity formalism, the following equation examines the variability in the organism fugacity:

$$f_a = \frac{D_u f_w}{D_e} \left(1 - e^{-\frac{D_e t}{V_a Z_a}} \right) \tag{14}$$

where f_a is the fugacity in the organism (pascals), D_u is the uptake transfer coefficient (moles per hour per pascal), D_e is the elimination transfer coefficient (moles per hour per pascal), V_a is the volume of the organism (m^3), Z_a is the fugacity capacity of the organism (moles per cubic meter per pascal), f_w is the fugacity of the water (pascals), and t is time (h). Thus, at steady state, $D_u f_w = D_e f_a$ and the ratio of the fugacities $f_a/f_w = D_u/D_e$. If both sides of the equation are multiplied by Z_a/Z_w then BCF $= f_a Z_a/f_w Z_w = C_a/C_w = D_u Z_a/D_e Z_w$. From the above, the equivalency of the various formalisms is clear. However, as the complexity of the model increases, the equivalency becomes more difficult to depict; however, each of the approaches provides a good fit and accurate description of the experimental data (Landrum et al., 1992a). The selection of one model formalism over another depends on the experimentalist's experience and the ease of data collection. In general, the simplest model that will adequately address the question should minimize the errors associated with parameter estimation and, thus, result in the most precise estimates (Landrum et al., 1992a).

Despite the general utility of each of the above formalisms, the critical issue to the application of such models and their extensions to more complex conditions than water-only exposures is to ensure that all the assumptions of the model are explicitly described. For instance, in the above simple models, the water concentration (fugacity) must remain constant, and there should be no biotransformation. These two assumptions are easily recognized and are usually explicitly stated. What is less easily recognized is that the uptake clearances, rate constants, and transfer coefficients do not change over the course of the study. These assumptions are rarely recognized with the implications that go with them. For instance, if the coefficients change, then the integrated forms shown above are not correct. The fact is, these coefficients do change with the physiology of the organism and with a variety of environmental factors. Thus, data used to generate such models are conditional, and the coefficients determined are conditional, based on the environmental and physiological conditions of the organisms under the conditions of the study.

To utilize a model that is useful for predicting accumulation in the field, the major factors affecting the coefficients for accumulation and loss must be identified. In the case of *Diporeia*, accumulation kinetics and the factors affecting kinetics were determined from a series of studies (Landrum, 1988,1989). These values and factors were combined into a seasonal model for field validation (Landrum et al., 1992b). Even when several of the factors that affect the coefficients were known, predictability was not accurate for the less hydrophobic compounds. Thus, the model still did not contain the appropriate coefficients needed to accurately estimate the accumulation of these less hydrophobic compounds.

Another rarely recognized assumption is that the systems must be homogeneous. If the organisms contact patches of contaminant, then the model assumption of a constant source would not hold. This may occur when organisms are initially added to sediments in typical sediment bioassays. It appears, particularly for some water-soluble compounds, that the initial accumulation is much faster than later in the experiment (Landrum, 1989). This was originally thought to be driven by sediment aging (Landrum et al., 1992c). However, more recent work with environmentally resident contaminants exhibits the same phenomenon of initial rapid accumulation followed by slower accumulation (Kukkonen et al., 1993). Accumulation of contaminants for organisms introduced into sediments initially occurs primarily from pore water, but the pore water is rapidly depleted of contaminant. Subsequently, accumulation occurs as a balance between pore water, replenished through contaminant desorption from particles, and assimilation of contaminant from ingested particles. The balance between which source dominates will depend on desorption kinetics for replenishing the pore water, the organism's selectivity, the organism's ingestion rate of particles, and the associated contaminant assimilation efficiency. Such nonhomogeneity invalidates the current models, i.e., models that assume homogeneity within a compartment, for sediment accumulation. New mathematical formalisms will need to be determined. The important point is that assumptions of the models need to be explicitly stated and care taken to ensure that there are no hidden assumptions for a particular application.

In addition to the classical compartment-based models described above, both bioenergetics and physiologically based pharmacokinetic models are other approaches for predicting concentrations in the organism and even concentrations in specific tissues (Landrum et al., 1992a). Bioenergetics models rely on the gross physiological functions of the organism, e.g., respiration and feeding, to provide encounter rates for the various source compartments. These physiological rates also are used to estimate the elimination and metabolism rates for compound removal and must be coupled to efficiency terms for compound transfer and transformation to produce the net compound accumulation or loss. Physiologically based pharmacokinetic models also include the details of transfer among organism tissues to produce concentrations in the tissue containing the receptor. Either the compound concentrations or the fugacities among the various compartments can be used to track the compound in the two model approaches. For internal transfer among organs, the transfer coefficients between the circulating fluid, e.g., blood, and the tissues needs to be known. Both models can respond to the physiological state of the organism, so that as this changes, the net transfer to the organism can be estimated. In the aquatic environment, most of the models have been made for fish; in only a few cases (Boese et al., 1990) have these models been considered for lower organisms. The main advantage of these models lies in their strong tie to physiological function and their adaptability to organisms of various sizes, assuming that the routes of transfer do not change.

The assumptions in these physiologically based pharmacokinetic models are as important as in the compartment models. For example, it is important to understand the routes of accumulation for physiologically based models. With lower organisms such as amphipods, the accumulation from the aqueous phase may not be limited to the accumulation across the respiratory membrane. For *Diporeia*, the accumulation of nonpolar organic contaminants was found to exceed that of oxygen and depends on the surface area or surface area-to-volume ratio of the organism (Landrum and Stubblefield, 1991). Thus, to expand beyond simple equilibrium models for biomagnification, routes of transfer and loss that are often ignored in equilibrium models should be considered. The question of which model to use for predicting hazard is often difficult to answer, and a comparison of the various models needs to be thoroughly considered by the researcher (Table 2). The simpler the model used to fit the data, the more likely the accuracy of prediction under the same assumptions and conditions. It is important to use the models within the limits of their assumptions. As the model structure is changed, the inherent assumptions will likely change and must be evaluated and compared with the conditions under which the data are gathered.

VIII. BIOAVAILABILITY

Whether thermodynamic or kinetic models are employed, all the contaminant in a particular compartment may not be readily available for bioaccumulation. The fraction of material available in a compartment for uptake by biota is the bioavailable fraction. In water, the freely dissolved contaminant is generally accepted as the bioavailable fraction. No such simple definition can be applied for food or sediment sources.

The difficulty in applying the bioavailability concept is that first, a quantitative measure to define the bioavailability in a compartment needs to be made, and second, a compound that is temporarily unavailable can, after desorption or other change in chemical speciation, become available. One approach to addressing the first issue is a recent attempt to define the uptake efficiency as a measure of environmental bioavailability (Landrum et al., 1994a). Thus, environmental bioavailability (EBA) from a compartment can be defined:

Table 2 Comparison of Models Used in Exposure Assessment

Model/attribute	Model					
	Equilibrium	Rate coefficient	Fugacity	Clearance volume	Physiological-based pharmacokinetic	Bioenergetic
Requires assumption of equilibrium	Yes	No	No	No	No	No
Models multiple compartments	No	Yes	Yes	Yes	Yes	Yes
Models multiple uptake routes	No	Yes	Yes	No	Yes	Yes
Can be used to model internal distribution of toxicants	No	Yes	Yes	Yes	Yes	No
Potential to scale to other species	Yes (by lipid content)	Some	Some	No	Yes	Yes
Data requirements	Low	Moderate	Moderate-high	Moderate	High	High

Table modified from Landrum et al., 1992a.

$$\text{EBA} = \frac{\text{uptake clearance (ml source compartment/g organisms/h)}}{\text{encounter rate (volume/mass; ml/g/h)}} \quad (15)$$

The encounter mass for a solid phase may be substituted for encounter volume; the uptake clearance would then have units of g source/g organism/h, so that EBA would remain a fraction and reflect accumulation efficiency. While this approach presents a quantitative value that can be compared among different exposure conditions such as exposure in different sediments, the term is subject to both chemical and physiological factors that alter the uptake clearance. Further, measuring encounter mass or volume is difficult at best. Thus, EBA may not be easily determined in most situations. For intercomparison of a single organism and compound, comparative bioavailability can be described by comparing the uptake clearances from exposures to different sources, e.g., different sediments (Lee, 1991).

The second issue, temporal changes in bioavailability, from a risk assessment perspective will likely require that total contaminant entering a system be considered bioavailable. The issue is not whether a compound is bioavailable, but on what time scale it will be available. Because the prefix *bio* specifies an interaction with biota, the term bioavailable then will have temporal limits associated with the temporality of particular organisms. The time frames that are important change with organism size and behavior (Landrum et al., 1994a). While all of a compound entering a system may eventually become available for some species at some point in time, the focus for practical purposes must limit such discussions to time frames associated with a particular species of interest.

IX. UTILITY AND ASSESSMENT

The purpose behind estimating bioaccumulation is to determine the accumulated dose to a particular trophic level for modeling accumulation through the next trophic level, or to estimate the impact of that internal dose on the organism. The primary regulatory use of this type of data has been for human health risk assessment and not for assessing the impact either on the food chain or the effect of contaminants. Food chain models, as discussed earlier, often employ equilibrium partitioning approaches, particularly for the lower end of the food web. The use of more realistic kinetic models should help bring the data into line with that observed through field monitoring studies. They should also be useful for explaining seasonal variation, such as observed in the PCB accumulation in mussels with changes in lipid content (Capuzzo et al., 1989).

Evaluating the effect of contaminants on various levels of the aquatic food chain has traditionally used concentrations in the external environment. When multiple sources are involved and the exposure becomes complicated due to significant bioavailability limitations such as exposures in sediments, then assessing effects based on the external environment may not be very predictable.

Rather, there is a body of knowledge that is developing to evaluate the effect of chemicals based on the internal concentration in organisms (McCarty and Mackay, 1993). This is analogous to utilizing blood levels in mammals to predict drug effects and behavior. The complication in using this approach is that most researchers who have measured toxicity have not measured the organism's internal dose. Despite the lack of internal dose data for fish, concentration ranges exist for several mechanisms of action (McCarty and Mackay, 1993). These ranges lead directly to the application of data collected from bioaccumulation bioassays that are not usually used to assess environmental impact. Additionally, this approach is readily applied to mixtures and in the study of compound interactions (Landrum et al., 1989, 1991; McCarty and Mackay, 1993).

The range of concentrations producing effects varies with both mechanism of action and duration of exposure. Mortality due to narcosis, for instance, ranges from 2 to 8 mmol/kg for acute responses to 0.2 to 0.8 mmol/kg for chronic exposures in fish. Similar values for mortality have been observed for fish (McCarty and Mackay, 1993) and for selected invertebrates depending on the mechanism of action, e.g., narcosis by PAH in *Diporeia* (Landrum et al., 1991) and in *Daphnia magna* (Pawlisz and Peters, 1993a,b), decoupling of oxidative phosphorylation by pentachlorophenol in *Diporeia* and *Mysis relicta* (Landrum and Dupuis, 1990), and neurotoxicity by tri-*n*-butyltin in amphipods (Meador et al., 1991). The use of internal concentrations and their resultant effects has been applied to scope for growth in mussels (Widdows and Donkin, 1989). Thus, the approach may well be useful for other effects, although such applications remain to be revealed.

The ability to predict the accumulation of compounds will aid in the ability to predict effects. The total effect may not always depend on the total parent compound in the organism, but at some point the expected flux into the organism may need to serve as the dose for readily metabolized compounds. An alternative measure may incorporate the concentration of the metabolites for a specific chemical relating the effect to metabolite concentration, particularly for compounds that are activated by biotransformation, e.g., parathion to paraoxon. Thus, the use of internal dose may well take some refinement before it can be applied to a wide range of organisms and contaminant classes. The utility of the approach is, however, obvious. When organisms are exposed to multiple sources where none are dominant or where simple equilibrium models do not effectively reflect the concentration, then prediction of body burdens through kinetic models and assessment of effects based on internal dose should provide better estimates of environmental hazard.

REFERENCES

Adams, W.J., R.A. Kimerle and R.G. Mosher, 1984. Aquatic safety assessment of chemicals sorbed to sediments. In: Aquatic Toxicology and Hazard Assessment: Seventh Symposium, ASTM STP 854, edited by R.D. Cardwell, R. Purdy and R.C. Bahner, American Society for Testing and Materials, Philadelphia, PA, pp. 429–453.

Aller, R.C., 1983. The importance of the diffusive permeability of animal burrow linings in determining marine sediment chemistry. J. Mar. Res. 41, 299–322.

Aller, R.C., J.Y. Yingst and W.J. Ullman, 1983. Comparative biogeochemistry of water in interstitial *Onuphis* (polychaeta) and *Upogebia* (crustecea) burrows: temporal patterns and causes. J. Mar. Res. 41, 571–604.

Ankley, G.T., P.M. Cook, A.R. Carlson, D.J. Call, J.A. Swenson, H.F. Corcoran and R.A. Hoke, 1992. Bioaccumulation of PCBs from sediments by oligochaetes and fishes: comparison of laboratory and field studies. Can. J. Fish. Aquat. Sci. 49, 2080–2085.

Aubert, M., J. Aubert, H. Augier and C. Guillemaut, 1985. Study of the toxicity of some silicone compounds in relation to marine biological chains. Chemosphere 14, 127–138.

Banerjee, S. and G.L. Baughman, 1991. Bioconcentration factors and lipid solubility. Environ. Sci. Technol. 25, 536–539.

Barron, M.G., 1990. Bioconcentration. Environ. Sci. Technol. 24, 1612–1618.

Biddinger, G.R. and S.P. Gloss, 1984. The importance of trophic transfer in the bioaccumulation of chemical contaminants in aquatic ecosystems. Residue Rev. 91,103–145.

Bierman, V.J. II, 1990. Equilibrium partitioning and biomagnification of organic chemicals in benthic animals. Environ. Sci. Technol. 24, 1407–1412.

Black, M.C. and J.F. McCarthy, 1988. Dissolved organic macromolecules reduce the uptake of hydrophobic contaminants by the gills of rainbow trout (*Salmo giardneri*). Environ. Toxicol. Chem. 7, 593–600.

Boese, B.L., H. Lee II, D.T. Specht, R.C. Randall and M. Winsor, 1990. Comparison of aqueous and solid phase uptake for hexachlorobenzene in the tellinid clam, *Macoma nasuta* (Conrad): a mass balance approach. Environ. Toxicol. Chem. 9, 221–231.

Boryslawskyj, M., A.C. Garrood, J.T. Pearson and D. Woodhead, 1987. Rates of accumulation of dieldrin by freshwater filter feeder: *Sphaerium corneum*. Environ. Pollut. 43, 3–13.

Brannon, J.M., C.B. Price, F.J. Reilly, Jr., J.C. Pennington and V.A. McFarland, 1993. Effects of sediment organic carbon on distribution of radiolabeled fluoranthene and PCBs among sediment, interstitial water, and biota. Bull. Environ. Contam. Toxicol. 51, 873–880.

Branson, D.R., G.E. Blau, H.C. Alexander and W.B. Neely, 1975. Bioconcentration of 2,2'4,4'–tetrachlorobiphenyl in rainbow trout as measured by an accelerated test. Trans. Am. Fish. Soc. 4, 785–792.

Broman, D., C. Näf, C. Zebühr, B. Fry and J. Hobbie, 1992. Using ratios of stable nitrogen isotopes to estimate bioaccumulation and flux for polychlorinated dibenzo-p-dioxins (PCDDs) and dibenzofurans (PCDFs) in two food chains from the northern Baltic. Environ. Toxicol. Chem. 11, 331–345.

Brown, V.M., D.H.M. Jordon and B.A. Tiller, 1967. The effect of temperature on the acute toxicity of phenol to rainbow trout in hard water. Water Res. 1, 587–594.

Bruggeman, W.A., A. Opperhuizen, A. Wijbenga and O. Hutzinger, 1984. Bioaccumulation of super-lipophilic chemicals in fish. Toxicol. Environ. Chem. 7, 173–189.

Bruner, K.A., 1993. Bioconcentration and trophic transfer of lipophilic contaminants by the zebra mussel, *Dreissena polymorpha*: the role of lipid, body size and route of exposure. Ph.D. Thesis, Ohio State University, Columbus, OH.

Capuzzo, J.M., J.W. Farrington, P. Rantamaki, C.H. Clifford, B.A. Lancaster, D.F. Leavitt and X. Jia, 1989. The relationship between lipid composition and seasonal differences in the distribution of PCBs in *Mytilus edulis*. Mar. Environ. Res. 28, 259–264.

Carlberg, G.E., K. Martinsen, A. Kringstad, E.T. Gjessing, M. Grande, T. Källqvist and J.U. Skåre, 1986. Influence of aquatic humus on the bioavailability of chlorinated micropollutants in Atlantic salmon. Arch. Environ. Contam. Toxicol. 15, 543–548.

Carlberg, G.E. and K. Martinsen, 1982. Adsorption/complexation of organic micropollutants to aquatic humus. Sci. Total Environ. 25, 245–254.

Carter, C.W. and I.H. Suffet, 1982. Binding of DDT to dissolved humic materials. Environ. Sci. Technol. 16, 735–740.

Chiou, C.T., R.L. Malcolm, T.I. Brinton and D.E. Kile, 1986. Water solubility enhancement of some organic pollutants and pesticides by dissolved humic and fulvic acids. Environ. Sci. Technol. 20, 502–508.

Choudhry, C.G., 1983. Humic substances. III. Sorptive interactions with environmental chemicals. Toxicol. Environ. Chem. 6, 127–171.

Clark, K.E. and D. Mackay, 1991. Dietary uptake and biomagnification of four chlorinated hydrocarbons by guppies. Environ. Toxicol. Chem. 10, 1205–1217.

Clark, K.E., F.A.P.C. Gobas and D. Mackay, 1990. Model of organic chemical uptake and clearance by fish from food and water. Environ. Sci. Technol. 24, 1203–1213.

Coats, J.T. and A.W. Elzerman, 1986. Desorption kinetics for selected PCB congeners from river sediments. J. Contamin. Hydrol. 1, 191–210.

Connell, D.W. 1988. Bioaccumulation behavior of persistent organic chemicals with aquatic organisms. Rev. Environ. Contam. Toxicol. 101, 117–147.

Connell, D.W. and G. Schüürmann, 1988. Evaluation of various molecular parameters as predictors of bioconcentration in fish. Ecotoxicol. Environ. Saf. 15, 324–335.

Connor, M.S., 1984. Fish/sediment concentration ratios for organic compounds. Environ. Sci. Technol. 18, 31–35.

Connolly, J.P., 1991. Application of a food chain model to polychlorinated biphenyl contamination of the lobster and winter flounder food chain in New Bedford Harbor. Environ. Sci. Technol. 25, 760–770.

Connolly, J.P. and C.J. Pedersen, 1988. A thermodynamic-based evaluation of organic chemical accumulation in aquatic organisms. Environ. Sci. Technol. 22, 99–103.

Dabrowska, H. and S.W. Fisher, 1993. Environmental factors affecting the accumulation of sediment-sorbed hexachlorobiphenyls by channel catfish. Aquat. Toxicol. 27, 179–198.

DeWitt, T.H., R.J. Ozretich, R.C. Swartz, J.O. Lamberson, D.W. Schults, G.R. Ditsworth, J.K.P. Jones, L. Hoselton and L.M. Smith, 1992. The influence of organic matter quality on the toxicity and partitioning of sediment-associated fluoranthene. Environ. Toxicol. Chem. 11, 197–208.

de Wolf, W., J.H.M. de Brujin, W. Seinen and J.L.M. Hermens, 1992. Influence of biotransformation on the relationship between bioconcentration factors and octanol-water partition coefficients. Environ. Sci. Technol. 26, 1197–1201.

Di Toro, D.M., C.S. Zarba, D.J. Hansen, W.J. Berry, R.C. Swartz, C.E. Cowen, S.P. Pavlou, H.E. Allen, N.A. Thomas and P.R. Paquin, 1991. Technical basis for establishing sediment quality criteria for nonionic organic chemicals by using equilibrium partitioning. Environ. Toxicol. Chem. 12, 1541–1583.

Di Toro, D.M., L.M. Horzempa, M.M. Casey and W. Richardson, 1982. Reversible and resistant components of PCB adsorption-desorption: adsorption concentration effects. J. Gt. Lakes Res. 8, 336–349.

Eadie, B.J., N.R. Morehead, J. V. Klump and P.F. Landrum, 1992. Distribution of hydrophobic organic compounds between dissolved and particulate organic matter in Green Bay waters. J. Gt. Lakes Res. 18, 91–97.

Eadie, B.J., N.R. Morehead and P.F. Landrum, 1990. Three-phase partitioning of hydrophobic organic compounds in Great Lakes waters. Chemosphere 20, 161–178.

Ewald, G. and P. Larsson, 1994. Partitioning of ^{14}C-labelled 2,2′,4,4′-tetrachlorobiphenyl between water and fish lipids. Environ. Toxicol. Chem. 13, 1577–1580.

Ferraro, S.P., H. Lee II, L. Smith, R. Ozretich and D. Specht, 1990a. Accumulation factors for eleven polychlorinated biphenyl congeners. Bull. Environ. Contam. Toxicol. 46, 276–283.

Ferraro, S.P., H. Lee II, R.J. Ozretich and D.T. Specht, 1990b. Predicting bioaccumulation potential: a test of a fugacity-based model. Arch. Environ. Contam. Toxicol. 46, 276–283.

Fisher, D.J. and J.R. Clark, 1990. Bioaccumulation of kepone by grass shrimp (*Palaemontes pugio*): importance of dietary accumulation and food ration. Aquat. Toxicol. 17, 167–186.

Fisher, S.W., 1991. Changes in the toxicity of three pesticides as a function of environmental pH and temperature. Bull. Environ. Contam. Toxicol. 46, 197–202.

Fisher, S.W., 1990. The pH dependent accumulation of PCP in aquatic microcosms with sediment. Aquat. Toxicol. 18, 199–218.

Fisher, S.W. and T.W. Lohner, 1986. Changes in the aqueous behavior of parathion under varying conditions of pH. Arch. Environ. Contam. Toxicol. 16, 79–84.

Fisher, S.W. and R.W. Wadleigh, 1986. Effects of pH on the acute toxicity and uptake of [^{14}C]pentachlorophenol in the midge, *Chironomus riparius*. Ecotoxicol. Environ. Saf. 11, 1–8.

Fisher, S.W., 1985. Effects of pH on the toxicity and uptake of [^{14}C]lindane in the midge, *Chironomus riparius*. Ecotoxicol. Environ. Saf. 10, 202–208.

Fu, G., A.T. Kan and M. Tomson, 1994. Adsorption and desorption hysteresis of PAHs in surface sediment. Environ. Toxicol. Chem. 13, 1559–1567.

Gauthier, T.D., W.R. Seitz and C.L. Grant, 1987. Effects of structural and compositional variations of dissolved humic materials on pyrene K_{oc} values. Environ. Sci. Technol. 21, 243–248.

Gerould, S., P.F. Landrum and J. P. Giesy, 1983. Anthracene bioconcentration and biotransformation in chironomids: effects of temperature and concentration. Environ. Pollut. 30A, 175–188.

Gjessing, E.T. and L. Bergling, 1981. Adsorption of PAH to aquatic humus. Arch. Hydrobiol. 92, 24–30.

Gobas, F.A.P.C., J.R. McCorquodale and G.D. Haffner, 1993. Intestinal absorption and biomagnification of organochlorines. Environ. Toxicol. Chem. 12, 567–576.

Gobas, F.A.P.C., E.J. McNeil, L. Lovett-Doust and G.D. Haffner, 1991. Bioconcentration of chlorinated aromatic hydrocarbons in aquatic macrophytes. Environ. Sci. Technol. 25, 924–929.

Gobas, F.A.P.C., K.E. Clark, W.Y. Shiu and D.Mackay, 1989a. Bioconcentration of polybrominated benzenes and biphenyls and related superhydrophobic chemicals in fish: role of bioavailability and elimination into feces. Environ. Toxicol. Chem. 8, 231–245.

Gobas, F.A.P.C., D.C. Bedard, J.J.H. Ciborowski and G.D. Haffner, 1989b. Bioaccumulation of chlorinated hydrocarbons by the mayfly (*Hexagenia limbata*) in Lake St. Clair. J. Gt. Lakes Res. 15, 581–588.

Gobas, F.A.P.C., D.C.G. Muir and D. Mackay, 1988a. Dynamics of dietary bioaccumulation and fecal elimination of hydrophobic organic chemicals in fish. Chemosphere 17, 943–962.

Gobas, F.A.P.C., J.M. Lahittete, G. Garofalo, W.Y. Shiu and D. Mackay, 1988b. A novel method for measuring membrane-water partition coefficients of hydrophobic organic chemicals: comparison with 1-octanol-water partitioning. J. Pharm. Sci. 77, 265–272.

Gobas, F.A.P.C., A. Opperhuizen and O. Hutizinger, 1986. Bioconcentration of hydrophobic chemicals in fish: relationship with membrane permeation. Environ. Toxicol. Chem. 5, 637–646.

Grovers, H., C. Ruepert and H. Aking, 1984. Quantitative structure-activity relationships for polycyclic aromatic hydrocarbons: correlation between molecular connectivity, physico-chemical properties, bioconcentration and toxicity in *Daphnia pulex*. Chemosphere 13, 227–236.

Hamelink, J.L. and A. Spacie, 1977. Fish and chemicals: the process of accumulation. Annu. Rev. Pharmacol. Toxicol. 17, 167–177.

Harkey, G.A., 1993. Investigation of the bioavailability of sediment-associated hydrophobic organic contaminants via laboratory bioassays. Ph.D. Thesis, Clemson University, Clemson, SC, 158 pp.

Harkey, G.A., M.J. Lydy, J. Kukkonen and P.F. Landrum, 1994a. Feeding selectivity and assimilation of PAH and PCB in *Diporeia* spp. Environ. Toxicol. Chem. 13, 1445–1455.

Harkey, G.A., P.F. Landrum and S.J. Klaine, 1994b. Partition coefficients of hydrophobic contaminants in natural water, porewater, and elutriates obtained from dosed sediment: a comparison of methodologies. Chemosphere 28, 583–596.

Harkey, G.A., P.F. Landrum and S.J. Klaine, 1994c. Comparison of whole sediment, elutriate, and porewater for use in assessing sediment-associated organic contaminants in bioaccumulation assays. Environ. Toxicol. Chem. 13, 1315–1329.

Hassett, J.P. and E. Milicic, 1985. Determination of equilibrium and rate constants for binding of a polychlorinated biphenyl congener by dissolved humic substances. Environ. Sci. Technol. 19, 638–643.

Hassett, J.P. and M.A. Anderson, 1979. Association of hydrophobic organic compounds with dissolved organic matter in aquatic systems. Environ. Sci. Technol. 13, 1526–1529.

Hayton, W.L. and M.G. Barron, 1990. Rate-limiting barriers to xenobiotic uptake by the gill. Environ. Toxicol. Chem. 9, 151–157.

Hickey, J.P. and D.R. Passino-Reader, 1991. Linear solvation energy relationships: "Rules of thumb" for estimation of variable values. Environ. Sci. Technol. 25, 1753–1760.

Hobbs, E.J., M.L. Keplinger and J.C. Calandra, 1975. Toxicity of polydimethylsiloxanes in certain environmental systems. Environ. Res. 10, 397–406.

Hoke, R.A., G.T. Ankley, A.M. Cotter, T. Goldenstein, P.A. Kosian, G.L. Phipps and F.M. Van der Meiden, 1994. Evaluation of equilibrium partitioning theory for predicting acute toxicity of field-collected sediments contaminated with DDT, DDE, and DDD to the amphipod *Hyalella azteca*. Environ. Toxicol. Chem. 13, 157-166.

Howe, G.E., L.L. Marking, T.D. Bills, J.J. Rach and F.L. Mayer, Jr., 1994a. Effects of water temperature and pH on toxicity of terbofos, trichlorfon, 4-nitrophenol and 2,4-dinitrophenol to the amphipod *Gammarus pseudolimnaeus* and rainbow trout (*Oncorhynchus mykiss*). Environ. Toxicol. Chem. 13, 51–66.

Howe, G.E., L.L. Marking, T.D. Bills, M.A. Boogaard and F.L. Mayer, Jr., 1994b. Effects of water temperature on the toxicity of 4-nitrophenol and 2,4-dinitrophenol to developing rainbow trout (*Oncorhynchus mykiss*). Environ. Toxicol. Chem. 13, 79–84.

Ikemoto, Y., K. Motoba, T. Suzuki and M. Uchida, 1992. Quantitative structure-activity relationships of nonspecific and specific toxicants in several organism species. Environ. Toxicol. Chem. 11, 931–939.

Kalsch, W., R. Nagel and K. Urich, 1991. Uptake, elimination and bioconcentration of ten anilines in zebrafish (*Brachydanio rerio*). Chemosphere 22, 351–363.

Kamlet, M.J., R.M. Doherty, M.H. Abraham and R.W. Taft, 1988. Linear solvation energy relationships: an improved equation for correlation and prediction of octanol/water partition coefficients of organic nonelectrolytes (including strong hydrogen bond donor solutes). J. Phys. Chem. 92, 5244–5255.

Kamlet, M.J., R.M. Doherty, J.L.M. Abboud and R.W. Taft, 1986. Solubility — a new look. Chemtech 16, 566–576.

Kango, R.A. and J.A. Quinn, 1992. A combined reverse-phase and purge and trap chromatographic method to study the interaction of volatile organic compounds with dissolved humic acid in aqueous solutions. Environ. Sci. Technol. 26, 163–165.

Karickhoff, S.W. and K.R. Morris, 1985. Impact of tubificid oligochaetes on pollutant transport in bottom sediments. Environ. Sci. Technol. 19, 51–56.

Karickhoff, S.W., 1980. Sorption kinetics of hydrophobic pollutants in natural sediments. In: Contaminants and Sediments, Vol. 2, edited by R.A. Baker, Ann Arbor Science Publishers, Ann Arbor, MI, pp. 193–206.

Keilty, T.J., D.S. White and P.F. Landrum, 1988a. Sublethal responses to endrin in sediment by *Stylodrilius heringianus* (Lumbriculidae) as measured by a [137]cesium marker layer technique. Aquat. Toxicol. 13, 251–270.

Keilty, T.J., D.S. White and P.F. Landrum, 1988b. Sublethal responses to endrin in sediment by *Lumbriculus hoffmeisteri* (Tubificidae) and in mixed culture with *Stylodrillus heringianus* (Lumbriculidae). Aquat. Toxicol. 13, 227–250.

Keilty, T.J., D.S. White and P.F. Landrum, 1988c. Short-term lethality and sediment avoidance assays with endrin-contaminated sediment and two oligochaetes from Lake Michigan. Arch. Environ. Contam. Toxicol. 17, 95–101.

Kier, L.B. and L.H. Hall, 1976. Molecular Connectivity in Chemistry and Drug Research, Academic Press, New York.

Kier, L.B. and L.H. Hall, 1986. Molecular Connectivity in Structure-Activity Analysis, Research Studies Press, Letchworth, U.K., 262 pp.

Kenega, E.E. and C.A.I. Goring, 1980. Relationship between water solubility, soil sorption, octanol-water partitioning, and concentration of chemicals in biota. In: Aquatic Toxicology, ASTM STP 707, edited by J.G. Eaton, P.R. Parrish and A.C. Hendricks, American Society for Testing and Materials, Philadelphia, PA, pp. 78–115.

Klaassen, C.D., 1986. Distribution, excretion and absorption of toxicants. In: Casarett and Doull's Toxicology, III, edited by C.D. Klaassen, M.O. Amdur and J. Doull, Macmillan, New York, pp. 33–63.

Klump, J.V., J.R. Kresoski, M.E. Smith and J.L. Kaster, 1987. Dual tracer studies of the assimilation of an organic contaminant from sediments by deposit feeding oligochaetes. Can. J. Fish. Aquat. Sci. 44, 1574–1583.

Kucklick, J.R., T.F. Bidleman, L.L. McConnell, M.D. Walla and G.P. Ivanov, 1994. Organochlorines in the water and biota of Lake Baikal, Siberia. Environ. Sci. Technol. 28, 31–37.

Kukkonen, J. and P.F. Landrum, 1995. Measuring assimilation efficiencies for sediment-bound PAH and PCB congeners by benthic organisms. Aquat. Toxicol. 32, 75–92.

Kukkonen, J. and P.F. Landrum, 1994a. Effects of sediment-bound PDMS on the bioavailability and distribution of benzo(a)pyrene in lake sediment. Environ. Toxicol. Chem. 14, 523–531.

Kukkonen, J. and P.F. Landrum, 1994b. Toxicokinetics and toxicity of sediment-associated pyrene to *Lumbriculus variegatus* (Oligochaeta). Environ. Toxicol. Chem. 13, 1457–1468.

Kukkonen, J., P.L. Van Hoof and P.F. Landrum, 1983. Bioavailability of sediment-associated PAHs by *Lumbriculus variegatus*: comparison of laboratory-dosed with environmentally resident PAH. Abstract 277. 14th Annual Meeting of the Society of Environmental Toxicology and Chemistry, November 14–18, 1993, Houston, TX.

Kukkonen, J., 1991. Effects of pH and natural humic substances on the accumulation of organic pollutants into two freshwater invertebrates. In: Substances in the Aquatic and Terrestrial Environment, edited by B. Allard, H. Borén and A. Grimvall, Lecture Notes in Earth Sciences, Vol. 33, Spring-Verlag, New York, pp. 413–422.

Kukkonen, J. and A. Oikari, 1991. Bioavailability of organic pollutants in boreal waters with varying levels of dissolved organic material. Water Res. 25, 455–463.

Kukkonen, J., J.F. McCarthy and A. Oikari, 1990. Effects of XAD-8 fractions of dissolved organic carbon on the sorption and bioavailability of organic micropollutants. Arch. Environ. Contam. Toxicol. 19, 551–557.

Kukkonen, J., A. Oikari, S. Johnsen and E. Gjessing, 1989. Effects of humus concentrations on benzo(a)pyrene accumulation from water to *Daphnia magna*: comparison of natural waters and standard preparations. Sci. Total Environ. 79, 197–207.

Lake, J.L., N.I. Rubinstein, H. Lee II, C.A. Lake, J. Heltshe and S. Pavignano, 1990. Equilibrium partitioning and bioaccumulation of sediment-associated contaminants by infaunal organisms. Environ. Toxicol. Chem. 9, 1095–1106.

Lake, J.L., N. Rubenstein and S. Pavignano, 1987. Predicting bioaccumulation: development of a simple partitioning model for use as a screening tool for regulating ocean disposal of wastes. In: Fate and Effects of Sediment-Bound Chemicals in Aquatic Systems, edited by K.L. Dickson, A.W. Maki and W. A. Brungs, Pergamon Press, Elmsford, NY, pp. 151–166.

Landrum, P.F., W.L. Hayton, H.Lee II, L. McCarty, D. Mackay and J. McKim, 1994a. Kinetics behind environmental bioavailability. In: Bioavailability: Physical, Chemical and Biological Interactions, edited by J.L. Hamelink, P.F. Landrum, W.H. Benson and H.L. Bergman, Lewis Publishers, Boca Raton, FL, pp. 203–219.

Landrum, P.F., W.S. Dupuis and J. Kukkonen, 1994b. Toxicity and toxicokinetics of sediment-associated pyrene in *Diporeia* spp.: examination of equilibrium partitioning theory and residue effects for assessing hazard. Environ. Toxicol. Chem. 13, 1769–1780.

Landrum, P.F., H. Lee II and M. Lydy, 1992a. Toxicokinetics in aquatic systems: model comparisons and use in hazard assessment. Environ. Toxicol. Chem. 11, 1709–1725.

Landrum, P.F., T.D. Fontaine, W.R. Faust, B.J. Eadie and G.A. Lang, 1992b. Modeling the accumulation of polycyclic aromatic hydrocarbons by the amphipod *Diporeia* (spp.). In: Chemical Dynamics in Fresh Water Ecosystems, edited by F.A.P.C. Gobas and J.A. McCorquodale, Lewis Publishers, Boca Raton, FL, pp. 111–128.

Landrum, P.F., B.J. Eadie and W.R. Faust, 1992c. Variation in the bioavailability of polycyclic aromatic hydrocarbons to the amphipod *Diporeia* (spp.) with sediment aging. Environ. Toxicol. Chem. 11, 1197–1208.

Landrum, P.F., B.J. Eadie and W.R. Fasut, 1991. Toxicokinetics and toxicity of a mixture of sediment-associated polycyclic aromatic hydrocarbons to the amphipod *Diporeia* spp. Environ. Toxicol. Chem. 10, 35–46.

Landrum, P.F. and W.R. Faust, 1991. Effect of variation in sediment composition on the uptake rate coefficient for selected PCB and PAH congeners by the amphipod *Diporeia* spp. In: Aquatic Toxicology and Risk Assessment XIV, ASTM STP 1124, edited by M.A. Mayes and M.G. Barron, American Society for Testing and Materials, Philadelphia, PA, pp. 263–279.

Landrum, P.F. and C.R. Stubblefield, 1991. Role of respiration in the accumulation of organic xenobiotics by the amphipod *Diporeia* spp. Environ. Toxicol. Chem. 10, 1019–1028.

Landrum, P.F. and W.S. Dupuis, 1990. Toxicity and toxicokinetics of pentachlorophenol and carbaryl to *Pontoporeia hoyi* and *Mysis relicta*. In: Aquatic Toxicology and Risk Assessment XIII, ASTM STP 1096, edited by W.G. Landis and W.H. Van der Schalie, American Society for Testing and Materials, Philadelphia, PA, pp. 278–289.

Landrum, P.F. and J.A. Robbins, 1990. Bioavailability of sediment-associated contaminants to benthic invertebrates. In: Sediments: Chemistry and Toxicity of In-Place Pollutants, edited by R. Baudo, J.P. Giesy and H. Muntau, Lewis Publishers, Ann Arbor, MI, pp. 237–263.

Landrum, P.F., W.R. Faust and B.J. Eadie, 1989. Bioavailability and toxicity of a mixture of sediment-associated chlorinated hydrocarbons to the amphipod *Pontoporeia hoyi*. In: Aquatic Toxicology and Hazard Assessment XII, ASTM STP 1027, edited by U.M. Cowgill and L.R. Williams, American Society for Testing and Materials, Philadelphia, PA, pp. 315–329.

Landrum, P.F., 1989. Bioavailability and toxicokinetics of polycyclic aromatic hydrocarbons sorbed to sediments for the amphipod *Pontoporeia hoyi*. Environ. Sci. Technol. 23, 588–595.

Landrum, P.F., 1988. Toxicokinetics of organic xenobiotics in the amphipod, *Pontoporeia hoyi*: role of physiological and environmental variables. Aquat. Toxicol. 12, 245 271.

Landrum, P.F. and R. Poore, 1988. Toxicokinetics of selected xenobiotics in *Hexagania limbata*. J. Gt. Lakes Res. 14, 427–437.

Landrum, P.F., S.R. Nihart, B.J. Eadie and L.R. Herche, 1987. Reduction in bioavailability of organic contaminants to the amphipod *Pontoporeia hoyi* by dissolved organic material of sediment interstitial waters. Environ. Toxicol. Chem. 6, 11–20.

Landrum, P.F., M.D. Reinhold, S.R. Nihart and B.J. Eadie, 1985. Predicting the bioavailability of organic xenobiotics to *Pontoporeia hoyi* in the presence of humic and fulvic materials and natural dissolved organic matter. Environ. Toxicol. Chem. 4, 459–467.

Landrum, P.F., S.R. Nihart, B.J. Eadie and W.S. Gardner, 1984. Reverse phase separation method for determining pollutant binding to Aldrich humic acid and dissolved organic carbon of natural waters. Environ. Sci. Technol. 18, 187–192.

Landrum, P.F., 1983. The effect of co-contaminants on the bioavailability of polycyclic aromatic hydrocarbons to *Pontoporeia hoyi*. In: Polynuclear Aromatic Hydrocarbons. Seventh International Symposium on Formation, Metabolism and Measurement, edited by M.W. Cooke and A.J. Dennis, Battelle Press, Columbus, OH, pp. 731–743.

Landrum, P.F. and D.G. Crosby, 1981. The disposition of p–nitroanisole by the sea urchin, *Strongylocentrotus purpuratus*. Ecotoxicol. Environ. Saf. 5, 240–254.

Lara, R. and W. Ernst, 1989. Interaction between polychlorinated biphenyls and marine humic substances: determination of association coefficients. Chemosphere 19, 1655–1664.

Larsson, P., L. Collvin, L. Okla and G. Meyer, 1992. Lake productivity and water chemistry as governors of the uptake of persistent pollutants in fish. Environ. Sci. Technol. 26, 346–352.

Lee D.Y. and W.J. Farmer, 1989. Dissolved organic matter interaction with napropamide and four other nonionic pesticides. J. Environ. Qual. 18, 468–474.

Lee II, H. and R. Swartz, 1980. Biological processes affecting the distribution of pollutants in marine sediments. II. Biodeposition and bioturbation. In: Contaminants and Sediments, Vol. II, edited by R.A. Baker, Ann Arbor Science, Ann Arbor, MI, pp. 555–606.

Lee II, H., 1992. Models, muddles, and mud: predicting bioaccumulation of sediment-associated pollutants. In: Sediment Toxicity Assessment, edited by G.A. Burton, Lewis Publishers, Ann Arbor, MI, pp. 267–293.

Lee II, H., 1991. A clam's eye view of the bioavailability of sediment-associated pollutants. In: Organic Substances and Sediments in Water; Vol. III, Biological, edited by R.A. Baker, Lewis Publishers, Ann Arbor, MI, pp. 73–93.

Leo, A., C. Hansch and D. Elkins, 1971. Partitioning coefficients and their uses. Chem. Rev. 71, 525–616.

Leversee, G.J., P.F. Landrum, J.P. Giesy and T. Fannin, 1983. Humic acids reduce bioaccumulation of some polycyclic aromatic hydrocarbons. Can. J. Fish. Aquat. Sci. 40(Suppl 2), 63–69.

Lewis, P.A. and W.B. Horning II, 1991. Differences in acute toxicity test results of three reference toxicants on *Daphnia* at two temperatures. Environ. Toxicol. Chem. 10, 1351–1357.

Lohner, T.W. and S.W. Fisher, 1990. Effects of pH and temperature on the acute toxicity and uptake of carbaryl in the midge, *Chironomus riparius*. Aquat. Toxicol. 16, 335–354.

Long, E.R. and L.G. Morgan, 1990. The potential for biological effects of sediment-sorbed contaminants tested in the National Status and Trends Program, NOAA Technical Memorandum NOS OMA 52. National Oceanic and Atmospheric Administration, Washington, D.C.

Lydy, M.J., T.W. Lohner and S.W. Fisher, 1990. Influence of pH, temperature and sediment type on the toxicity, accumulation and degradation of parathion in aquatic systems. Aquat. Toxicol. 17, 27–44.

Lyman, W.J., W.F. Reehl and D.H. Rosenblatt, 1990. Handbook of Chemical Property Estimation Methods. American Chemical Society, Washington, D.C.

Macek, K.J. and S. Korn, 1970. Significance of the food chain in DDT accumulation by fish. J. Fish. Res. Board Can. 27, 1496–1498.

Macek, K.J., C.R. Rodgers, D.L. Stalling and S. Korn, 1970. The uptake, distribution and elimination of dietary ^{14}C-DDT and ^{14}C-dieldrin in rainbow trout. Trans. Am. Fish. Soc. 99, 689–695.

Mackay, D. 1991. Multimedia Environmental Models: The Fugacity Approach. Lewis Publishers, Ann Arbor, MI, 257 pp.

Mackay, D., 1982. Correlation of bioconcentration factors. Environ. Sci. Technol. 16, 274–278.

McCarthy, J.F., L.E. Roberson and L.W. Burris, 1989. Association of benzo(a)pyrene with dissolved organic matter: prediction of K_{dom} from structural and chemical properties of the organic matter. Chemosphere 19, 1911–1920.

McCarthy, J.F. and M.C. Black, 1988. Partitioning between dissolved organic macro-molecules and suspended particulates: effects on bioavailability and transport of hydrophobic organic chemicals in aquatic systems. In: Aquatic Toxicology and Hazard Assessment X, ASTM STP 971, edited by W.J. Adams, G.A. Chapman and W.G. Landis, American Society for Testing and Materials, Philadelphia, PA, pp. 233–246.

McCarthy, J.F. and B.D. Jimenez, 1985a. Interactions between polycyclic aromatic hydrocarbons and humic material: binding and dissociation. Environ. Sci. Technol. 19, 1072–1076.

McCarthy, J.F. and B.D. Jimenez, 1985b. Reduction in bioavailability to bluegills of polycyclic aromatic hydrocarbons bound to dissolved humic material. Environ. Toxicol. Chem. 4, 511–521.

McCarthy, J.F., B.D. Jimenez and T. Barbee, 1985. Effect of dissolved humic material on accumulation of polycyclic aromatic hydrocarbons: structure-activity relation-ship. Aquat. Toxicol. 7, 15–24.

McCarty, L.S. and D. Mackay, 1993. Enhancing ecotoxicological modeling and assess-ment. Environ. Sci. Technol. 27, 1719–1728.

McCarty, L.S., G.W. Ozburn, A.D. Smith and D.G. Dixon, 1992. Toxicokinetic mod-eling of mixtures of organic chemicals. Environ. Toxicol. Chem. 11, 1037–1047.

McElroy, A.E., J.W. Farrington and J.M. Teal. 1989. Bioavailability of polycyclic aromatic hydrocarbons in the aquatic environment. In: Metabolism of Polycyclic Aromatic Hydrocarbons in the Aquatic Environment, edited by U. Varanasi, CRC Press, Boca Raton, FL, pp. 1–39.

McFarland, V.A. and J.U. Clarke, 1989. Environmental occurrence, abundance, and potential toxicity of polychlorinated biphenyl congeners: considerations for a congener-specific analysis. Environ. Health Perspect. 81, 225–239.

McFarland, V.A., 1984. Activity-based evaluation of potential bioaccumulation from sediments. In: Dredging and Dredged Material Disposal. Proceedings of the Con-ference Dredging '84, , edited by R.L. Montgomery and J.W. Leach, American Society of Civil Engineering, New York, pp. 461–466.

McGowan, J.C. and A. Mellors, 1986. Molecular Volumes in Chemistry and Biology: Applications Including Partitioning and Toxicity. Ellis Horwood, Chichester, U.K., 259 pp.

McKim, J., P. Schmieder and G. Veith, 1985. Absorption dynamics of organic chemical transport across trout gills as related to octanol-water partition coefficient. Toxicol. Appl. Pharmacol. 77, 1–10.

McLeay, D.J., C.C. Walden and J.R. Munro, 1979a. Influence of dilution water on the toxicity of kraft pulp and paper mill effluent, including mechanisms of effect. Water Res. 13, 151–158.

McLeay, D.J., C.C. Walden and J.R. Munro, 1979b. Effect of pH on toxicity of kraft pulp and paper mill effluent to salmonid fish in fresh and seawater. Water Res. 13, 249–254.

Meador, J.P., U. Varanasi and C.A. Krone, 1991. Differential sensitivity of marine infaunal amphipods to tributylin. Mar. Biol. 116, 231–239.

Meyer, C.L., B.C. Suedel, J.H. Rodgers II and P.B. Dorn, 1993. Bioavailability of sediment-sorbed chlorinated ethers. Environ. Toxicol. Chem. 12, 493–505.

Morehead, N.R., B.J. Eadie, B. Lake, P.F. Landrum and D. Berner, 1986. The sorption of PAH onto dissolved organic matter in Lake Michigan waters. Chemosphere 15, 403–412.

Muir, D.C.G., W.K. Marshall and G.R.B. Webster, 1985. Bioconcentration of PCDD's by fish: effects of molecular structure and water chemistry. Chemosphere 14, 829–833.

Murry, W.J., L.H. Hall and L.B. Kier, 1978. Molecular connectivity. III. Relationship to partition coefficients. J. Pharm. Sci. 64, 1978–1981.

Nebeker, A.V., G.S. Schuytema, W.L. Griffis, J.A. Barbitta and L.A. Carey, 1989. Effect of sediment organic carbon on survival of *Hyalella azteca* exposed to DDT and endrin. Environ. Toxicol. Chem. 8, 705–718.

Neely, W.B., 1979. Estimating rate constants for the uptake and clearance of chemicals by fish. Environ. Sci. Technol. 13, 1506–1510.

Neely, W.B., D.R. Branson and G.E. Blau, 1974. Partition coefficient to measure bioconcentration potential of organic chemicals in fish. Environ. Sci. Technol. 8, 1113–1115.

Neff, J.M.. B.W. Cornaby, R.M. Vaga, T.C. Gulbransen, J.A. Scanlon and D.J. Bean, 1988. An evaluation of the screening level concentration approach for validation of sediment quality criteria for freshwater and saltwater ecosystems. In: Aquatic Toxicology and Hazard Assessment X, ASTM STP 971, edited by W.J. Adams, G.A. Chapman and W.G. Landis, American Society for Testing and Materials, Philadelphia, PA, pp. 115–127.

Neff, J.M., 1979. Polycyclic aromatic hydrocarbons in the marine environment: sources, fate and biological effects. Applied Science Publishers, London, U.K., 262 pp.

Newman, M.C., 1993. Regression analysis of log-transformed data: statistical bias and its correction. Environ. Toxicol. Chem. 12, 1129–1133.

Niimi, A.J. and V. Palazzo, 1985. Temperature effect on the elimination of pentachlorophenol, hexachlorobenzene and mirex by rainbow trout (*Salmo gairdneri*). Water Res. 19, 205–297.

Nkedi-Kizza, P., P.S.C. Rao and A.G. Hornsby, 1985. Influence of organic co-solvents on sorption of hydrophobic organic chemicals by soils. Environ. Sci. Technol. 19, 975–979.

Ogner, G. and M. Schnitzer, 1970. Humic substances: fulvic acid-dialkyl phthalate complexes and their role in pollution. Science 170, 317–318.

Oliver, B.G. and A.J. Niimi, 1988. Trophodynamics analysis of polychlorinated biphenyl congeners and other chlorinated hydrocarbons in Lake Ontario ecosystem. Environ. Sci. Technol. 22, 388–397.

Oliver, B.G., 1987. Biouptake of chlorinated hydrocarbons from laboratory-spiked and field sediments by oligochaete worms. Environ. Sci. Technol. 21, 785–790.

Oliver, B.G. and A. J. Niimi, 1983. Bioconcentration of chlorobenzenes from water by rainbow trout: correlation with partition coefficients and environmental residues. Environ. Sci. Technol. 17, 287–291.

Opperhuizen, A., 1991. Bioaccumulation kinetics: experimental data and modeling. In: Organic Micropollutants in the Aquatic Environment, Proceedings of the Sixth European Symposium, edited by G. Angeletti and A. Bjoresth, Kluwer Academic Publishers, Dordrecht, Netherlands, pp. 61–70.

Opperhuizen, A. and S.M. Schrap, 1988. Uptake efficiencies of two polychlorinated biphenyls in fish after dietary exposure to five different concentrations. Chemosphere 17, 253–262.

Opperhuizen, A., P. Serné and J.M.D. Van der Steen, 1988. Thermodynamics of fish/water and octan-1-ol/water partitioning of some chlorinated benzenes. Environ. Sci. Technol. 22, 286–292.

Opperhuizen, A., H.W.J. Damen, G.M. Asyee, J.M.D. van der Steen and O. Hutzinger, 1987. Uptake and elimination by fish of polydimethylsiloxanes (silicones) after dietary and aqueous exposure. Toxicol. Environ. Chem. 13, 265–285.

Opperhuizen, A., E.W. Velde, F.A.P.C. Gobas, D.A.K. Liem, J.M.D. Steen and O. Hutzinger, 1985. Relationship between bioconcentration in fish and steric factors of hydrophobic chemicals. Chemosphere 14, 1871–1896.

Owens, J.W., S.M. Swanson and D.A. Birkholz, 1994. Bioaccumulation of 2,3,7,8-tetrachlorodibenzo-p-dioxin, 2,3,7,8-tetrachlorodibenzofuran and extractable organic chlorine at a bleached-kraft mill site in northern Canadian river system. Environ. Toxicol. Chem. 13, 343–354.

Park, J.H. and H.J. Lee, 1993. Estimation of bioconcentration factor in fish, adsorption coefficient for soils and sediments and interfacial tension with water for organic nonelectrolytes based on the linear solvation energy relationships. Chemosphere 26, 1905–1916.

Pärt, P., 1989. Bioavailability and uptake of xenobiotics in fish. In: Chemicals in the Aquatic Environment. Advanced Hazard Assessment, edited by L. Landner, Springer, Berlin, pp. 113–127.

Pawlisz, A.V. and R.H. Peters, 1993a. A radioactive tracer technique for the study of lethal body burdens of narcotic organic chemicals in *Daphnia magna*. Environ. Sci. Technol. 27, 2801–2806.

Pawlisz, A.V. and R.H. Peters, 1993b. A test of the equipotency of internal burdens of nine narcotic chemicals using *Daphnia magna*. Environ. Sci. Technol. 27, 2801–2806.

Quigley, M.A., 1988. Gut fullness of the deposit-feeding amphipod, *Pontoporeia hoyi*, in southeastern Lake Michigan. J. Gt. Lakes Res. 14, 178–187.

Randall, R.C., H. Lee II, R.J. Ozretich, J.L. Lake and R.J. Purell, 1991. Evaluation of selected lipid methods for normalizing pollutant bioaccumulation. Environ. Toxicol. Chem. 10, 1431–1436.

Rhead, M.M. and J.M. Perkins, 1984. An evaluation of the relative importance of food and water as sources of p,p'-DDT to goldfish, *Carassius auratus* (L.). Water Res. 18, 719–725.

Robbins, J.A., 1986. A model for particle-selective transport of tracers in sediments with conveyor belt deposit feeders. J. Geophys. Res. 91, 8542–8558.

Rowan, D.J. and J.B. Rasmussen, 1992. Why don't Great Lakes fish reflect environmental concentration of organic contaminants? — An analysis of between-lake variability in the ecological partitioning of PCBs and DDT. J. Gt. Lakes Res. 18, 724–741.

Rubenstein, N.I., J.L. Lake, R.J. Pruell, H. Lee II, B. Taplin, J. Heltshe, R. Bowen and S. Pavignano, 1987. Predicting bioaccumulation of sediment-associated organic contaminants: development of a regulatory tool for dredged material evaluation. Internal report. U.S. Environmental Protection Agency, 600/x-87/368, Narragansett, RI.

Rutherford, D.W., C.T. Chiou and D.E. Kile, 1992. Influence of soil organic matter composition on the partition of organic compounds. Environ. Sci. Technol. 26, 336–340.

Saarikoski, J. and M. Viluksela, 1981. Influence of pH on the toxicity of substituted phenols to fish. Arch. Environ. Contam. Toxicol. 10, 747–753.

Saarikoski, J. and M. Viluksela, 1982. Relation between physicochemical properties of phenols and their toxicity and accumulation in fish. Ecotoxicol. Environ. Saf. 6, 501–512.

Saarikoski, J., R. Lindström, M. Tyynelä and M. Viluksela, 1986. Factors affecting the absorption of phenolics and carboxylic acids in the guppy (*Poecilia reticulata*). Ecotoxicol. Environ. Saf. 11, 158–173.

Sabljic, A. and M. Protic, 1982. Molecular connectivity: a novel method for prediction of bioconcentration factor of hazardous chemicals. Chem.-Biol. Interact. 42, 301–310.

Sagiura, K., N. Ito, N. Matsumoto, Y. Mihara, K. Murata, Y. Tsukakoshi and M. Goto, 1978. Accumulation of polychlorinated biphenyls and polybrominated biphenyls in fish: correlation between partition coefficients and accumulation factors. Chemosphere 9, 731–736.

Saito, S., C. Tateno, A. Tanoue and T. Matsuda, 1990. Electron microscope autoradiographic examination of uptake behavior of lipophilic chemicals into fish gill. Ecotoxicol. Environ. Saf. 19, 184–191.

Schlautman, M.A. and J.J. Morgan, 1993. Effects of aqueous chemistry on the binding of polycyclic aromatic hydrocarbons by dissolved humic materials. Environ. Sci. Technol. 27, 961–969.

Schüürmann, G. and W. Klein, 1988. Advances in bioconcentration prediction. Chemosphere 17, 1551–1574.

Servos, M.R., D.C.G. Muir and G.R.B. Webster, 1989. The effect of dissolved organic matter on the bioavailability of polychlorinated dibenzo-*p*-dioxins. Aquat. Toxicol. 14, 169–184.

Servos, M.R. and D.C.G. Muir, 1989. Effect of dissolved organic matter from Canadian shield lakes on the bioavailability of 1,2,6,8-tetrachlorodibenzo-*p*-dioxin to the amphipod *Crangonyx laurentianus*. Environ. Toxicol. Chem. 8, 141–150.

Shaw, G.R. and G.W. Connell, 1984. Physicochemical properties controlling polychlorinated biphenyl (PCB) concentrations in aquatic organisms. Environ. Sci. Technol. 18, 18–23.

Sijm, D.T.H.M., W. Seinen and A. Opperhuizen, 1992. Life-cycle biomagnification study in fish. Environ. Sci. Technol. 26, 2162–2174.

Sijm, D.T.H.M. and A. Opperhuizen, 1988. Biotransformation, bioaccumulation and lethality of 2,8,-dichlorodibenzo-*p*-dioxin: a proposal to explain the biotic fate and toxicity of PCDDs and PCDFs. Chemosphere 17, 83–99.

Sithole, B.B. and R.D. Guy, 1985, Interactions of secondary amines with bentonite clay and humic materials in dilute aqueous systems. Environ. Int. 11, 499–504.

Spacie, A. and J.L. Hamelink, 1985. Bioaccumulation. In: Fundamentals of Aquatic Toxicology, edited by G.M. Rand and S.R. Petrocelli, Hemisphere Publishing, New York, pp. 495–525.

Spehar, R.L., H.P. Nelson, M.J. Swanson and J.W. Renoos, 1985. Pentachlorophenol toxicity to amphipods and fathead minnows at different test pH values. Environ. Toxicol. Chem. 4, 389–397.

Stehly, G.R. and W.L. Hayton, 1990. Effect of pH on the accumulation kinetics of pentachlorophenol in goldfish. Arch. Environ. Contam. Toxicol. 19, 464–470.

Stehly, G.R. and W.L. Hayton, 1983. pH control of weak electrolyte toxicity to fish. Environ. Toxicol. Chem. 2, 325–328.

Stemmer, B.L., G.A. Burton, Jr. and S. Leibfritz-Frederick, 1990. Effect of sediment test variables on selenium toxicity to *Daphnia magna*. Environ. Toxicol. Chem. 9, 381–389.

Suedel, B. C., J.H. Rodgers II and P.A. Clifford, 1993. Bioavailability of fluoranthene in freshwater sediment toxicity tests. Environ. Toxicol. Chem. 12, 155–165.

Suter, G.W., 1993. Ecological Risk Assessment, Lewis Publishers, Ann Arbor, MI, 538 pp.

Swackhamer, D.L. and R.S. Skoglund, 1993. Bioaccumulation of PCBs by algae: kinetics versus equilibrium. Environ. Toxicol. Chem. 12, 831–838.

Swackhamer, D.L. and R.S. Skoglund, 1991. The role of phytoplankton in the partitioning of hydrophobic organic contaminants in water. In: Organic Substances and Sediments in Water: Processes and Analytical, Vol. II, edited by R.A. Baker, Lewis Publishers, Ann Arbor, MI, pp. 79–105.

Swackhamer, D.L. and R.A. Hites, 1988. Occurrence and bioaccumulation of organochlorine compounds in fishes from Siskiwit Lake, Isle Royal, Lake Superior. Environ. Sci. Technol. 22, 543–548.

Swartz, R.C., D.W. Schultz, T.H. Dewitt, G.R. Ditsworth and J.O. Lamberson, 1990. Toxicity for fluoranthene in sediment to marine amphipods: a test of the equilibrium partitioning approach to sediment quality criteria. Environ. Toxicol. Chem. 9, 1071–1080.

Thomann, R.V., 1981. Equilibrium model of fate of microcontaminants in diverse aquatic food chains. Can. J. Fish. Aquat. Sci. 38, 280–296.

Thomann, R.V., 1989. Bioaccumulation model of organic chemical distribution in aquatic food chains. Environ. Sci. Technol. 23, 699–707.

Thomann, R.V. and J.P. Connolly, 1984. Model of PCB in the Lake Michigan lake trout food chain. Environ Sci. Technol. 18, 65–71.

Thomann, R.V., J.P. Connolly and T.F. Parkerton, 1992. An equilibrium model of organic chemical accumulation in aquatic food webs with sediment interaction. Environ. Toxicol. Chem. 11, 615–629.

U.S. EPA, 1989. Briefing report to the EPA science advisory board on the equilibrium partitioning approach to generating sediment quality criteria. U.S. Environmental Protection Agency Office of Water Regulations and Standards, EPA 440/5-89-002, Washington, DC, 160 pp.

Veith, G.D., D.L. DeFoe and B.V. Bergstedt, 1979. Measuring and estimating the bioconcentration factor of chemicals in fish. J. Fish Res. Board Can. 36, 1040–1048.

Veith, G.D., K.J. Macek, S.R. Petrocelli and J. Carroll, 1980. An evaluation of using partition coefficients and water solubility to estimate bioconcentration factors for organic chemicals in fish. In: Aquatic Toxicology, ASTM STP 707, edited by J.G. Eaton, P.R. Parrish and A.C. Hendricks, American Society for Testing and Materials, Philadelphia, PA, pp. 116–129.

Weston, D.P., 1990. Hydrocarbon bioaccumulation from contaminated sediment by the deposit-feeding polychaete *Abarenicola pacifica*. Mar. Biol. 107, 159–170.

Widdows, J. and P. Donkin, 1989. The application of combined tissue residue chemistry and physiological measurements of mussels (*Mytilus edulis*) for the assessment of environmental pollution. Hydrobiologia 188/189, 455–461.

Witkowski, P.J., P.R. Jaffe and R.A. Ferrara, 1988. Sorption and desorption dynamics of Aroclor 1242 to natural sediment. J. Contam. Hydrol. 2, 249–269.

Word, J.Q., J.A. Ward, L.M. Franklin, V.I. Cullinan and S.L. Kiesser, 1987. Evaluation of the equilibrium partitioning theory for estimating the toxicity of the nonpolar organic compound DDT to the sediment dwelling amphipod *Rhepoxynius abronius*. Report WA56-1. Battelle Marine Research Laboratory, Sequim, WA.

Ziegenfuss, P.S., W.J. Renaudette and W.J. Adams, 1986. Methodology for assessing the acute toxicity of chemicals sorbed to sediments: testing the equilibrium partitioning theory. In: Aquatic Toxicology and Environmental Fate, IX, ASTM STP 921, edited by T.M. Poston and R. Purdy, American Society for Testing and Materials, Philadelphia, PA, pp. 479–493.

CHAPTER 5

Molecular Markers to Toxic Agents

Lee R. Shugart

I. THE PROBLEM

Ecotoxicology is a multidisciplinary science dealing with the adverse effects of toxic agents on living systems. At first glance, this appears to be a simple definition, but one must remember that living systems include a continuum from single organisms to multiple species, and toxic agents may be something rather innocuous, such as an uncomplicated organic molecule like carbon monoxide, or a complex ecological phenomenon, such as global climate change that has discrete aspects of toxicity. The ecotoxicologist attempts to bring some semblance of understanding to his discipline by performing manageable research. His efforts are often directed towards distinct endpoints, such as identifying the toxic agent, understanding its movement in the biosphere, revealing the mechanism of toxic action, or by defining the limits of safe exposure. For many in this field, the primary concern of exposure to environmental contamination is the effects that may occur at the population, community, or ecosystem level (Forbes and Forbes, 1994). However, with increasing concerns about complex environmental issues, such as global climate change, stratospheric ozone depletion, and species extinction, today's ecotoxicologist finds him/herself limited in his ability to adequately assess the status of the environment. There are many reasons for this, but mainly it is the ineffectiveness and/or inappropriateness of traditional toxicological approaches to problems of these magnitudes where environmental species are chronically exposed to pollutants at low levels.

In an attempt to simplify the extent of these problems, there has been a shift in emphasis toward understanding sublethal effects of long-term exposure to contamination at the level of biological organization where exposure can be adequately described and assessed. This, in turn, has led to greater attention to metabolic pathways, induction of resistance mechanisms, and environmental monitoring employing physiological and biochemical criteria, which in essence is the philosophy embodied in the Biological Markers approach to evaluating the status of the environment under stress (Shugart et al., 1992). To understand the significance of toxic effects at higher levels of biological organization (i.e., population), it has become necessary to conduct studies on individual organisms at the biochemical and molecular levels where toxicant-induced responses are initiated.

It is the purpose of this chapter to discuss the current use of molecular markers to determine the relationships between effects observed at the biochemical level and the ultimate response(s) of the organism to its environment. In this regard, it is important to recognize that many of the techniques and methods employed by the ecotoxicologist for studying environmental problems were borrowed and/or modified from mammalian toxicology investigations. For example, the detection and quantitation of DNA adducts as a measure of the biologically significant dose of a genotoxicant was developed from toxic exposure studies with rodents (Ehrenberg et al., 1983). This is just one of many examples that could be cited, but underlines the significance of ecotoxicologists maintaining a strong connection with other fields of toxicology. Furthermore, this chapter has a bias towards genetic ecotoxicology that reflects the research interests of the author.

II. BIOLOGICAL MARKERS

A. CONCEPT

It is sometimes difficult to demonstrate the effect of environmental pollution at the ecosystem level, where populations and communities are studied. Although pollution can produce stress at the ecosystem level, the response observed is latent and so far removed from the initial event of exposure that causality is almost impossible to establish. Risk assessment to date has relied on models that use toxicity data and physical properties of chemicals. The rationale for chemical monitoring is our concern for the potential threat that chemicals can pose to biological resources. This approach has been ineffective at the ecosystem level because of the complexity of pollutants in the environment and the variability in susceptibility of the numerous species present. The environment is subjected to a complex and dynamic array of physical, chemical, and biological interactions, and there is only a limited understanding of how pollution will influence these interactions or how perturbations are expressed through various levels of physical and biological organization.

The monitoring of biological responses (biological markers or biomarkers) for assessing the biological and ecological significance of contaminants present in the environment is a complementary approach to chemical monitoring that is becoming a reasonable and necessary component of many environmental monitoring programs. The rationale behind the concept of biomarkers follows directly from previous discussions on toxic responses. Living organisms integrate exposure to contaminants in their environment and respond in some measurable and predictable way. Responses can be observed at several levels of biological organization from the biomolecular level, where pollutants can cause damage to critical cellular macromolecules and elicit defensive strategies such as detoxication and repair mechanisms, to the organismal level, where severe disturbances are manifested as impairment in growth, reproduction, developmental abnormalities, or decreased survival. Biomarkers can provide not only evidence of exposure to a broad spectrum of anthropogenic chemicals, but also a temporally integrated measure of bioavailable contaminant levels. Preferably a suite of biomarkers are evaluated over time to determine the magnitude of the problem at the individual level and possible consequences at the population or community levels (Cormier and Daniel, 1994).

The "biomarker approach" has recently received considerable attention in ecotoxicology as a new and potentially very powerful and informative tool for detecting and documenting exposure to, and effects of, environmental contamination (Shugart et al., 1989, 1992; McCarthy and Shugart, 1990; Huggett et al., 1992; Peakall and Shugart, 1992; Fossi and Leonzio, 1993; Travis, 1993). The detection of biomarkers in organisms present in a polluted environment is an approach that may resolve some of the problems outlined. Chemicals are known to elicit measurable and characteristic biological responses in exposed organisms. Such evidence can provide a link between exposure and effect. The primary use of biomarkers in environmental monitoring is to assess the health of the species present in order to detect and identify potential problems so that unacceptable and irreversible effects at high levels of biological organization such as disease in humans, mass mortality, and loss of commercially or ecologically important species, can be avoided. It is important to recognize that our current understanding of biomarker responses in environmental species is limited and to achieve their full potential as a tool for environmental protection, a great deal of research will be needed to develop, validate, and interpret biomarker-based monitoring.

B. APPROACH

i. Introduction

Biomarkers can be a cost-effective tool for a number of applications, most of which are motivated, directly or indirectly, by regulatory concerns. These numerous regulatory applications devolve into a more limited and workable number of objectives that provide the basis for focusing strategies for development and application of biomarkers. The objectives of these applications may include documenting exposure, determining the geographic extent of bioavailable contaminants, distinguishing relative degrees of contaminant-related stress at different sites, or prediction of ecological risks to an environment. To design an appropriate and focused study, it is essential to consider

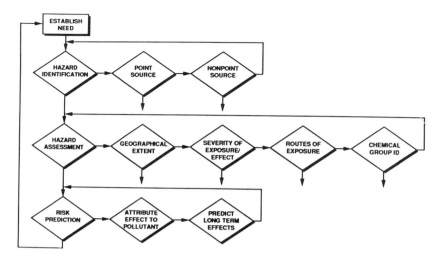

Figure 1 Three-tiered hierarchy of environmental monitoring objectives for which biomarker responses can be a meaningful tool (McCarthy et al., 1992).

different applications of biomonitoring and identify specific objectives for the study. Most of the information provided in this section is taken from a strategy document (McCarthy et al., 1992) on biomarkers prepared for a NATO Advanced Research Workshop on this topic (Peakall and Shugart, 1992).

ii. Objectives

Before any study is initiated, an objective or need for the study must be established. Generally the objective is derived from a question (e.g., Has exposure occurred? What is the extent of exposure? What are the effects?) and the results of the proposed study should provide information needed to answer that question. Often, the objectives of the study are derived from regulatory issues.

Once a need has been established, an evaluation of information already available can be used to establish and define the objective of the proposed study. Will the study be intended for hazard identification, hazard assessment, or risk assessment? Figure 1 describes a hierarchy comprised of three tiers defined by increasing levels of information about the existence and severity of an environmental problem, and by the need for additional information to evaluate the existence, extent, and severity of the problem. Although many components or elements of a properly designed monitoring study will be influenced by the specific objectives motivating the research, many elements and considerations in study design are generally applicable to a broad range of situations.

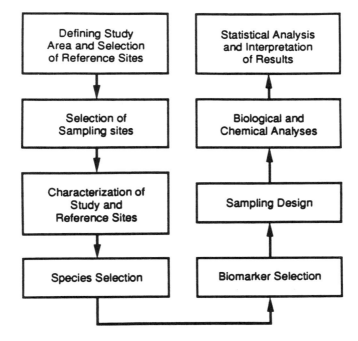

Figure 2 General elements of a biomarker-based environmental monitoring study (McCarthy et al., 1992).

iii. General Elements

Regardless of the specific objectives or motivation for an environmental monitoring study, many elements of the study design remain constant. The principal elements of the design and implementation of a study in a logical pathway are illustrated in Figure 2. The major issues within each element, along with a checklist of specific considerations to be addressed in arriving at a final study design, are outlined in detail by Peakall and Shugart (1992). While the general elements, issues, and considerations are largely unchanged by the study objectives, the answers to the checklist queries, and the weight or importance to be assigned to that issue in the final study design will vary with specific objectives, resources available, types and sources of pollution, and accumulated field experience.

iv. Research Component

Development of a biomarker-based capability will require a close integration of controlled laboratory studies in conjunction with ongoing field studies. Many fundamental research questions can be probed by straightforward experimental design. For example, dose-response studies, or questions related to the effect of single variables on specific biomarker responses (e.g., effect of a species reproductive status on contaminant-related induction of detoxication activity) can be designed and executed based on classic approaches in the toxicological literature. The goals of the research component are to (1) improve basic understanding of biomarker responses; (2) elucidate the role of environmental toxicological stresses on biomarker responses; and (3) confirm relationships between exposure, biomarker responses, and long-term adverse effects to individuals and populations.

More importantly, there is a need for more work demonstrating that biological responses to pollutants in the laboratory are comparable to field data. This need arises for several reasons. First, most studies that examine the effects of anthropogenic contaminants on aquatic organisms focus either on laboratory or field research, with only a few investigations directly linking the two. Second, a large proportion of laboratory studies report results obtained from a single time point or from short-term exposures, with relatively little information on long-term development of biological aberrations in response to toxicants. More detailed information is needed to understand the mechanisms of action of pollutants on aquatic animals, and is necessary to properly interpret laboratory findings with respect to field data, and vice versa. Results obtained in the laboratory may be contradictory to field data if the improper length of time of exposure is chosen (e.g., Shugart, 1988). Third, in the field, it is difficult to control environmental variables that may affect or modulate toxicological responses. Therefore, field data must be validated with appropriate laboratory studies in which these environmental correlates can be controlled.

Many studies, both in the laboratory and in the field, focus on only one or, at most, a few endpoints. Although focusing on a small number of endpoints may be adequate for monitoring exposure to a particular compound, it has limited utility in predicting the effect of pollution, which often occurs as a complex mixture in the environment, on the overall health of the organism. Therefore, there is a need for more studies examining multiple biological endpoints in response to toxic stress. For example, a list of several biological endpoints that are measures of genotoxic responses is found in Table 1 (Shugart, 1993a). These biological indicators of genotoxic stress can be much more sensitive and ecologically relevant than simply measuring the levels of genotoxicant in the environment (McCarthy and Shugart, 1990).

It is often difficult to relate effects observed in the field to the contaminants themselves or their source found in the environment because of the influence of noncontaminant-mediated factors. In such instances, laboratory studies may sometimes be important for establishing a chain of causality. Recently, Theodorakis et al. (1992) examined the temporal expression of various biomarkers in sunfish exposed under controlled laboratory conditions to sediment containing PAHs, PCBs, and heavy metals. This sediment was taken from a known contaminated site and was used to determine whether it was the major

Table 1 Biomarkers of Genotoxicity

Biomarker	Biological response	Level of biological organization	Temporal occurrence[a]	Restriction[b] Detection level	Restriction[b] Limitations
DNA damage	Adducts	Molecular	Early	Low	Repair/analysis
DNA damage	Strand breaks	Molecular	Early	Moderate	Repair
DNA damage	Repair	Molecular	Early	Low	Analysis
Mixed function oxidase	Enzyme induction	Molecular	Early	Moderate	Species variability
DNA damage	Minor nucleoside	Biochemical	Early/middle	High	
Chromosomal aberrations	Abnormal DNA	Subcellular	Middle/late	High	

[a] Temporal occurrence subsequent to exposure: early, hours to days; middle, days to weeks/months; late, weeks/months to years.

[b] Constraints on biomarker application: (1) detection level — anticipated probability of detecting biomarker, (2) limitation — factor(s) contributing to detection or affecting use of biomarker.

source of genotoxicants for the native population of fish. Their results are summarized in Table 2 and show that laboratory data on biological responses associated both with phase I detoxication processes and structural damage to DNA parallel those recorded in fish collected from the site of the sediment's origin. Because results obtained from laboratory-exposed fish were similar to the field-sampled fish, it was concluded that the sediment was the most likely source of genotoxicant for the biological responses observed *in situ*. The fact that the response levels observed between field and laboratory studies were not the same magnitude could have been due to individual differences, lifetime toxicant exposures, natural selection, population and community interactions, or abiotic mediating factors associated with field exposure.

v. Limitations

The exposure of a living organism to pollution may result in the induction of a cascade of cellular events starting with an initial insult at the molecular level and culminating eventually in the appearance of an overt pathological disease. The rationale for the selection of any given biomarker to assess this exposure or the consequence thereof is dictated and limited by two important considerations: (1) the present state of our knowledge concerning those cellular mechanisms involved in the metabolism, distribution, and detoxication of a deleterious substance or its metabolites within the organism; and (2) the analytical technology available to define the biomarker response. The success of current *in situ* efforts in the field of environmental monitoring has been due mainly to the incorporation of analytical techniques that are extremely selective and/or sensitive for detecting and quantitating biological responses (McCarthy and Shugart, 1990; Peakall, 1992; Huggett et al., 1992). The full

Table 2   Summary of Values for Biomarkers from Redbreast
           Sunfish Living in East Fork Poplar Creek (EFPC)

Marker (units)	Field data[a]		Laboratory data[b]	
	Reference stream	EFPC	Control sediment	EFPC sediment
Microsomal cytochrome P_{450} content (pmol/mg protein)	400	800	80	120
Microsomal cytochrome b_5 content (pmol/mg protein)	40	100	275	325
Microsomal NADH-cytochrome b_5 reductase activity (nmol/mg protein/min)	230	380	60	90
Microsomal NADPH-cytochrome c reductase activity (nmol/mg protein/min)	90	170	110	120
EROD activity (pmol/mg protein/min)	2	120	5	50
Liver somatic index (% body mass)	1.0	1.6	1.0	2.0
DNA strand breaks (N value)	—	4.0	—	1.7

[a] Source: Loar et al., 1988; Jimenez et al., 1990; Shugart, 1990.
[b] Source: This study at longest exposure period.
From Theodorakis, C.W. et al., 1992. Ecotoxicology 1, 45–73. With permission.

potential of this approach in future studies will be realized only if continued effort and emphasis is directed towards the development of new and sensitive biomarkers.

C. DOSE-RESPONSE AND MULTIPLE-RESPONSE PARADIGMS

In recent years, technological advances have made it possible to detect and quantify molecular and biochemical responses in biota to environmental pollutants that were previously unknown or below the levels of detection. Although strategies for research and evaluation of these responses are being developed in various fields of study (e.g., environmental exposure/effects, chemical risk assessment, and human epidemiological studies), discussions and debates concerning the proper criteria to use in evaluating and interpreting the results have not kept pace with these technological advances. Presented here is a discussion of the dose-response concepts in relation to an emerging concept based on multiple responses (McCarthy et al., 1992; Peakall and Shugart, 1992).

Many molecular and biochemical responses are measurements of cellular or physiological processes that are normal components of an organism's attempt to maintain a constant internal balance or to deal with normal metabolic processes. The measured values of these processes will fluctuate within some normal range for an organism. If the organism is exposed to a xenobiotic contaminant, it may respond to this exposure by compensatory increases or decreases in one or several of these cellular or physiological processes. The field of biomarker research, for example, attempts to measure and interpret this abnormal increase or decrease in the cellular or physiological process.

Dose-response is a concept formulated for controlled laboratory studies in which an organism is exposed to a single chemical and a response is measured over a range of doses. Information obtained from these laboratory responses are then extrapolated to organisms living under natural environmental conditions. The accuracy of this extrapolation has been the subject of much concern. Natural populations are not usually exposed to a single chemical but rather to mixtures of chemicals and are subjected to many stresses not normally encountered in laboratory studies (e.g., heat and cold stress, nutritional and energy deficiency, and predator/prey stress). These variables greatly confound the extrapolation of laboratory studies to field situations. An organism under natural field conditions may respond quite differently to exposure than an organism in a laboratory situation. In fact, individuals within the same population may respond differently to the same exposure scenario, depending on individual idiosyncracies, genetic variability, differing states of nutrition, and local microenvironments. Also it is clear that interacting stresses and multiple exposures as well as natural physiological processes will affect the dose-response.

The multiple-response concept is an expansion of ideas published by Depledge (1989). The basic scenario is that an organism in its natural environment maintains homeostasis. As that organism is exposed to xenobiotic contaminants, physiological compensatory mechanisms become active, an increase/decrease in one or more physiological processes or functions occurs, and/or structural changes occur and, as such, indicate that exposure has occurred. If the exposure continues or increases, the organism's compensatory mechanisms become overwhelmed, damage occurs, and physiological repair mechanisms become active. At this point and with increased exposure or dose, additional responses (e.g., reduced body weight, increased liver, spleen, and adrenal weights) are more prevalent and the susceptibility to disease and/or decimating factors (predation, freezing, starvation, and stress) increases. Under natural environmental conditions, as an organism progresses through these phases (homeostasis, compensatory response, and repair response), the energy required for normal maintenance is reduced as energy is needed for compensatory response and repair. The organism becomes progressively "weaker" and will most likely be quickly eliminated from the population. Therefore, *in situ* surveys of animal populations may not normally detect abundant organisms in a diseased condition even though exposure and effects have occurred.

In this multiple-response paradigm, the goal is not to quantitatively define the dose of different toxicants to which an organism is exposed but rather to determine where a population of organisms is located on the continuum between normal homeostasis and disease. Responses indicate whether the animal is challenged but readily coping with toxicant stress (compensatory phase) or is being damaged and is having to expend resources to repair that damage. The goal of research in this arena is to identify which responses correspond with different levels of departure from normal homeostasis.

Both the dose-response and the multiple-response paradigms can provide useful information for evaluating exposure and effects of contaminants, and research needs to provide data that can be interpreted within the context of both models. Alternative hypotheses can be evaluated in laboratory and field research to determine which paradigm is most appropriate and useful. The dose-response paradigm would predict that molecular and biochemical responses will be well correlated with quantitative exposure (dose) and that dose will be predictive of some ultimate disease state. In contrast, the multiple-response paradigm would predict that these responses, while correlated to at least some extent with dose, will also be correlated with other stressors (multiple contaminants or environmental stress) and will have a much higher correlation with expression of some disease state or functional impairment than does the dose of a toxicant to which the organism is exposed.

Currently, data are not available to judge which paradigm is preferable, and information to make quantitative assessments of environmental impact based on biological responses is lacking. The challenge is to build from existing approaches and develop a data base that enables us to evaluate these and other paradigms and to parameterize the more effective model to provide a more quantitative and predictive assessment of environmental contamination.

III. MOLECULAR MARKERS TO TOXICANTS

A. INTRODUCTION

For a toxicant to elicit a response at the molecular or biochemical level, it is necessary that the toxicant becomes bioavailable to the exposed organism. Bioaccumulation and bioavailability of toxicants are addressed in Chapter 4. There are numerous biological processes that affect the absorption, distribution, and metabolism of a toxicant and determine the chemical form that interacts with the primary target or receptor in the cell and ultimately result in a biochemical or molecular response.

Important cellular targets for toxicant interaction and observed responses are

1. Membranes — disruption of permeability (please note the more detailed discussion on the role of the cellular membrane as barrier to toxicants in Chapter 3)
2. Enzymes — loss of biological activity
3. Protein biosynthesis — dysfunction
4. DNA — structural damage

Toxic responses may be caused by irreversible covalent binding between the chemical and the biological substrate or receptor. For example, the metabolic activation of polycyclic aromatic hydrocarbons (PAHs) can result in the production of activated intermediates that can bind covalently to DNA or proteins forming adducts, as shown in Figure 3. Also the covalent binding of other chemicals (heavy metals such as cadmium, lead, and mercury) to -SH groups in proteins often leads to the loss of enzymatic activity. Toxic responses may be caused by oxidative stress through the formation of free radicals. The microsomal metabolism of some hydrocarbons produces free radicals, including oxyradicals, that are thought to play important roles in chemical carcinogenesis. The biochemical mechanisms whereby fluxes of these radicals are produced in aquatic animals have been reviewed by DiGiulio et al. (1989). Both the superoxide radical and the hydroxyl radical are known to damage DNA directly through strand scission and oxidation of the bases of DNA, in particular guanine. Free radicals can initiate autoxidation reactions of lipoproteins of membranes, resulting in loss of their viability. Metabolism of halogenated hydrocarbons in the liver can produce reactants that bind to cell components and induce lipid peroxidative processes which are detrimental to the cell (i.e., liver necrosis). It is evident from this brief discussion that it is important for ecotoxicologists to have an understanding of the metabolic processes that form reactive metabolites if they are to comprehend how toxicants interact with receptors, elicit biochemical and molecular responses, and cause toxic responses.

Two processes of detoxication are biotransformation and bioactivation. Biotransformation is a part of a process called detoxication which is a mechanism for the elimination of xenobiotics (foreign chemicals) that have become bioavailable (Sipes and Gandolfi, 1986). Bioactivation is the conversion of the xenobiotic by cellular enzymes to a reactive intermediate that may damage cellular macromolecules. This relationship is depicted in Figure 4. Biotransformation is biphasic. It is caused by enzymatic reactions that involve oxidation, reduction, and hydrolysis (phase I), and those that consist of conjugation (phase II). During phase II reactions, the foreign chemical or its product produced by phase I reactions is covalently linked to a normal cellular constituent such as glutathione, glucuronic acid, or sulfate. Both phase I and phase II reactions render the initial xenobiotics more water soluble and permit their excretion from the organism. Phase I enzymes are found in the endoplasmic reticulum of the cell, a lipoprotein matrix that facilitates the partitioning of the xenobiotic to the site of enzymatic biotransformation. Phase I enzymes are composed of the mixed function oxygenase system (also referred to as the cytochrome P450 system) and the mixed function amine oxidase system. Many of the phase II enzymes are found in the cytosol fraction of the cell. Bioactivation is the process whereby some of the metabolites that are formed enzymatically during biotransformation (either by phase I or phase II processes) become chemically more reactive than the parent compound. Reactive intermediates have the potential to interact with any nucleophilic sites within the cell, including those on lipids, proteins, and DNA, and induce a toxic response. It should be noted that the formation of reactive intermediates may not lead to cellular toxicity if there is a balance between the rates of their formation and detoxication.

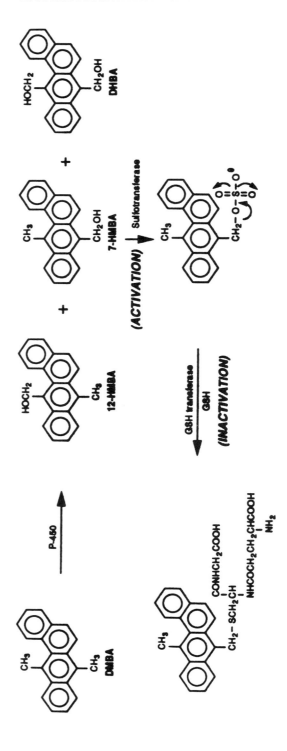

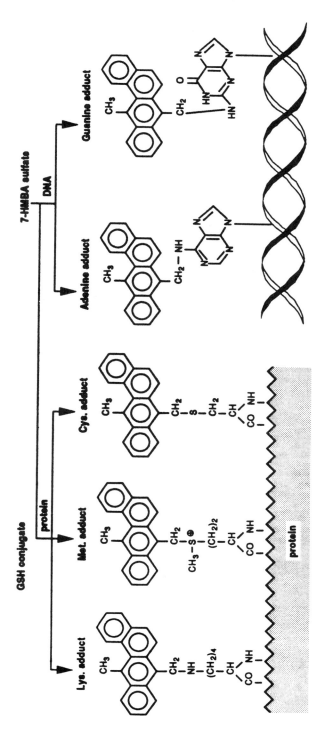

Figure 3 Metabolic conversion of chemical carcinogen 7,12-dimethyl-benz[a]anthracene to activated products and subsequent adduction of cellular macromolecules (DNA and protein).

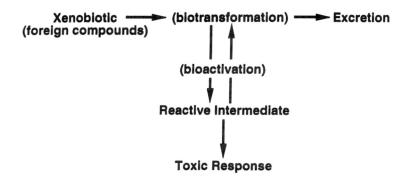

Figure 4.   Components of the detoxication process.

The preceding discussion explains, admittedly in an abbreviated manner, some of the expected detoxication processes that occur in a living organism once a toxicant has become bioavailable to that organism. In essence, these processes as they are understood to date provide the rationale for the selection and examination of molecular and biochemical responses that may be used to determine whether or not an individual has been exposed to a toxicant and the consequences of that exposure (Walker, 1992). Scientists in many disciplines are now applying this approach to investigations both in the laboratory with experimental species under controlled conditions, with indigenous species *in situ*, or with selected species placed in an impacted environment.

Molecular and biochemical responses can usually be placed in one of several general categories. Classification depends upon the endpoints under examination, but may consist of the following: (1) protein induction or expression — P450, metallothionein, stress proteins; (2) detection of metabolites and excretory products; and (3) DNA damage — adducts, strand breakage, clastogenic events, protein dysfunction.

The subsequent sections will discuss these categories in more detail. It would be an impossible task to cover all the work currently undertaken in these areas; therefore, the focus will be to provide insight into approaches that scientists are using to study relevant ecotoxicological problems within the listed categories. It should be noted that the success of current efforts to detect and quantify molecular and biochemical responses in environmental species is due mainly to the incorporation of analytical techniques that are extremely selective and sensitive for the endpoint being examined.

B. PROTEIN INDUCTION

i. Cytochrome P450 Enzymes

The cytochrome P450 enzymes are the major monooxygenase detoxication system of phase I metabolism in living organisms (Guengerich, 1993) including both eukaryotes and prokaryotes. Not only do they detoxify thousands of hazardous substances, but they are also involved in the metabolism of natural endogenous substances such as steroids. Numerous cytochrome P450 genes code for different versions of the enzymes. For convenience, these genes have been grouped into families (numbers 1, 2, etc.), subfamilies (letters A, B, etc.), and individual enzymes (number 1, 2, etc.). For example, the cytochrome P450 enzyme induced by 2,3,7,8-tetrachlorodibenzo-*p*-dioxin (dioxin) is given the designation P4501A1.

Studies on the molecular mechanisms that regulate the induction of the cytochrome P450 genes can provide useful knowledge and insight on the various molecular and chemical events that define the detoxication process. The induction of a cytochrome P450 enzyme is thought to occur through a series of molecular events. For example, with the P4501A system, a toxic chemical binds initially to the aryl hydrocarbon hydroxylase (Ah) receptor in the cytoplasm of a cell and causes the release of a heat shock protein (hsp 90). The heat shock protein is replaced by another protein, the aryl hydrocarbon receptor nuclear transferase. This new complex enters the nucleus of the cell where it binds to a specific part of the DNA molecule, which in turn permits transcription factors access to the promoter region of the cytochrome P450 gene and synthesis of the messenger ribonucleic acid (mRNA) transcript of the cytochrome P450 gene (Goksoyr and Forlin, 1992; Guengerich, 1993; Whitlock, 1993). This leads to the synthesis of a protein with specific catalytic activity which is incorporated into the endoplasmic reticulum of the cell and thus facilitates the metabolism of the toxic chemical.

As noted above, the inductive process is composed of a series of steps. Two major products of these steps, the P450 protein and its mRNA transcript, are routinely analyzed to determine the status of the cytochrome P450 detoxifying enzymes in environmental species exposed to toxicants. The P450 protein can be detected by its catalytic activity or conversely by an immunochemical assay specific for a particular P450 protein while the mRNA transcript can be determined with a cDNA probe.

Although the induction response has not been fully characterized in all cytochrome P450 families, the P4501A subfamily has been studied extensively because of its role in the metabolism of aromatic hydrocarbon carcinogens and other toxic chemicals of environmental relevance. In mammals, such chemicals as dioxin, PAHs, and beta-naphthoflavone are exogenous inducers. Two catalytic activities specific for the P4501A isoenzymes are benzo[*a*]pyrene hydroxylation (AHH) and ethoxyresorufin *O*-deethylation (EROD). Antibodies to purified P4501A1 isoenzyme have been made to detect this protein using the Western blotting technique or indirect enzyme-linked immunosorbent assay (ELISA) with anti-fish P4501A1 (Goksoyr and Forlin, 1992). Cloning of the P4501A1 gene has resulted in the production of a cDNA probe for the determination of the mRNA transcript of the gene.

As previously discussed, many of the inducers of the P450 system, especially the P4501A1 activity, are known environmental pollutants (aromatic and chlorinated hydrocarbons). Thus, the analysis of protein levels by catalytic or immunochemical methods provide a convenient way to monitor pollution effects at the molecular level. In this regard, considerable attention has been given to aquatic species (Payne et al., 1987; Stegeman and Kloepper-Sams, 1987; Goksoyr and Forlin, 1992; Curtis et al., 1993) and to other species, especially birds (Fossi et al., 1990; Peakall, 1992). Cytochrome P4501A gene expression was shown to be environmentally induced in feral populations of tomcod (*Microgadus tomcod*); however, mRNA levels rapidly declined when the fish were introduced into clean water (Kreamer et al., 1991). Also, probes to the mRNA and the genomic DNA detected polymorphism in the tomcod gene for the cytochrome P4501A protein (Wirgin et al., 1991).

The ecotoxicologist should be aware of factors, both exogenous and endogenous to the organism and its environment, that may affect the interpretation of data generated by the various methods that analyze the cytochrome P450 system (Goksoyr and Forlin, 1992). Paramount among these is the fact that certain inducers (mainly organochlorines) can be inhibitors of catalytic activity and that normal physiological states of the organism (i.e., reproduction) produce endogenous compounds such as steroids which in turn may modulate catalytic activity (Jimenez et al., 1990). Furthermore, other chemicals present, in addition to the inducer, may affect in some subtle way the overall processes of biotransformation and in turn the interaction of the toxicant under consideration with the cytochrome P450 system.

ii. Stress Proteins

Cells respond to numerous types of stressors by synthesizing a set of proteins described originally as heat shock proteins (hsps). Each stress protein is comprised of a multigene family in which some are constitutively expressed and others are highly inducible (Lindquist,1986). Stress proteins have been grouped into several classes: (1) those related to the heat shock phenomenon, with variants between 100 and 20 kDa molecular weight (including ubiquitin at 7 kDA); (2) glucose-regulated proteins; and (3) stressor-specific proteins, such as metallothionein. The reader is directed to the excellent review by Sanders (1990) on the subject of heat shock protein classification.

Collectively, the first two groups of proteins appear to be involved in the protection, enhanced survival, and restoration of normal cellular activities in stressed cells. As described above, hsps or stress proteins are an integral component of the Ah receptor, which is involved in the induction and synthesis of cytochrome P450 enzymes when a cell is exposed to a toxicant (chemical stressor). The induction of these proteins by mild stress may confer acquired tolerance to more severe stress situations. Because they are highly conserved they can be detected immunologically with antibodies not necessarily from the same species. The response is quantitative and the variety of proteins synthesized makes possible the identification of effects of specific contaminants or sets of contaminants. Thus, the stress protein response becomes an integrated signal (or set of signals) for environmental stress (Sanders, 1990; Bradley, 1990; Bradley et al., 1994).

iii. Metallothionein

Metallothionein, although not usually classified with hsps or stress proteins (i.e., it is not related structurally or functionally to them, nor is it composed of a homologous group of proteins), is nevertheless induced by toxic chemicals. It is a low-molecular-weight protein (~7 kDa) with about one-third of its amino acids as cysteine residues which can bind up to seven metal atoms and is found in most living organisms. The primary metabolic role for metallothionein appears to be the maintenance of homeostasis of zinc and copper which along with cadmium and mercury induce its synthesis. These metals are ubiquitous environmental pollutants and can be extremely toxic to living organisms. Therefore, another possible metabolic role for metallothionein is to protect cells from toxic metal ions by lowering their concentration at critical intracellular sites (Garvey, 1990; Petering et al., 1990). In Chapter 2, Campbell and Tessier detail the usefulness of measuring metallothionein content in the gill tissue of mollusks to track heavy metal gradients in Canadian lakes. Also, the role of metallothionein in carcinogenesis is currently under intensive study, particularly in cancer cell pathobiology, drug resistance, and toxicity (Cherian et al., 1994).

C. METABOLITES AS INDICATORS

The induction of a specific protein or set of proteins as a biological response to toxicant exposure was discussed previously. Protein expression is a biological phenomenon intimately associated with exposure to a toxicant and occurs usually via the direct interaction of the toxicant with a cellular receptor (e.g., cytochrome P450 protein synthesis). During the processes of biotransformation and bioactivation of a toxicant, however, products such as metabolites are produced that may be considered biological responses to toxicant stress. The metabolism of many chemicals (e.g., PAHs) results in the formation of numerous conjugated and unconjugated metabolites that are excreted and have been used as biochemical indicators of exposure of environmental species to these hazardous chemicals (Varanasi et al., 1989). In addition, metabolites may be breakdown products of cellular catabolism. For the purpose of this discussion, metabolite indicators are defined as small molecules such as amino acids, nucleosides, organic acids, and porphyrins that are associated with physiological fluids (i.e., urine, bile, milk, or blood) and whose presence or amounts are indicative of cellular detoxication processes or dysfunctions.

There are two important drawbacks to the use of metabolites as biological indicators for assessing toxicant stress in environmental species. The first is the temporal occurrence of the indicator. Excretion and/or partitioning of the biochemical into the physiological fluid of concern is usually a rapid process that occurs shortly after exposure of the organism to the toxicant and the process ceases upon its removal. Thus, in the case of acute or intermittent exposures, it may not be possible to perform the analysis in the proper time frame. Second, many of these metabolites are present in the tissue or fluid at extremely low concentrations and, therefore, are not easily detected. Several analytical techniques are currently available that possess the sensitivity to overcome this hindrance. Among these are gas chromatography-mass spectrometry, high-performance liquid chromatography, and capillary electrophoresis (DeAntonis and Brown, 1993).

Many metabolite markers are present in physiological fluids under normal cellular conditions, but altered profiles of these compounds can be indicative of a toxic response (Flynn et al., 1994) or diseased state by the organism. Two examples are nucleosides and porphyrins. Catabolism of nucleic acids produces nucleosides. In certain types of human cancer, viral infections, and organ failures (pathological dysfunctions), the concentration and distribution of these compounds change in such a manner that the altered profile becomes a marker for the disorder. Nucleosides have not generally been examined as biochemical indicators in environmental species, but their usefulness should be investigated. Porphyrins are cyclic tetrapyrroles that are intermediate metabolites of heme biosynthesis. Disturbance of heme biosynthesis can lead to an excess of porphyrins in the blood and other tissues. Many environmental chemicals (e.g., lead and aromatic hydrocarbons) deregulate heme synthesis and cause increases in liver concentrations of several individual species of porphyrins. The cytochrome P4501A1 monooxygenases appear to be involved in the development of these chemically induced hepatic porphyria. Sensitive cell culture bioassays are currently being employed to detect chemicals in tissues of exposed environmental species that disrupt normal porphyrin synthesis (Kennedy et al., 1993).

D. DNA MARKERS

i. DNA Damage

A model that reflects events and cellular processes related to DNA integrity is depicted in Figure 5 (Shugart, 1990a; Shugart et al., 1992). DNA is present in cells as a functionally stable, double-stranded entity without discontinuity (strand breaks) or abnormal structural modifications (adducts or chemically altered bases). As such, it is considered to have high integrity. The rigid maintenance of this integrity is important for survival and is reflected in the low mutational rate observed in living organisms, which has been estimated to be on the order of one mutation per average gene per 200,000 years (Alberts et al., 1989). DNA damage can occur as the result of:

1. Wear and tear by normal cellular events, such as metabolism and random thermal collisions (pathway 1)
2. Interaction with physical agents, such as ultraviolet light and ionizing radiation (pathway 2)
3. Interaction with chemical agents (pathway 3)

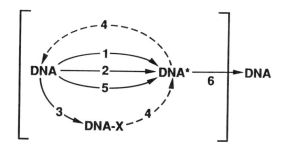

INSULT	REPAIR	SYNTHESIS
1. Normal wear and tear	4. Incision	5. Replication
2. UV and ionizing radiation (γ and X irradiation)	excision resynthesis	6. Postreplication modification
3. Chemical	ligation	

DNA: Normal double-stranded DNA with no strand breaks.
DNA-X: Chemically modified DNA.
DNA*:   DNA with strand breaks.

Figure 5 A schematic representation of the status of DNA in relation to insults that disrupt DNA integrity and cellular processes that maintain DNA integrity. (Reprinted from Shugart, L.R. 1990a. Biological Markers of Environmental Contaminants, ©Lewis Pub., an imprint of CRC Press.)

These various processes give rise to structural alterations, which are usually rapidly repaired, but in the process produce a transient population of DNA with strand breaks and low integrity. Some chemicals work through free-radical mechanisms and cause strand breaks directly, whereas other genotoxic agents can interfere with the fidelity of normal repair or DNA modifications (pathway 6). The loss of bases from the DNA molecule (abasic sites) creates frequent lesions that occur as a result of random thermal collisions or from breakdown of chemically unstable adducts. Even the normal cellular process of replication (pathway 5) produces DNA with strand breaks. Therefore, at any one time, a background level of DNA with low integrity (DNA with various types of structural alterations) may exist in the cell. Fortunately, most cells have DNA repair mechanisms (Sancar and Sancar, 1988) that under normal circumstances efficiently eliminate DNA of low integrity (pathway 4).

Structural modifications of DNA are probably the best understood genotoxic event for which analytical techniques with the appropriate selectivity and sensitivity are available (Wogan and Gorelick, 1985; Bartsch et al., 1988; Santella, 1988; Shugart, 1988; Jones and Parry, 1992) and can take the form of adducts (where the chemical or its metabolite becomes covalently attached to the DNA), of strand breakage, or of chemically altered bases. For example, cellular metabolism of genotoxic chemicals, once they become bioavailable, can be a relatively complex phenomenon and the lack of complete detoxication sometimes lead to the formation of highly reactive electrophilic metabolites

(Phillips and Sims, 1979; Harvey, 1982; Phillips, 1983). These intermediates can form adducts when they undergo attack by nucleophilic centers in macromolecules such as lipids, proteins, DNA, and RNA, which often results in cellular toxicity. Metabolic activation (bioactivation) of the carcinogenic chemical, 7,12-dimethyl-benz[a]anthracene (DMBA) to several hydroxymethylated intermediates (12-HMBA, 7-HMBA, and DHBA) is depicted in Figure 3. Subsequent enzymatic activation may proceed to a sulfate derivative (7-HMBA sulfate) that can become covalently attached to bases on the DNA molecule. Note also that other cellular macromolecules such as proteins may also become adducted with this chemical. This latter situation and its diagnostic importance to DNA adduct formation will be discussed subsequently. Methods of varying sensitivity have been devised to measure DNA adducts, including physiochemical (Rahn et al., 1982), immunological (Santella et al., 1987), and ^{32}P-postlabeling (Randerath et al., 1981). The methods most commonly employed are ^{32}P-postlabeling and immunoassays using adduct-specific antibodies.

Strand breakage and modification of bases as a result of genotoxic insult (i.e., free radial formation) are other forms of structural modification that can occur to the DNA molecule and are being investigated as responses at the molecular level that are indicative of exposure as well as effect. Shugart (1993a) recently discussed current methodologies for detecting these types of genotoxic insult as well as examples of their use in environmentally exposed animals.

ii. Surrogates to DNA Damage

Damage to ancillary molecules may serve as a surrogate (protein adducts for example) for estimating exposure to genotoxic agents. Blood proteins, such as hemoglobin and serum albumin, have proved useful in this context (Ehrenberg et al., 1983; Shugart and Kao, 1985). This concept is illustrated in Figure 3 and shows that in addition to DNA bases, amino acids and proteins may also be modified. In the example provided, it should be noted that it is the same metabolite of DMBA that becomes adducted to the two different cellular macromolecules (protein and DNA). For this reason, protein adducts may serve as surrogates for DNA adducts. Furthermore, since protein adducts are not subject to repair, they persist for the lifetime of the protein and accumulate in a dose-related manner. The detection and measurement of adducts to hemoglobin have been seriously studied in humans exposed to hazardous occupational chemicals for some time and a considerable scientific literature exists on methodologies and approaches in this area (Wogan and Gorelick, 1985; Bartsch et al., 1988). Recently Shugart (1993b) discussed the approach of using protein adducts to assess the status of genotoxicant impact on biota in a contaminated environment.

iii. Consequences of Structural Perturbations

The consequence of the structural perturbations discussed above can be either innocuous due to repair of the damage or lethal by causing death of the

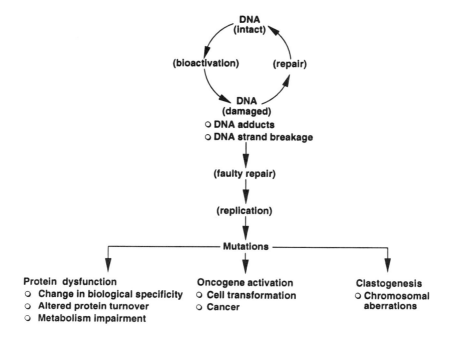

Figure 6   Consequences of genotoxic effects to DNA.

cells containing the damaged DNA. However, those lesions not properly repaired may result in alterations that become fixed and are eventually transmitted into daughter cells. It has been postulated that the reaction of chemicals with DNA and ensuing changes that result may cause deleterious, pathological conditions (Harvey, 1982). Affected cells often exhibit altered cellular functions indicative of subclinical manifestation of genotoxic disease. For example, structural alterations to DNA caused by genotoxic chemicals may potentiate irreversible changes to the DNA molecule and result in the expression of cellular responses such as chromosomal aberrations and oncogene activation. Such responses have been associated with the latent events of chemical carcinogenesis (Weinstein, 1978).

From the preceding discussion, it is evident that, upon exposure to genotoxicants, a cascade of cellular events may be induced that can change the integrity of an organism's DNA and result in various types of damage. This flow of genotoxic stress within a somatic cell is depicted in Figure 6 and the mechanisms involved have been reviewed (Brusick, 1980; Thilly and Call, 1986); however, several salient points need to be reiterated. Effects to the cell such as the occurrence of chromosomal aberrations, oncogene activation, and protein dysfunction are not usually caused by the direct interaction of the genotoxicant with DNA but rather as the result of a subsequent series of separate and very complex processes, of which there is presently only a rudimentary understanding. These processes are affected differently in different species and may depend upon, for example, the type or class of chemical and the reactivity of its metabolite(s), the capacity of the cell to repair DNA damage, and the ability of the cell to recognize and suppress the multiplication of cells with aberrant properties (Thilly and Call, 1986). Effects expressed in somatic cells can be detrimental to exposed individuals, whereas mutational events in germ cells will affect subsequent generations. Extrapolation of observations made at the somatic cell level of biological organization to predict effects at the germ cell level of biological organization is difficult. This is due to the inherent difference in sensitivity of these types of cells to genotoxicants (Clive, 1987). Furthermore, establishing a causal relationship between a genotoxic agent in the environment and an effect in subsequent generations of that organism is also highly unlikely because individuals carrying harmful mutations are eliminated from the population due to a strong selection against less fit and less well-adapted individuals (Wurgler and Kramers, 1992).

If the structural damage to DNA is not repaired and persists or if genotoxic stress is expressed at other levels of biological organization, then an opportunity to test for genotoxicity is provided (Shugart, 1990a; Jones and Parry, 1992). Early attempts at assessing the effects of genotoxicants on species present in the environment usually involved direct observation such as the visual occurrence of neoplasms and chromosomal aberrations in various plants, wild terrestrial mammals, and aquatic vertebrates (Sandhu and Lower, 1989). Current attempts use more sensitive and sophisticated analytical techniques. The ecotoxicological goals of these investigations are to link contaminant-induced genetic effects to physiological effects and the identification of causal relationships between exposure to genotoxicants with increased risk of effects on individuals and populations that can lead to decreased ecological integrity (Wurgler and Kramers, 1992; Karr, 1993).

IV. GENETIC ECOTOXICOLOGY

As mentioned previously, it is usually difficult to demonstrate the direct effect of environmental pollution at the ecosystem level. To circumvent this deficiency, measurement of biological responses at various levels of biological organization within the individual have been proposed to demonstrate relatedness, that is, exposure to deleterious substances may induce a cascade of biological events that lead to some dysfunction. Once impairment at the individual level is observed, it becomes less difficult to extrapolate the potential for harm to the population and community levels (Cormier and Daniel, 1994). However, this is a tenuous argument at best and the reader is referred to other discussions on the subject (e.g., Gooch and van Veld, 1990).

Another way to approach this problem is to ask mechanistic questions about how an organism relates to other organisms and, in turn, to its physical environment. Ecosystems result from the dynamic interactions of living and inert matter where the living material acclimates and adapts to environmental change. These processes are physiological and have a genetic basis; therefore, understanding change at the genetic level should help define the more complex

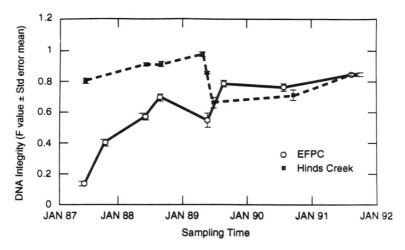

Figure 7   Temporal status of double-stranded (*F* value) DNA in liver samples of sunfish from East Fork Poplar Creek (contaminated stream) and Hinds Creek (reference stream) over a 4-year period (Shugart and Theodorakis, 1994).

changes at the ecosystem level. For example, analyzing for strand breaks in the DNA of sunfish has been employed as a biological marker for environmental genotoxicity as part of the Biological Monitoring and Abatement Program at East Fork Poplar Creek (Shugart, 1990b). This creek is the receiving stream for industrial effluent from the U.S. Department of Energy (DOE) reservation in Oak Ridge, TN. Water and sediments downstream contain metals, organic chemicals, and radionuclides discharged over many years of operation.

DNA strand break data (*F* values), measured in sunfish from the head waters of the creek (near the U.S. DOE reservation) and at Hinds Creek (reference stream) over a period of 4 years, are presented in Figure 7. *F* values are a measure of the relative double-strandedness of a particular DNA preparation which in turn can be related to the number of strand breaks present. *F* values are determined under *in vitro* conditions by the alkaline unwinding assay (Shugart, 1988) where the rate of conversion of the DNA from double-stranded to single-stranded structures is proportional to the number of strand breaks present. Thus, large *F* values are indicative of DNA with few strand breaks. Two points are clear: (1) DNA structural integrity of the sunfish from the reference stream is high and relatively constant (large *F* value); and (2) DNA structural integrity of the sunfish from East Fork Poplar Creek improved during the study period to reach levels similar to those for Hinds Creek. In all probability, the large genotoxic response observed in sunfish from East Fork Poplar Creek during the years 1987 and 1988 (small *F* value) was related to the release of chemicals from the U.S. DOE reservation. Diminution of this response in subsequent years may be due to the remedial activities that occurred on the U.S. DOE reservation to attenuate the release of pollutants. However, the possibility that there has been an adaptive response over time by the resident population of sunfish to their environment cannot be excluded.

Genetic ecotoxicology can be defined as the study of pollutant-induced changes in the genetic material of biota in nature and has two aspects: (1) genotoxicity of pollutants, and (2) long-term heritable effects (i.e., changes in gene frequency within exposed populations, mutational events, etc.). An example of the first aspect of this definition was illustrated in the studies on the occurrence of DNA strand breaks in sunfish where the central dogma is that exposure to genotoxicants may result in modification of cellular DNA. It should be noted that it is now possible to identify molecular targets of genotoxicants with extreme sensitivity and, furthermore, to determine how chemical modification affects function at a precise molecular level (Marnett, 1993).

Approaches and studies related to the second aspect of genetic ecotoxicology are not as far advanced. The challenge will be to develop assays with the sensitivity to demonstrate those subtle changes in the genetic material of organisms exposed to genotoxicants that are responsible for heritable effects (Dieter, 1993; Thaler, 1994). In this regard it should be remembered that changes in gene frequencies can operate on a relatively short time scale and, furthermore, that toxic chemicals can exert a strong selective pressure. For example, the reader is directed to Dr. Michael Newman's discussion in this book on the sensitivity of genotype to heavy metal exposure (Chapter 8) and the recent work of Shugart and Theodorakis (1994) that examines the effect on genetic diversity from exposure to genotoxicants in the environment. Finally, research efforts must be broadened to cover pathophysiological phenomena associated with reproduction and development in wildlife (Colborn et al., 1993) that may be manifestations of the genotoxic disease syndrome (Kurelec, 1993).

Recent advances in the discipline of molecular biology provide the experimental tools with which to precisely investigate key biological mechanisms at the genetic level that regulate and limit the response of individual organisms to genotoxicants in their environment (Chasan, 1991). This is a fruitful area for new research initiatives in genetic ecotoxicology, as it offers an opportunity to rapidly advance our knowledge and understanding of the effect of pollution on ecosystems.

V. CONCLUSIONS

The ecotoxicologist in today's world faces a difficult challenge: how to assess the long-term impacts of low levels of contaminants that enter ecosystems. Traditional approaches used in the field of environmental toxicology principally include the ranking of relative toxicities of chemicals, the detection and quantification of toxic chemicals in the environment, the study of mechanisms of toxicity of chemicals, and the determination of no observable effects levels in the laboratory. While these approaches are useful, they are unfortunately too narrow in scope and their validity is being seriously questioned. Current approaches in ecotoxicology examine molecular and biochemical responses in exposed individuals that are elicited by toxicants in an effort to assess the status of an impacted environment. As a result, the field of biomarker research and investigation evolved to address the objections raised against the traditional approaches. Although contributing significantly to our maturing concepts of the effects of pollutants, this approach has not provided a complete picture of the issues. What is needed are new and innovative approaches that integrate effects across different levels of biological complexity and provide a clear understanding of all the danger posed by environmental pollution, not only to ecological systems but for human health as well.

ACKNOWLEDGMENTS

The Oak Ridge National Laboratory is managed by Martin Marietta Energy Systems, Inc., for the U.S. Department of Energy under contract No. DE-AC05-84OR21400. This is Environmental Sciences Division publication number 4403.

REFERENCES

Alberts, B., D. Bray, J. Lewis, M. Raff, K. Roberts and J.D. Watson, 1989. Molecular Biology of the Cell, 2nd ed., Garland Publishing, New York, pp. 220–227.

Bartsch, H., K. Hemminki and I. O'Neill (Eds.), 1988. Methods for Detecting DNA Damaging Agents in Humans: Application in Cancer Epidemiology and Prevention. IARC Scientific Publication No. 89, IARC, Lyon, France.

Bradley, B.P., 1990. Stress proteins: their detection and uses in biomonitoring. In: Aquatic Toxicology and Risk Assessment: Sublethal Indicators of Toxic Stress, Vol. 13, STP 1096, edited by W.G. Landis and W.H. van der Schalie, American Society for Testing and Materials, Philadelphia, PA, pp. 338–347.

Bradley, B.P., C.M. Gonzales, J.-A. Bond and B.E. Tepper, 1994. Complex mixture analysis using protein expression as a qualitative and quantitative tool. Environ. Toxicol. Chem. 13, 1043–1050.

Brusick, D., 1980. Principles of Genetic Toxicology. Plenum Press, New York.

Chasan, R., 1991. Molecular biology and ecology: a marriage of more than convenience. Plant Sci. News, Nov. 1143–1145.

Cherian, M.G., S.B. Howell, N. Imura, C.D. Klaassen, J. Koropatnick, J.S. Lazo and M.P. Waalkes, 1994. Role of metallothionein in carcinogenesis. Toxicol. Appl. Pharmacol. 126, 1–5.

Clive, D., 1987. Genetic toxicology: from theory to practice. Clin. Res. Drug Dev. 1, 11–41.

Colborn, T., F.S. vom Sall and A.M. Soto, 1993. Development effects of endocrine-disrupting chemicals in wildlife and humans. Environ. Health Perspect., 101, 378–384.

Cormier, S.M. and F.B. Daniel, 1994. Biomarkers: taking the science forward. Environ. Toxicol. Chem. 13, 1011–1012.

Curtis, L.R., H.M. Carpenter, R.M. Donohoe, D.E. Williams, O.R. Hedstrom, M.L. Deinzer, M.A. Bellstein, E. Foster and R. Gates, 1993. Sensitivity of cytochrome P4501A1 induction in fish as a biomarker for distribution of TCDD and TCDF in the Willamette River, Oregon. Environ. Sci. Technol. 27, 2149–2157.

DeAntonis, K.M. and P.R. Brown, 1993. Determination of biochemical markers by HPLC and CE for the diagnosis of disease. Am. Lab. Oct., 62–71.

Depledge, M., 1989. The rational basis for detection of the early effects of marine pollutants using physiological indicators. Ambio 18, 301–302.

Dieter, M.P., 1993. Identification and quantification of pollutants that have the potential to affect evolutionary processes. Environ. Health Perspect. 101, 278.

Di Giulio, R.T., P.C. Washburn, R.J. Wenning, G.W. Winston and C.S. Jewell, 1989. Biochemical responses in aquatic animals: a review of determinants of oxidative stress. Environ. Toxicol. Chem. 8, 1103–1123.

Ehrenberg, L.E., S. Moustacchi, G. Osterman-Golkar and G. Ekman, 1983. Dosimetry of genotoxic agents and dose/response relationship of their effects. Mutat. Res. 123, 121–182.

Flynn, K.J., K.J. Jones, R. Raine, J. Richard and K. Flynn, 1994. Use of intracellular amino acid analysis as an indicator of the physiological status of natural dinoflagellate populations. Mar. Ecol. Prog. Ser. 103, 175–186.

Forbes, V.E. and T.L. Forbes, 1994. Ecotoxicology in Theory and Practice. Chapman and Hall, London, p. 13.

Fossi, C., C. Leonzio, S. Focardi and D.B. Peakall, 1990. Avian mixed function oxidase induction as a monitoring device: the influence of normal physiological function. In: Biomarkers of Environmental Contamination, edited by J.F. McCarthy and L.R. Shugart, Lewis Publishers, Boca Rotan, FL, pp. 143–149.

Fossi, M.C. and C. Leonzio (Eds.), 1993. Nondestructive Biomarkers in Vertebrates. Lewis Publishers, Boca Raton, FL.

Garvey, J.S., 1990. Metallothionein: a potential biomonitor of exposure to environmental toxins. In: Biomarkers of Environmental Contamination, edited by J.F. McCarthy and L.R. Shugart, Lewis Publishers, Boca Raton, FL, pp. 267–287.

Goksoyr, A. and L. Forlin, 1992. The cytochrome P450 system in fish, aquatic toxicology and environmental monitoring. Aquat. Toxicol. 22, 287–311.

Gooch, J.W. and P. van Veld, 1990. Methodologies for suborganismal (biochemical and cellular) toxicity testing. In: Chesapeake Bay Ambient Toxicity Assessment Workshop Report, edited by E.C. Krome, printed by U.S. Environmental Protection Agency, Washington, D.C., for the Chesapeake Bay Program, pp. 41–52.

Guengerich, F.P., 1993. Cytochrome P450 enzymes. Am. Sci. 81, 440–447.

Harvey, R.C., 1982. Polycyclic hydrocarbons and cancer. Am. Sci. 70, 386–393.

Huggett, R.J., R.A. Kimerle, P.M. Mehrle and H.L. Bergman (Eds.), 1992. Biomarkers: Biochemical, Physiological and Histological Markers of Anthropogenic Stress, Lewis Publisher, Boca Raton, FL.

Jimenez, B.D., A. Oikari, S.M. Adams, D.E. Hinton and J.F. McCarthy, 1990. Hepatic enzymes as biomarkers: interpreting the effects of environmental, physiological and toxicological variables. In: Biomarkers of Environmental Contamination, edited by J.F. McCarthy and L.R. Shugart, Lewis Publishers, Boca Raton, FL, pp. 123–142.

Jones, N.J. and J.M. Parry, 1992. The detection of DNA adducts, DNA base changes and chromosome damage for the assessment of exposure to genotoxic pollutants. Aquat. Toxicol. 22, 323–344.

Karr, J.R., 1993. Defining and assessing ecological integrity: beyond water quality. Environ. Toxicol. Chem. 12, 1521–1531.

Kennedy, S.W., A. Lorenzen, C.A. James and B.T. Collins, 1993. Ethoxyresorufin-O-deethylase and porphyrin analysis in chicken embryo hepatocyte cultures with a fluorescence multiwell plate reader. Anal. Biochem. 211, 102–112.

Kreamer, G.-L., K. Squibb, D. Gioeli, S.J. Garte and I. Wirgin, 1991. Cytochrome P4501A mRNA expression in feral Hudson River tomcod. Environ. Res. 55, 64–78.

Kurelec, B., 1993. The genotoxic disease syndrome. Mar. Environ. Res. 35, 341–348.

Lindquist, S., 1986. The heat shock response. Annu. Rev. Biochem. 55, 1151–1191.

Loar, J.M., S.M. Adams, H.L. Boston, B.D. Jimenez, J.F. McCarthy, J.G. Smith, G.R. Southworth and A.J. Stewart, 1988. First annual report on the Y–12 plant biological monitoring and abatement program. Oak Ridge: ORNL/TM, Oak Ridge National Laboratory, Oak Ridge, TN.

Marnett, L.J., 1993. Frontiers in molecular toxicology. Chem. Res. Toxicol. 6, 739–740.

McCarthy, J.F. and L.R. Shugart (Eds.). 1990. Biomarkers of Environmental Contamination, Lewis Publishers, Boca Raton, FL.

McCarthy, J.F., R.S. Halbrook and L.R. Shugart, 1992. Conceptual Strategy for Design, Implementation and Validation of a Biomarker-Based Biomonitoring Capability, ORNL/TM-11783, Oak Ridge National Laboratory, Oak Ridge, TN.

Payne, J.F., L.L. Fancey, A.D. Rahimtula and E.L. Porter, 1987. Review and perspective on the use of mixed-function oxygenase enzymes in biological monitoring. Comp. Biochem. Physiol. 86C, 233–245.

Peakall, D.B., 1992. Animal Biomarkers as Pollution Indicators. Chapman and Hall, London, pp. 86–107.

Peakall, D.B. and L.R. Shugart (Eds.), 1992. Strategy for Biomarker Research and Application in the Assessment of Environmental Health, Springer-Verlag, Heidelberg.

Petering, D.H., M. Goodrich, W. Hodgman, S. Krezoski, D. Weber, C.G. Shaw III, R. Spieler and L. Zettergren, 1990. Metal-binding proteins and peptides for the detection of heavy metals in aquatic organisms. In: Biomarkers of Environmental Contamination, edited by J.F. McCarthy and L.R. Shugart, Lewis Publishers, Boca Raton, FL, pp. 239–254.

Phillips D. and P. Sims, 1979. PAH metabolites: their reaction with nucleic acids. In: Chemical Carcinogens and DNA, Vol. 2, edited by P.L. Grover, CRC Press, Boca Raton, FL, pp. 9–57.

Phillips, D., 1983. Fifty years of benzo[a]pyrene. Nature 303, 478–482.

Rahn R.O., S.S. Chang, J.M. Holland and L.R. Shugart, 1982. A fluorometric HPLC assay for quantitating the binding of benzo[a]pyrene metabolites to DNA. Biochem. Biophys. Res. Commun. 109, 262–268.

Randerath, K., M.V. Reddy and R.C. Gupta, 1981. 32P-Labeling test for DNA damage. Proc. Natl. Acad. Sci. USA 78, 6126–6129.

Sancar A. and G. G. Sancar, 1988. DNA repair enzymes. Annu. Rev. Biochem. 57, 29–67.

Sanders, B., 1990. Stress proteins: potential as multitiered biomarkers. In: Biomarkers of Environmental Contamination, edited by J.F. McCarthy and L.R. Shugart, Lewis Publisher, Boca Raton, FL, pp. 165–191.

Sandhu, S.S. and W.R. Lower, 1989. *In situ* assessment of genotoxic hazards of environmental pollution. Toxicol. Ind. Health 5, 73–83.

Santella, R.M., F. Gasparo and L. Hsieh, 1987. Quantitation of carcinogen-DNA adducts with monoclonal antibodies. Prog. Exp. Tumor. Res. 31, 63–75.

Santella, R.M., 1988. Application of new techniques for the detection of carcinogen adducts to human population monitoring. Mutat. Res. 205, 271–282.

Shugart, L.R. and J. Kao, 1985. Examination of adduct formation in vivo in the mouse between benzo[a]pyrene and DNA of skin and hemoglobin of red blood cells. Environ. Health Perspect. 62, 223–226.

Shugart, L.R., 1988. Quantitation of chemically induced damage to DNA of aquatic organisms by alkaline unwinding assay. Aquat. Toxicol. 13, 43–52.

Shugart, L.R., S.M. Adams, B.D. Jimenez, S.S. Talmage and J.F. McCarthy, 1989. Biological markers to study exposure in animals and bioavailabilty of environmental contaminants. In: Biological Monitoring for Pesticide Exposure: Measurement, Estimation and Risk Reduction, edited by R.G.M. Wang, C.A. Franklin, R.C. Honeycutt and J.C. Reinert, ACS Symposium Series 382, American Chemical Society, Washington, D.C., pp. 86–97.

Shugart, L.R., 1990a. Biological monitoring: testing for genotoxicity. In: Biological Markers of Environmental Contaminants, edited by J. F. McCarthy and L. R. Shugart, Lewis Publishers, Boca Raton, FL, pp. 205–216.

Shugart, L.R., 1990b. DNA damage as an indicator of pollutant-induced genotoxicity. In: Aquatic Toxicology and Risk Assessment: Sublethal Indicators of Toxic Stress, Vol. 13, STP 1096, edited by W.G. Landis and W.H. van der Schalie, American Society for Testing and Materials, Philadelphia, PA, pp. 348–355.

Shugart, L.R., J. Bickham, G. Jackim, G. McMahon, W. Ridley, J. Stein and S. Steinert, 1992. DNA alterations. In: Biomarkers: Biochemical, Physiological and Histological Markers of Anthropogenic Stress, edited by R.J. Huggett, R.A. Kimerle, P.M. Mehrle and H.L. Bergman, Lewis Publishers, Boca Raton, FL, pp. 125–153.

Shugart, L.R., J.F. McCarthy and R.S. Halbrook. 1992. Biological markers of environmental and ecological contamination: an overview. J. Risk Analysis 12, 352–360.

Shugart, L.R., 1993a. Genotoxic responses in blood. In: Nondestructive Biomarkers in Vertebrates, edited by M.C. Fossi and C. Leonzio, Lewis Publishers, Boca Raton, FL, pp. 131–145.

Shugart, L.R., 1993b. Hemoglobin adducts. In: Nondestructive Biomarkers in Vertebrates, edited by M.C. Fossi and C. Leonzio, Lewis Publishers, Boca Raton, FL, pp. 159–168.

Shugart, L.R. and C.W. Theodorakis, 1994. Environmental genotoxicity: probing the underlying mechanisms. Environ. Health Perspect. 102, 13–17.

Sipes, I.G. and A.J. Gandolfi, 1986. Biotransformation of toxicants. In: Casarett and Doull's Toxicology, 3rd edition, edited by C.D. Klaassen, M.O. Amdur and J.Doull, Macmillan, New York, pp. 64–98.

Stegeman, J.J. and P.J. Kloepper-Sams, 1987. Cytochrome P450 isoenzymes and monooxygenase activity in aquatic animals. Environ. Health Perspect. 71, 87–95.

Thaler, D.S., 1994. The evolution of genetic intelligence. Science 264, 224–225.

Theodorakis, C.W., S.J. D'Surney, J.W. Bickham, T.B. Lyne, B.P. Bradley, W.E. Hawkins, W.L. Farkas, J.F. McCarthy and L.R. Shugart, 1992. Sequential expression of biomarkers in bluegill sunfish exposed to contaminated sediment. Ecotoxicology 1, 45–73.

Thilly, W.G. and K.M. Call, 1986. Genetic toxicology. In: Casarett and Doull's Toxicology, 3rd edition, edited by C.D. Klaassen, M.O. Amdur and J. Doull, Macmillan, New York, pp. 174–194.

Travis, C.C. (Ed.), 1993. Use of Biomarkers in Assessing Health and Environmental Impacts of Chemical Pollutants, NATO ASI Series A: Life sciences Vol. 250, Plenum Press, New York.

Varanasi, U., J.E. Stein and M. Nishimoto, 1989. Biotransformation and disposition of polycyclic aromatic hydrocarbons (PAH) in fish. In: Metabolism of Polycyclic Aromatic Hydrocarbons in the Aquatic Environment, edited by U. Varanasi, CRC Press, Boca Raton, FL, pp. 93–149.

Walker, C.H., 1992. Biochemical responses as indicators of toxic effects of chemicals in ecosystems. Toxicol. Lett. 64/65, 527–533.

Weinstein, I.B., 1978. Current concepts on mechanism of chemical carcinogenesis. Bull. N.Y. Acad. Med. 54, 336–383.

Whitlock, J.P. 1993. Mechanistic aspects of dioxin action. Chem. Res. Toxicol. 6, 754–763.

Wirgin, I., G.-L. Kreamer and S.J. Garte, 1991. Genetic polymorphism of cytochrome P4501A in cancer-prone Hudson River tomcod. Aquat. Toxicol. 19, 205–214.

Wogan, G.N. and N.J. Gorelick, 1985. Chemical and biochemical dosimetry to exposure to genotoxic chemicals. Environ. Health Perspect. 62, 5–18.

Wurgler, F.E. and P.G.N. Kramers, 1992. Environmental effects of genotoxins (ecogenotoxicology). Mutagenesis 7, 321–327.

Responses at the Tissue Level: Quantitative Methods in Histopathology Applied to Ecotoxicology

Charles H. Jagoe

I. CONCEPTUAL BACKGROUND

Pollutants affect biological systems at many levels, but all chemical pollutants must initially act by changing structural and/or functional properties of molecules essential to cellular activities. These molecular changes then affect the structure and function of organelles and cells, which alters the physiological state of the organism. Changes in physiological status influence energy allocation, which can affect growth and reproduction (Calow, 1991). Altered patterns of growth and reproduction affecting a number of individuals can produce effects in populations, and changes in populations can influence community structure. This hierarchical concept serves as the organizing theme of this volume, and provides a valuable framework for envisioning the effects of pollutants at different levels of biological organization (see Adams, 1990). Of course, the cascade of responses among various levels is not unidirectional, and there will be feedback among the levels in this type of conceptual model.

While the concept of a hierarchy of responses to pollutant stress has considerable utility, extrapolation of observations made at one level to predict effects several levels higher or lower in the hierarchy is generally not possible. However, effects observed at one level can provide useful information about processes occurring a level or two away. To illustrate this, consider that by studying cell- and tissue-level effects, it is often possible to learn something about mechanisms at the biochemical and molecular levels. For example, Lanno et al. (1987) fed rainbow trout (*Oncorhynchus mykiss*) a diet enriched in copper, and observed electron-dense granules in hepatocytes using transmission electron microscopy. Electron microprobe analysis showed that these particles were rich in copper and sulfur, leading to the conclusion that excess copper was bound to the sulfur-rich protein metallothionein and packaged into discrete granules for elimination in the bile. As another example, polychlorinated biphenyl (PCB) exposure caused apparent proliferation of smooth endoplasmic reticulum in fish hepatocytes; this ultrastructural response was associated with the induction of microsomal mixed-function oxidase enzymes (Hinton et al., 1978; Klaunig et al., 1979).

The cell and tissue levels lie between the biochemical-molecular level and the organismal level in this conceptual hierarchy. Thus, one might expect that responses to pollutants at the cell and tissue levels would be useful in predicting effects on the neighboring level of whole organisms. As an example, the development of a liver neoplasm after exposure to some chemical might lead to the prediction of premature death for an exposed individual. For a more subtle illustration of the relationship between these levels, pollutant-induced histopathological changes in sensory organs (Gardner, 1975; DiMichele and Taylor, 1978; Solangi and Overstreet, 1982) might infer interference with normal activities, including detection of food or predator avoidance.

Histological responses can also be useful in predicting or interpreting responses at the population level. As an example, changes in gonads, such as atresia of oocytes, have been reported in fish exposed to contaminated sediments (Johnson et al., 1988), petroleum production wastes (Stott et al., 1981), and low environmental pH (McCormick et al., 1989; Leino et al., 1990). In the latter example, fathead minnows (*Pimephales promelas*) exposed to low pH and elevated concentrations of dissolved aluminum had dose-related abnormalites of ovaries and testes that were associated with poor reproductive success. Declines in fish populations in acidified waters have been well documented, and are thought to be related to reproductive failure (Haines, 1981; Baker and Schofield, 1985; Mills et al., 1987). Thus, a population-level response to a pollutant (decline in number of individuals) is closely associated with a histological effect (gonadal abnormalities).

Carcinogenesis may occur in fishes exposed to some pollutants, and this response is best detected by studies focused at the cell and tissue levels. For example, the process of neoplasia in livers of several fish species exposed to carcinogens in the laboratory has been studied in considerable detail (Hinton et al., 1984, 1988a, 1988b; Hinton and Lauren, 1990). Other studies document increased frequencies of tumors, neoplasms, and other abnormalities in feral populations inhabiting polluted sites, and these clearly demonstrate a bridge between the cell/tissue level and the population level. Brown bullheads (*Ictalurus nebulosus*) inhabiting industrially polluted rivers in Ohio had increased frequencies of liver tumors (Baumann et al., 1987;1991). Skin and liver neoplasms occurred at increased frequency in brown bullhead and white sucker (*Catostomus commersoni*) living over contaminated sediments in Lake Ontario (Hayes et al., 1990). Recently, Myers et al. (1994) documented associations between hepatic lesions, including neoplasms, and the concentrations of chemical contaminants in sediments and stomach contents of bottom-dwelling fishes along the Pacific coast.

In addition to cancerous and precancerous lesions, a variety of other histopathologic conditions occur at increased frequency in populations exposed to pollutants. Sindermann (1979) provided an early catalog and overview of the tissue abnormalities associated with pollutant exposure. More recently, Overstreet (1988) surveyed histopathological indicators of pollutant exposure along the southeastern U.S. coast, and noted correlations of pathological lesions with polluted locations. Khan et al. (1994) captured winter flounder from sites near a pulp mill and reference locations, and showed that lesions, including gill hyperplasia, focal vacuolation in the liver, and hemosiderosis in the spleen, were more prevalent in fish taken near the mill.

A considerable body of evidence has accumulated linking structural lesions in cells and tissues with pollutant exposure in feral populations, especially in fishes. There are several reviews on this topic (Hinton and Lauren, 1990; Hinton et al., 1992; Hinton, 1993), as well as a number of works detailing normal and abnormal histology of fishes (Grizzle and Rogers, 1976; Yasutake and Wales, 1983; Meyers and Hendricks, 1984; Ferguson, 1989; Sindermann, 1990). Histopathological responses can indicate potential problems before the effects appear at higher organizational levels, so alterations at the cell and tissue levels provide early warnings of pollution effects. In other words, structural changes in cells and tissues are excellent biomarkers of pollutant exposure and effects, as has been persuasively argued in previous reviews (Hinton et al., 1992; Hinton, 1993).

McCarthy and Shugart (1990) defined biological markers as "measurement of body fluids, cells or tissues that indicate . . . the presence of contaminants or the magnitude of the host response." Chapter 5 of this volume reviews some biochemical indices of pollutant exposure, and discusses their relevance to ecotoxicology. However, assessment of biochemical responses requires homogenization of tissue samples, so that spatial relationships among tissue components and other structural information are lost. Bolender (1979) argued that morphological data, especially quantitative (morphometric) information, provides essential information needed to interpret biochemical measurements, and to understand the impact of chemicals on tissues or organisms. Hinton (1993) has correctly pointed out that interdisciplinary integration, such as incorporation of biochemical, morphological, and physiological measurements to determine pollutant effect, strengthens the biomarker approach, and assists in interpreting causes and effects.

Regardless of whether histological methods are used alone or in concert with physiological and biochemical indices to evaluate pollutant effects, methods of quantifying changes in structure are readily available. However, these methods are not as widely applied as they might be, perhaps because their power is unappreciated, or because they are viewed as difficult and tedious. Histopathology, as currently applied to the study of the effects of environmental pollutants, is for the most part a qualitative endeavor. For example, in a recent comprehensive text often used in aquatic toxicology courses, the histopathology chapter contains an extensive review of terminology and laboratory techniques along with an organ by organ description of lesions resulting from exposure to various toxicants (Meyers and Hendricks, 1984). However, no mention is made in the chapter of any quantitative methods to detect or measure toxicant responses. Lesions are discussed only in qualitative terms (e.g., present or absent), or, at best, semiquantitative terms (e.g., severe hyperplasia or decreased numbers of mucous cells). My intent is not to criticize this particular text; instead, I have cited it as an example of the widespread bias toward qualitative description in this field. Indeed, given that quantitative studies constitute only a small fraction of the primary literature, it is not surprising that qualitative results are so commonly cited in review and synthetic papers. For example, Mallat (1985) reviewed the literature on the effects of chemical and physical irritants on fish gill structure. Of the 130 studies he cited, most contained no quantitative histopathological results. Less than one-quarter of the studies cited provided numerical data such as epithelial thickness, gill surface area, or counts of different cell types.

In this chapter, it is not my intent to present an extensive survey of various histological responses to pollutants. There are a number of excellent reviews of various histopathological responses to environmental pollutants, as cited above, emphasizing the utility of these responses as biomarkers. Instead, in keeping with a major theme of this volume (quantitative ecotoxicology), I focus on the utility of quantitative measures of structural changes in response to pollutant exposure. I believe that more widespread incorporation of quantitative methods is essential to the continued development of histopathological indices of pollutant exposure, and to the interpretation of histological responses within the hierarchical framework outlined at the beginning of this section. A number of other workers have also called attention to the largely subjective methods commonly employed in histopathological studies of pollutant effects, and argued that quantitative methods would greatly increase the power of the histopathological approach (see Hinton et al., 1987). It may be that this field is ready for a major transformation into a quantitative science, especially given the rapid development of improved morphometric techniques and computer-assisted methods.

The remainder of this chapter is divided into three sections. First, the need for quantitative approaches in studies of pollutant effects at the tissue level is discussed, and the view is presented that these are necessary both to extract the maximum amount of information from experimental results, and to allow the rigorous hypothesis testing characteristic of a mature science. Second, brief reviews are provided of some of the methods that have been developed for quantifying histopathological effects. Finally, it is demonstrated how some of these methods have been applied to obtain a better understanding of a particular environmental problem, the effects of dissolved metals on fish.

II. THE NECESSITY OF QUANTITATION

Consider a study in which organisms were exposed to various concentrations of some pollutant and suffered "some" or "moderate" or "severe" mortality. Suppose these adjectives alone, without the mortality numbers or percentages, were the only results presented. This would seem strange indeed, because we expect studies of acute toxicity to yield quantitative results, such as estimates of LD50 with confidence limits. Similarly, a physiological study in which exposure to some pollutant was reported to produce, say, a decline in blood oxygen level, or inhibition of some enzyme would have little chance of publication unless numerical measurements were presented. In these examples, qualitative statements of results would be inadequate to demonstrate convincingly that responses had occurred. However, such results are commonplace in histopathology studies of pollutant effects, and these studies are usually profoundly qualitative and descriptive. As mentioned above, most papers cited by Mallat (1985) about effects of pollutants on gill structure do not contain quantitative results; studies cited in a later review (Evans, 1987) of the same topic are also primarily qualitative.

A. SITUATIONS IN WHICH QUALITATIVE
   OBSERVATIONS ARE SUFFICIENT

Qualitative studies are useful in many instances; I do not argue that such studies are without value. There are two general areas in histopathology as applied to ecotoxicology where quantitative measurements at the tissue level are probably unnecessary. First, in initial studies of some pollutant, there may be insufficient information to predict target organs or cell types, or to estimate the likely magnitude of any effect. In this case, a qualitative survey can be extremely useful. For example, early field and laboratory studies showed that acidification of aquatic systems from atmospheric acid deposition caused gill pathologies in fish (Daye and Garside, 1976; Muniz and Leivestad, 1980; Jagoe and Haines, 1983). Other studies have demonstrated increased incidences of liver neoplasms in fish living over sediments contaminated with polyaromatic hydrocarbons (PAHs) (Smith et al., 1979; Malins et al., 1984; Baumann et al., 1991). These studies suggested mechanisms of toxic action, and gave some indication of the magnitude of effects that could be expected at environmentally realistic pollutant levels. Another goal of this type of study is often to explore whether different substances have similar effects on tissues. For example, Jagoe et al. (1993) showed that dissolved Be and Al, whose aqueous speciation properties are similar, produced similar gill pathologies.

Second, especially in acute studies, the effects of a pollutant at the tissue level may be so profound that qualitative examination is often sufficient to detect response. There are numerous examples of this in the literature, and several are cited here to illustrate this point. Skidmore and Tovell (1972) exposed rainbow trout to extremely high levels of zinc (40 mg/l) and found severe edema in the gills, along with epithelial separation and increased infiltration of granulocytes into gill tissues. Daoust et al. (1984) found degeneration of epithelial cells, and swelling and fusion of lamellae in gills of rainbow trout exposed to mercury (0.35 mg/l) or copper (0.135 mg/l). Rojik et al. (1983) observed that hepatic cells of three species of fish were highly vacuolated, with swollen mitochondria and substantially decreased amounts of nuclear heterochromatin after exposure to pesticides (paraquat, $CuSO_4$, $ZnCl_2$; 100 mg/l). Conti (1987) observed damage to gills and epidermal sensory structures of the lugworm, *Arenicola marina* after 48 h exposure to detergents, Carbaryl and Parathion at concentrations up to 20 mg/l. It is important to note that in many acute toxicity studies, organisms are exposed to pollutants at concentrations far higher than would be encountered in the environment. Toxic effects resulting in histopathological changes are often pronounced and readily observable with high toxicant concentrations, but the responses found may or may not reflect those occurring under more realistic conditions.

B. PROBLEMS WITH PURELY QUALITATIVE OBSERVATIONS

When using purely qualitative observations, it may be difficult to assess interindividual variation. Organisms differ somewhat in their sensitivities to environmental stressors, which is often reflected in their responses at the cell and tissue levels. Interindividual variability is probably less of a problem in acute toxicity studies of the type discussed above. In such studies, a large amount of some pollutant is present, and all or most organisms exposed experience some toxic effect. With the toxicant concentration above the sensitivity threshold for all or most of the organisms, interindividual differences in sensitivity become academic. Under these conditions, it is not surprising that common histopathological responses can be detected in exposed organisms. The problem arises at lower concentrations, where some fraction of the exposed organisms show a reduced response, or no response at all. In most qualitative studies, common patterns of responses at subacute concentrations can be difficult to detect.

Variation within individuals can also confound qualitative studies. The detection of histopathological effects requires the collection, fixation, and microscopic examination of a small piece of tissue from some organism. This obviously allows examination of only a small part of the target organ or tissue. Admittedly, this is less of a problem when smaller organisms are used. Smaller fish such as medaka (*Oryzias latipes*) or Poeciliidae (mosquitofish, guppies) are often used in laboratory studies. Because these fish have smaller organs, proportionally more of the organ can be seen in a section or photograph. This argument also holds for larvae and juveniles of larger species. Nonetheless, it must be recognized that histological examination typically samples only a small part of a larger structure. The problem becomes worse with higher levels of magnification, because the area of the field decreases as magnification increases; using electron microscopy only a few cells (at most) can be examined simultaneously.

There may be considerable variation from field to field even in normal tissues, and some of the problems in experimental design associated with this variability are discussed in detail below. Because of this variability, it is necessary to look at multiple areas within a tissue to obtain an accurate picture of structural features, as well as to examine tissues from several organisms. Without quantitative measurements, the only ways to address inter- and intraindividual variation are to take photographs of each field, or to note normal or abnormal features of each field, and later compare observations or photos within and between organisms. I suspect that this is seldom done systematically, and the natural tendency is to note only features that are abnormal or unusual. Of course, whether done consciously or not, this produces a biased sample, from which it is impossible to judge the actual frequency or severity of a lesion.

Another consideration is that changes in tissues which are essentially quantitative in nature need to be fairly large before they can be detected by a qualitative examination. Reith and Mayhew (1988) noted that "volume or number increases must be on the order of 30–50 % to be recognizable, and increases in membrane, that is, surface density, have to be even higher. However, in toxicological studies, for example, a constant 10–20 % increase in endoplasmic reticulum may be an important observation, but this will be lost in any routine . . . investigation." By their estimation, qualitative examination is likely to overlook subtle, yet potentially important alterations in morphology. Thus, the chance of a type II error, the acceptance of a false null hypothesis, is greatly increased by relying only on qualitative examination. In his review of the literature documenting toxicant-induced structural changes in gills, Mallat (1985) explicitly noted that he found it impossible to assess how often subtle gill lesions were overlooked.

C. SEMIQUANTITATIVE APPROACHES

A number of investigators have employed semiquantitative techniques to avoid some of the shortcomings and potential problems of purely qualitative studies. In a typical semiquantitative approach, a number of slides are made from tissue of organisms exposed to various concentrations of the pollutant of interest. The slides are then coded, so that the person examining them cannot associate the slide with a particular experimental treatment. The examiner then looks at one or more fields per slide, and assigns each a score based on the presence or severity of some preselected abnormality. These scores are then treated statistically to determine if there is an association between the abnormality and pollutant exposure. This is obviously a more sophisticated approach than simply examining pollutant-exposed tissues and reporting unusual structural features. A formal semiquantitative approach overcomes many of the problems discussed in the previous section. However, there are also some potential difficulties inherent in this approach that warrant consideration.

Two recent studies of the effects of water quality on gill structure illustrate the application of this approach. In one, Coulillard et al. (1988) exposed rainbow trout to diluted bleached kraft mill effluent for up to 60 d, and examined their gills for lesions. Using three randomly selected gill regions from each fish, they noted the number of regions showing cellular hypertrophy or clubbed lamellae (0–3 per fish), and scored each region for severity of lamellar fusions. The severity of these fusions was ranked from 1 to 4 based on the number of lamellae affected (1 = 0 to 25% of the lamellae fused, 2 = 25 to 50% of the lamellae fused, and so forth). They also counted the number of cell layers at the base of one lamella in each region as an index of epithelial hypertrophy. Using the nonparametric Mann-Whitney test to compare the rankings among treatments, they found increased lamellar fusion with 96-h acute exposures to concentrated effluent. However, in 20- to-60 d exposures to less concentrated effluent, there was no relationship between effluent concentration and lamellar fusion. In the chronic experiments, there were some significant differences among treatments in cellular hypertrophy and hyperplasia, but these were not consistent over time.

In the second study, Leino et al. (1990) reared fathead minnows (*Pimephales promelas*) in soft water at pH levels ranging from 5.2 to 7.5, with added aluminum up to 60 µg/l. They examined 3 to 12 sections per fish, and rated preselected abnormalities on a scale of 1 to 3. They also measured the thickness of the lamellae to detect lamellar hyperplasia. For each type of abnormality (e.g., chloride cell proliferation, lamellar edema, chloride cell vacuolation), they calculated a mean severity score for each experimental treatment. These mean severity scores were then multiplied by the percentage of the fish in each treatment showing the particular abnormality. These products were then summed for each treatment, to calculate an overall pathology score. The overall pathology scores significantly increased as treatment pH decreased, but addition of aluminum did not produce significantly greater abnormalities (by paired t tests).

In both studies, application of semiquantitative estimates of the severity of various pathologies provided clear evidence of dose-response effects. While neither study explicitly addressed inter- versus intraindividual variation, each used multiple fields per fish, and several fish per treatment. Thus, it would be possible in principle to evaluate the contributions of these sources to the total variation between treatments. By using numerical values to describe the extent and severity of pathological responses, statistical comparisons among treatments were possible. The researchers were therefore able to test hypotheses about the differences in treatment effects and calculate the probabilities that the treatment means were different. Such analyses of data are obviously not possible in purely qualitative studies.

While semiquantitative approaches overcome many of the drawbacks inherent in purely qualitative studies, there are still some shortcomings inherent in these methods. First, rankings of the severity of abnormalities are likely to be somewhat observer specific. Similar abnormalities may be assigned different scores by different workers. This can be minimized by having the same individual assign all scores, or addressed statistically by comparing the scores assigned by several individuals to the same group of slides. However, because rankings are somewhat observer specific, the results of semiquantitative studies can be difficult to compare to studies of other pollutants, done in different laboratories at different times. Moreover, semiquantitative studies cannot address the concern raised by Reith and Mayhew (1988) and discussed above: that essentially quantitative changes, such as the increase in number of some cell type or organelle, need to be fairly large before they can be detected readily by qualitative examination. In other words, semiquantitative studies are likely to overlook smaller, but still potentially important, differences between treatment groups, and thus may be prone to type II errors to some unknown extent.

In addition to the semiquantitative studies described above, there are several other partially quantitative approaches to detect and describe responses to pollutants at the histopathological level. For example, a number of studies have associated liver tumors and other abnormalities with exposure to polluted sediments (Hayes et al., 1990; Baumann et al., 1987, 1991; Myers et al., 1994). In these studies, samples of fish from polluted or reference locations were collected and dissected, and the livers examined for visible tumors. A portion or slice of liver was fixed, and later examined for abnormalities. The frequency of abnormalities were then compared among sites, and correlated with known pollutant distributions, or other environmental or biological factors. This is defined as a partially quantitative approach because, although the frequency of some lesion within a population is quantified, the diagnosis or detection of the lesion is based on qualitative or semiquantitative criteria. In other words, there is no quantification of the severity of a lesion; only the frequency within a population is expressed numerically. This approach has considerable utility in detecting and documenting abnormalities associated with environmental contaminants. However, detection of the lesions suffers from the limitations of the qualitative approaches discussed above.

Another partially quantitative approach has been proposed by Adams and Greely (1991) and Adams et al. (1993), termed the health assessment index (HAI). Determining the HAI involves examining a suite of physiological and morphological features in groups of fish collected from polluted and reference locations. The morphological appearances of various organs are assigned a score based on visual examination, and an overall health score is calculated by summing the scores for the organs. Population means are then used to allow interlocation comparison of fish health. As is the case for the frequency-of-abnormality studies discussed above, this technique is partially quantitative. It considers frequency of occupance of lesions in different organs in fish within a population, but the lesions are detected by qualitative or semiquantitative examination. The limitations of qualitative approaches, as previously discussed, including observer specificity and difficulty in reliably detecting subtle changes, apply to this methodology.

D. RATIONALES FOR QUANTITATIVE STUDIES

Forbus (1952) hypothesized that disease resulted from the quantitative alteration, such as increase or decrease of flux rate, of existing metabolic processes. He rejected the notion that pathological responses resulted from the formation of new biological structures or functional pathways. Today, this interpretation seems somewhat simplistic, but there is much utility in the paradigm proposed by Forbus. The neoplastic transformation of a cell probably involves quantitative changes in preexisting processes, including enzyme phosphorylation and gene expression (Alberts et al., 1983). Besides carcinogenesis, many pathological states are the result of quantitative changes in existing tissue components and physiological processes. For example, inhalation of beryllium oxide causes thickening and edema of alveolar walls in mammalian lungs, affecting the capacity for gas exchange (Wilber, 1980).

Acutely lethal pollutant exposure often causes gross structural damage to target organs. However, exposure to pollutants at sublethal levels also may affect histology and ultrastructure. Exposure to a pollutant at a sublethal level produces physiological stress in organisms, and this stress, or the response to it, involves some change in metabolic function. Alterations in metabolic function are usually reflected as changes in tissue or cell structure. As an example, increased requirements for ATP would necessitate increasing the volume fraction of mitochondria within a cell. Likewise, alterations in metabolic function at the organismal level typically involve structural changes in the tissues performing that metabolic function.

In their seminal book, *Strategies of Biochemical Adaptation*, Hochachka and Somero (1973) observed that organisms living under different environmental conditions must adapt at the biochemical level to maintain the correct direction and rate of flux through metabolic pathways. Because the metabolic activities of organisms are entirely dependent on macromolecules such as enzymes and nucleic acids, adaptation must involve changes in these macromolecules. They postulated three major strategies by which metabolic processes could be modulated: (1) the types of macromolecules involved in some process could change, (2) the amounts or concentrations of the macromolecules could vary, and (3) the functions or capacities of the macromolecules could be altered. By one or more of these general approaches, homeostasis could be maintained at the biochemical levels as environmental conditions changed.

These basic biochemical strategies represent ways in which organisms might respond to environmental stimuli at the molecular level. By analogy, one can imagine three basic ways by which an organism's cells and tissues can respond to stresses associated with sublethal pollutant exposure. First, the types of cells present in the tissue may change. For example, these changes might include neoplastic transformations, or the infiltration of leukocytes into the parenchymal spaces of organs. Second, the numbers or distribution of cells normally present within a tissue may change. Examples include hyperplasia of epithelial cells, or proliferation of mucocytes associated with increased mucous secretion. Third, the function or capacity of cells within a tissue may change. Changes such as hypertrophy or increased volume of rough endoplasmic reticulum in hepatocytes, implying enhanced protein turnover, exemplify this category. It is important to note that these strategies are not exclusive, and combinations of these responses can occur as a consequence of exposure to a single pollutant. Moreover, not all histopathological changes fit into one of these categories. However, this concept of common strategies of response to environmental challenges is useful in understanding the responses of organisms to pollutants, and this utility is illustrated later in this chapter by consideration of some effects of dissolved metals on gill structure.

Of the general categories of response discussed above, only the first is essentially qualitative in nature. The others involve modulations in cell numbers or sizes, or in the number, size, or density of subcellular components. Because changes in tissue in the latter two categories are essentially quantitative in nature, they can best be detected and described using quantitative techniques. Although responses falling into the latter two categories can be described qualitatively, such description nonetheless involves basically quantitative concepts: more, less, larger, smaller, and so forth. In evaluating responses that fall into these quantitative categories, it is important to remember the argument of Lord Kelvin, " When you can measure what you are speaking about, and can express it in numbers, you know something about it; but when you can not express it in numbers, your knowledge is of a meager and unsatisfactory kind; it may be the beginning of knowledge, but you have scarcely, in your thoughts, advanced to the state of science, whatever the matter may be" (quoted in Elias and Hyde, 1983).

III. METHODS FOR QUANTIFYING EFFECTS AT THE CELL AND TISSUE LEVELS

Over the last 30 years or so, a number of methodologies have been devised to quantify structural features of three-dimensional objects from two-dimensional images. These methodologies, collectively termed either "morphometry" or "stereology" were originally developed for materials science (De Hoff and Rhines, 1968; Underwood, 1970). Weibel (1979) was one of the first to apply these techniques to problems in the biomedical sciences, and Elias and Hyde (1983) published a compendium of methods intended as a practical guide for biologists a few years later. Despite these efforts, the use of these techniques is still far from universal. Weibel (in Reith and Mayhew, 1988) has written that stereology and morphometry deal "with numbers derived from pictures . . . and biologists, particularly morphologists, commonly experience a deep sense of horror or shock when they are asked to think mathematically, that is to say, to make an effort to reduce part of the wealth contained in their pictures to some simple, quantitative bits of information. I daresay it is not beauty that is lost in this process. The picture goes undamaged and can safely be hung on the wall. It is rather beauty that is gained, but beauty of a different kind: one that allows us to discover some of the laws of nature that we . . . need to single out and study one by one before we can put them back together."

Given the number of books and papers detailing the mathematical foundations of stereological and morphometric techniques, they do not need to be repeated in detail here. Weibel (1979), Elias and Hyde (1983), and Reith and Mayhew (1988) give excellent descriptions of techniques and practical applications in biology. Gundersen and coworkers have documented a number of new approaches to stereological studies that eliminate problems with shape dependence of certain measurements (Gundersen et al., 1988a, 1988b). My intent here is to summarize a few basic principles of these methodologies, and show how they are relevant to questions in environmental toxicology. I also consider questions of sampling design and economy of effort in designing quantitative histopathology studies.

A. STEREOLOGY AND MORPHOMETRY

Morphometry is defined as the measurement of shape or form. There is a subdiscipline in statistics concerned with the mathematical analysis and description of shape (e.g., Bookstein, 1991) which will not be further considered here. As used in this chapter, morphometry simply means measurement of some feature, such as shape or size or surface area, of some biological structure. Stereology is the measurement of three-dimensional forms from information contained in two-dimensional samples of the form. Underwood (1970) correctly noted that this is essentially the opposite of photogrammetry as used by mapmakers; their concern is representing a three-dimensional object, such as the surface of the earth, as a two-dimensional map. The two-dimensional samples used in stereology can be physical sections of three-dimensional structures, or photos of the sections. They can also be photographs of the surface of a three-dimensional object. For example, measurements can be taken from a scanning electron micrograph of an epithelial surface. While no sectioning is involved, this is actually a two-dimensional projection of a three-dimensional object.

In most cases involving a three-dimensional structure, the measurement of interest involves volume. This may be either the actual volume of some structure, the fraction of the total volume of a structure occupied by some component, or the relationship between surface area and volume. In stereology, three-dimensional volume is estimated from two-dimensional area. Consider a three-dimensional object cut into a number of parallel slices of equal thickness. The volume of the object could be calculated by measuring the area of the profile of the object in a number of the sections. The true volume would be more accurately estimated by measuring a larger number of thinner sections, all other things being equal. Obviously, this approach would be prohibitively labor intensive if the object were relatively large and the section thickness relatively thin. Still, this type of sectioning, measuring, and reconstructing has been used in some instances; Elias and Hyde (1983) concluded that serial sectioning, or the measurement of repeated, sequential slices of some object, is the best method available when a structure is solitary, or has an especially complicated shape.

It is much simpler to estimate the volume of structures from a few sections (or even one). This is possible because in many instances a section contains many cells of a particular type, or in the case of electron microscopy, a cell contains many organelles of interest. By measuring the profiles of these, an average volume for the cells or organelles can be obtained. It is also relatively simple to estimate a volume fraction, that is, the portion of a larger structure occupied by one or more smaller structures. If the volume fraction for a particular component is known, and the volume of the larger tissue can be obtained, then the volume of the component can be obtained by simple arithmetic. As an example, Myking (1988) used a simple point counting method, as discussed below, to estimate the volume fraction of the paracortical region in mammalian lymph nodes. He then calculated the volume of individual lymph nodes from their mass, and estimated the volume of the paracortical region from these two measurements.

Weibel et al. (1966) outlined a number of simple techniques to estimate volume fractions. First, they noted that the volume fraction of some component in a tissue (V_V) could be estimated from the area fraction (A_A) of that component in random sections of the tissue. One method of determining A_A is by planimetry, or actually measuring the area of the components in the sections (Figure 1A). Because the early development of stereology occurred before computer-assisted methods for planimetric measurements were available, this option was viewed as overly cumbersome and laborious. An alternative method of determining A_A is by lineal integration, or measuring the length of test lines passing though the component of interest. This approach is illustrated in Figure 1B, where a grid of parallel lines is superimposed over the tissue. The sum of the lengths of the lines in the grid is known, so measurement consists of determining the lengths of the line segments overlaying the component of interest, shown in the figure as bold lines. The sum of the lengths of these segments, divided by the sum of the lengths of the lines forming the grid yields the length fraction (L_L). Elias and Hyde (1983) noted that this method is especially convenient for small details, such as the width of cristae within mitochondria or the cisternae of endoplasmic reticula in electron micrographs.

Probably the most widely used method of estimating V_V is based on point counting, to determine the point fraction, P_P. As shown in Figure 1C, a grid is superimposed over the tissue. Here, the points are the intersections of the horizontal and vertical lines of the grid. The number of points falling within the components of interest are simply counted (equal to 25 in Figure 1C), and divided by the total number of points in the grid (equal to 144), so $P_P = 25/144 = 0.174$. This is equivalent to stating that, by this estimate, the component of interest occupies 17.4% of the tissue in this example. Note that this compares well with L_L (= 0.173) as measured from Figure 1B and A_A (= 0.179) from Figure 1A. In fact, all of these measures of volume density are equivalent: $V_V = A_A = L_L = P_P$ (Underwood, 1970; Weibel, 1979; Elias and Hyde, 1983). Intuitively, one can picture that making the number of points or the number of

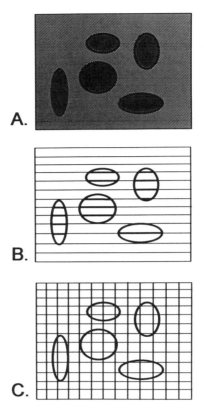

Figure 1   Schematic diagram of various methods for the volumetric analysis of tissue
composition. Ellipses represent a hypothetical tissue component measured by
(A) planimetry, (B) lineal integration, (C) differential point counting. (From
Weibel et al., 1966. J. Cell Biol. 30, 23–38. Copyright permission from the
Rockefeller University Press.)

lines in the grid larger and larger would improve the estimate of area; counting
a very large number of points would be equivalent to planimetrically measuring an area. In fact, this is the basis for much current image analysis software,
where the number of pixels within an image are automatically counted to
determine area. However, if the point counting is done manually, the added
labor of counting many additional points to achieve a slight increase in accuracy must be considered. Weibel et al. (1966) convincingly argued that increasing the number of points in the test grid is generally not an efficient way of
obtaining better estimates of stereological parameters.

In addition to volume of some tissue component, it is often desirable to
estimate surface area. Obviously, this parameter can be of critical importance
when considering diffusional fluxes or gas and ion exchanges. In stereology,
surface area is generally expressed per unit volume as surface density, S_V
(Weibel, 1966; Underwood, 1970; Elias and Hyde, 1983). If a grid of parallel
lines is placed over a tissue, as in Figure 1B, and the number of intersections

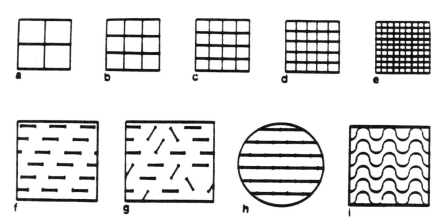

Figure 2 Nine representative test patterns or grids for differential point counting and other stereological methods. (From Elias, H. and D.M., Hyde, 1983. A Guide to Practical Stereology. S. Karger, New York, With permission.)

of the lines with the surfaces of the components of interest (N_L) are counted, then $S_V = 2 N_L$. For Figure 1B, S_V equals 56. Units depend on the length of the test lines. If each line on the grid in Figure 1B was 1 cm long, S_V here would be 56 cm²/cm³. Chalkey et al. (1949) reported a variant of this technique for estimating the surface-to-volume ratio of individual components within a tissue when multiple components are present. In other words, the technique presented here estimates the total surface-to-volume ratio for, say, all the mitochondria within some tissue volume, while the technique described by Chalkey et al. provides an estimate for each mitochondrion.

A number of grids have been devised to facilitate stereological measurements (Figure 2). These can be placed over a photograph or video monitor, or used as a reticle in the eyepiece of a microscope. Most of these employ squares as the basic geometrical shape, but some use equilateral triangles. Figure 2 (i) shows a Merz grid, usually recommended for regular, oriented structures (Elias and Hyde, 1983). To make stereological measurements, the grid is placed in a random orientation over the section, photo, or image and the appropriate counts or measurements taken. The grid is then moved randomly to another location and the process repeated. Because stereological estimates require random orientation of the grid relative to the tissue, grids based on straight lines work best on irregular, nonoriented tissues. However, for tissues such as fish gills with a repeating, oriented structure, a grid based on curved lines such as the Merz type avoids correlations between grid and tissue orientation.

In addition to measurements of the volume or surface area of various components, sometimes it is necessary to determine the number of some structures within a tissue. Counting is obviously a straightforward means of accomplishing this. For example, one may estimate the number of glomeruli in a kidney by counting them in a number of sections. Similarly, the number of chloride cells in a gill filament can be estimated from counts per unit area. Despite the ease of this approach, there are a number of potential problems to be aware of, mostly arising from the fact that sections are not infinitely thin. When the thickness of some section is an appreciable fraction of the structure being counted, the number of structures seen in the section becomes a function of the thickness (Elias and Hyde, 1983). Also, because many components of interest are irregular in shape and possibly oriented at an angle to the cutting plane, adjustment of the estimate of number for relative section thickness and mean diameter of the component of interest is necessary. These considerations, and the mathematical formulas for the appropriate adjustments, are further developed in Weibel et al. (1966), Underwood (1970), Weibel (1979), and Elias and Hyde (1983). For the purposes of this review, it is sufficient to note that in cases where the objects are roughly spherical and relatively large compared to the section thickness, counts of the objects in the sections generally provide a good estimate of N_V, the numerical density. Examples include glomeruli and chloride cells in plastic sections, as mentioned above. However, when the objects are highly irregular, and/or small compared to the section thickness, corrections to the raw count are necessary to accurately estimate N_V. Examples include mitochondria and endoplasmic reticula within cells. Recently, Gundersen et al. (1988a) reviewed a new technique, based on parallel sections, that allows unbiased estimation of N_V regardless of shape.

Most of the stereological methods discussed thus far are based on counting or measuring the lengths of line segments rather than measuring actual areas. This is partly for historical reasons, because techniques for rapid and simple planimetric measurements did not exist until fairly recently. At present, inexpensive personal computers can be equipped with pointing devices (mouse or digitizing tablet) and software allowing easy measurement by tracing around the perimeter of the object of interest. Software to measure digitized images based on differences in pixel intensity is also becoming commonplace, and this field is developing rapidly. (For an overview of computerized image analysis of electron micrographs, see Rigaut, 1988; although a bit dated, this is still a very good synopsis and survey of the literature in the field through the late 1980s.) It would be surprising if this trend towards more automated methods of image analysis did not continue, given the increasing computing power available at progressively lower costs. Several studies have compared computer-based planimetric measurements with simple point-counting methods (Gundersen and Osterby, 1981; Mathieu et al., 1981; Myking, 1988; Gundersen et al., 1988c). These all noted that while computerized planimetry was much more accurate than point-counting methods, it was also somewhat slower. They concluded that the increase in accuracy was not worth the decrease in speed of measurement. This echoes the earlier opinion of Weibel et al. (1966) that increasing the number of points in the test grid to improve the accuracy of P_P was usually inefficient. All argued that measuring larger numbers of fields with slightly lower accuracy was preferable to measuring smaller number of fields with high accuracy, because variation within and among individuals was more important than slight measurement errors. Gundersen and Osterby (1981) expressed this concept as "do more less well," maintaining that measuring more fields with lower precision allowed better estimation of true tissue dimensions than measuring a few fields with high precision. Given the rapid increase in the ease and accuracy of computer-based measurements, these conclusions may or may not hold for future studies. I suspect that those working in the field will need to evaluate the speed and ease of various techniques on a case by case basis before deciding which is best for a particular application.

B. SAMPLING DESIGN AND EFFICIENCY

Typically, to evaluate the effects of some pollutant, groups of organisms are exposed to different concentrations of the pollutant, usually for some fixed time. For acute toxicity studies the time of exposure is usually 96 h, and then some marker of response to the pollutant is measured. In the widely used, 96 h LC50 assay, the endpoint is death of some fraction of the exposed organisms. Dead and live organisms are tallied and used to estimate the concentration of the pollutant that will kill 50% of the exposed organisms in 96 h. In most cases, the organisms are discarded after the test. In considering sampling design and efficiency, it seems appropriate to begin by pointing out the basic inefficiency of this practice. Considering the widespread application of acute toxicity testing, if histopathologic or other biomarkers had been measured on even a fraction of the organisms used, we would have a much better knowledge of the mechanisms of action and target organs, tissue, and cells for most pollutants. This information could obviously be useful at other hierarchical levels as well.

It would be relatively simple to slightly modify future studies so that additional information, especially histopathological data, could be obtained. Tissue samples from surviving organisms (or whole organisms) could easily be fixed and archived. Because of rapid postmortem changes, histological studies on tissue fixed some time after the death of an organism can be difficult to interpret or of questionable value. However, to avoid this problem, moribund organisms could be scored as dead, removed, and fixed. After chemical fixation (usually in some mixture containing formaldehyde and/or glutaraldehyde), tissue samples have a very long shelf life. Even if funding is not available at the time of the experimental exposure for histopathological studies, such studies could be performed months or even years later if the tissues have been correctly fixed and archived. Although the focus has been on acute toxicity studies in the laboratory, this argument is also valid for longer-term studies or field studies, where more time and effort is involved. After the costs associated with experimental treatments, it is both cost-effective and efficient to preserve samples in case studies need to be expanded from the individual level in the hierarchy to the organ, tissue, or cell levels.

Once organisms have been exposed to some pollutant, an important question in histopathology arises: how to design a subsampling scheme to detect abnormalities, and to estimate how these abnormalities vary among and within individual organisms. In most studies, one or more tissue samples per individual are fixed, embedded, and sectioned. Generally, many sections are cut from each tissue block, representing slightly different regions of the tissue.

From each section, several fields may be examined. This is clearly a nested sampling design, with a nested hierarchy of fields within sections, sections within blocks, blocks within individuals, and individuals within treatments. Specific studies may have more or fewer levels within this hierarchy, but this type of nested design is widely applied. Nested analysis of variance is usually the appropriate way to analyze such results (Sokal and Rohlf, 1981), and methods are available to estimate the contribution of each level to the total variation observed (e.g., PROC VARCOMP, SAS, 1990).

Consider a nested sampling scheme where treatments (t) consist of a number of individuals (i), and a number of tissue blocks (b) are taken from each individual. Some number of sections (s) are prepared from each block, and some fields (f) are examined on each section. Thus, there are n_t treatments, $n_t \times n_i$ individuals, $n_t \times n_i \times n_b$ tissue blocks, $n_t \times n_i \times n_b \times n_s$ sections, and $n_t \times n_i \times n_b \times n_s \times n_f$ fields. Each sub-level contributes a component (s_i, s_b, s_s, s_f) to the total observed variance among treatments (Os_t). The contribution of these components to the observed variance among treatments is

$$Os_t = s_t \frac{s_i}{n_i} + \frac{s_b}{n_i n_b} + \frac{s_s}{n_i n_b n_s} + \frac{s_f}{n_i n_b n_s n_f} \qquad (1)$$

(Gundersen and Osterby, 1981; Gupta et al., 1983). Simple inspection of this equation suggests a straightforward way to decrease the overall variance, and thus increase our ability to detect "true" differences among treatments: increase n_i, because this term appears in all of the denominators. This would indeed be the best course of action if $s_i = s_b = s_s = s_f$, and if the costs of sampling at each level were roughly equivalent. In practice, the variance terms are not equal at all levels. In many cases, the differences can be substantial, and efforts should be concentrated at the most variable level or levels. Additionally, the costs of increasing sampling effort are not the same for all levels. For example, it may be much more economical, in terms of time and cost, to prepare another section from a tissue block than to add another animal to an experiment (Shay, 1975; Mayhew 1982).

A number of studies have suggested that interanimal variation is the largest contributor to the overall variance in this type of multilevel sampling design (Shay, 1975; Gundersen and Osterby, 1981; Mayhew, 1982; Gupta et al., 1983). These studies conclude that biological variation, or actual differences between individuals, is the major factor affecting the precision of stereological measurements. This implies that increasing the number of organisms per treatment is the best way to decrease treatment variances, and thus increase the probability of detecting actual differences among treatments. However, Egginton (1988) showed that intraindividual variation was greater than interindividual variation for S_V of mitochondria and V_V of myofibrils in red and white fish muscle. He argued that, in some tissues, strict functional requirements limit the amount of variation between individuals. His results also suggest that scale may be an important factor affecting the variation observed: in comparing biochemical measurements from whole tissue, stereological measurements of light micrographs and stereological measurements of electron micrographs, interanimal variation was most important in whole tissue samples, and least important in subcellular samples. By this argument, variation between fields may become more important as smaller fields are examined, because smaller fields are more likely to be heterogeneous, whereas larger fields integrate small-scale variations in tissue structure to yield a more homogeneous picture. Similarly, Jagoe and Haines (1990) showed that interindividual variation in chloride cell number in gills of Atlantic salmon (*Salmo salar*) was smaller than variation within individuals, and interpreted this as an example of functional requirements limiting interindividual variation.

The studies discussed above suggest that there is no universal answer to the question of the relative importance of variation at different sampling levels. In some instances, interindividual variation may be most significant, and in some, variation between tissue samples, sections, or fields may be more important. It is also important to note that the experimental treatments themselves may affect the variance at some levels. For example, Mayhew (1982) showed in a study of experimental carcinogenesis that induced hyperplasia in hamster cheek pouches increased the variance between fields, and decreased the interindividual variance. This also illustrates the importance of scale, in that individuals generally responded to the carcinogen in the same manner, but different areas of the tissue varied in magnitude of response. In sum, in applying stereology to studies of environmental toxicants, the investigator must consider the contributions of variation at each level of sampling. It may be desirable in many cases to do a small pilot study to estimate the contributions of inter- and intraindividual variation. This information, in concert with considerations of the costs of increasing replication at each level, can then determine the final sampling design. Obviously, it will always be necessary to examine more than one individual per treatment, because interindividual variation always exists. However, in many cases looking at relatively more sections or fields from relatively fewer individuals may be more economical, while still yielding a sufficiently low overall variance to allow a high probability of detecting true differences among treatments.

IV. AN EXAMPLE: EFFECTS OF LOW PH
AND DISSOLVED METALS ON FISH GILLS

Acidification of surface waters by atmospheric deposition of acids from fossil fuel usage was recognized as a potentially serious environmental problem in the 1970s. It was soon realized that low pH exposure alone was not the only threat to resources. Acid precipitation increases the mobility of metals such as aluminum, and acidified surface waters can contain toxic levels of aluminum leached from soils (Cronan and Schofield, 1979; Driscoll et al., 1980). Subsequent studies showed that other metals such as beryllium could be released from soils, resulting in increased concentrations in acidic waters

(Vesely et al., 1989). Increased concentrations of other trace metals, such as copper, lead, and zinc, may be associated with atmospheric deposition and/or acidification (Stennes, 1990). Most recently, atmospheric deposition of mercury to remote lakes (Mierle, 1990) and enhanced methylation of mercury, especially in low pH, colored waters (Winfrey and Rudd, 1990) has emerged as a new environmental concern.

It is known that exposure to low pH and metals such as aluminum cause ionoregulatory and osmoregulatory failure in fish (McDonald, 1983; Witters, 1986; Booth et al., 1988). Gas exchange and acid-base balance may also be disturbed at very low pH or high concentration of metals, including Al (Wood et al., 1988; Playle et al., 1989). Gills are the principal organ responsible for ion regulation, acid-base homeostasis, nitrogenous waste excretion, and gas exchange in fish. Thus, they are the main organ affected by exposure to low pH or elevated concentrations of dissolved metals in water. Gills are a multifunctional organ forming a complicated interface between the external environment and the internal composition of the organism. In addition to diffusional exchanges, the gill epithelium functions as an active transporting epithelium, moving ions including HCO_3^-, NH_4^+, Na^+, and Cl^- by ion exchange mechanisms. Because of the complex and interconnected functional and structural features of the gill epithelium, it is an excellent model system to examine the effects of various dissolved substances on tissues (Evans, 1987).

A number of earlier studies, mostly qualitative in approach, showed that changes in gill structure were associated with low pH (Daye and Garside, 1976; Matey et al., 1981; Jagoe and Haines, 1983) and dissolved metal exposure (Baker, 1969; Gardner and Yevich, 1970; Skidmore and Tovell, 1972). The responses described in these papers were generally nonspecific, including edema, swelling, necrosis, and proliferation of epithelial cells, but did suggest that gas exchange could be hampered, and that ionoregulatory cells were often involved. However, because these were mostly acute exposures, it was difficult to relate the often-severe morphological damage observed to the physiological responses observed with chronic exposure to more environmentally realistic conditions.

In the 1960s, Weibel and co-workers developed basic morphometric methods and applied them to the study of the diffusing capacity of mammalian lungs (e.g., Weibel and Knight, 1964). Hughes and co-workers adapted and modified these techniques, and applied them to develop quantitative descriptions of the structure of fish gills (Hughes, 1972; Hughes and Perry, 1976). While previous workers had provided some numerical descriptions of gill structure and its response to environmental challenge, these were among the first studies to systematically examine the response of various gill components in a reproducible and quantifiable manner. The techniques developed were aimed towards estimating the diffusing capacity of the gill. They were thus especially suited to studies of low pH and metal exposure which interfered with gas exchange.

Figure 3 is reproduced from Hughes and Perry (1976), and shows a test grid of the Merz type superimposed over secondary lamellae of the fish gill. Secondary lamellae are the major site of gas exchange in the gill, and thus

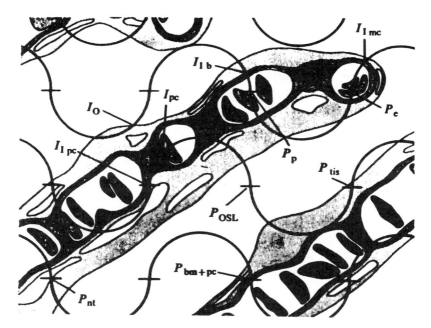

Figure 3   Portion of a Merz grid superimposed on the secondary lamellae of a fish gill. In practice, a finer grid is used which does not cover the entire field. Abbreviations: I, indicates intersections with structures; P, indicates points counted for stereological measures; OSL, outside secondary lamellae; O, outer surface; t, tissue space; nt, non-tissue space; pc, pillar cell; p, plasma; mc, marginal channel; bm, basement membrane; e, erythrocyte. (From Hughes, G.M. and S.F. Perry, 1976. J. Exp. Biol. 64, 447–460. With permission.)

alterations in their structure, especially the blood to water diffusion distance, affects oxygen uptake. Hughes and Perry (1976) subdivided the secondary lamellae into regions, including the interlamellar water space, the epithelial region exterior to the basement membrane, the pillar cell system, erythrocytes, and plasma. Points on the grid intersecting the various portions of the secondary lamellae were counted to calculate P_P, and obtain the volume density for each region. Intersections of the wavy lines of the grid with the outer epithelial surfaces served to randomly select points for the measurement of diffusion distances. From the point where the test line crossed the epithelium, the distance to the outer edge of the nearest erythrocyte was measured. This represented the minimum distance an oxygen molecule would need to diffuse to be taken up by a blood cell. The harmonic mean of a number of these measurements was calculated to estimate the diffusing capacity of the gill. Intersections of the test lines with the epithelial surfaces were also counted to obtain the surface density. Obviously, this parameter is directly related to the surface area of the gill, another critical factor influencing diffusional flux.

Hughes et al. (1979) applied this methodology to study the effects of cadmium, chromium, and nickel on gills. Figure 4A shows a portion of their results, the relative volumes of the various components of the gill lamellae of rainbow trout exposed to 0, 2.0, and 3.2 mg Ni/l for 3.5 d. The area outside the

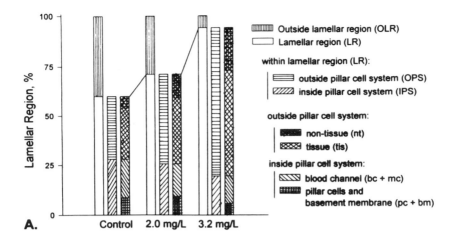

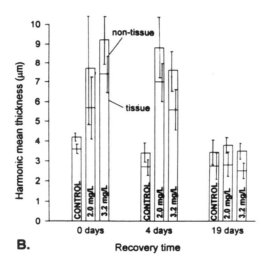

Figure 4   (A) Histograms of the relative volumes of different portions of the lamellar region of gills from fish held in water containing 0 to 3.2 mg Ni/l. Data are expressed as percent of the total volume. The first bar for each concentration shows the proportions of the tissue and interlamellar spaces. The next two bars for each concentration show the relative volumes of the various components within the tissue. (B) Harmonic mean thickness, with standard deviation, of the blood-water diffusion barrier in the gills of fish exposed to nickel. Effects of various concentrations, with recovery times of 0, 4, and 19 d in nickel-free water are shown. (From Hughes et al., 1979. Water Res. 13, 665–679. Copyright permission from Elsevier Science Ltd.)

lamellae, or the water space between the respiratory surfaces, became much smaller with increasing nickel. This was due to an increase in the volume of the epithelium outside the capillary-pillar cell area. Figure 4B, also from their results, shows that these changes in relative volume resulted in a significant increase in the harmonic mean thickness of the blood-water barrier. While

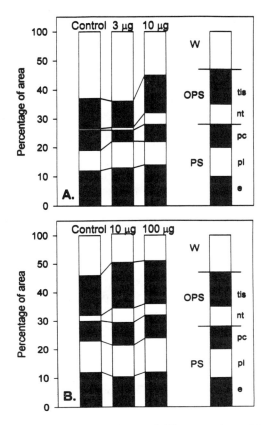

Figure 5 Histograms of the relative volumes of different portions of the lamellar region of gills from fish held in water containing cadmium. (A) Zebrafish, (B) rainbow trout, W, water (interlamellar space); OPS, outside pillar cell system; PS, within pillar cell system; tis, tissue; nt, non-tissue; pc, pillar cell; pl, plasma; e, erythrocytes. (Redrawn from Karlsson-Norrgren et al., 1985.)

these experiments can be criticized because of the high nickel concentrations and short exposure period, they clearly show quantitative changes in the relative volumes of the various tissue compartments consistent with impaired respiratory function.

Several years later, Karlsson-Norrgren et al. (1985) applied this same technique to study longer-term effects of dissolved cadmium on rainbow trout and zebrafish (*Brachydanio rerio*). The latter were exposed to up to 10 μg Cd/l in fresh water, while the trout were exposed to up to 100 μg Cd/l in brackish water for 6 weeks. Their results (Figure 5) showed similar trends to those of Hughes et al. (1979). The interlamellar water spaces decreased with increasing metal concentration, largely due to increases in the nontissue component of the epithelium. This nontissue space was filled with fluid, cellular debris, and myelin bodies, consistent with edema and cellular damage. In the zebrafish, the blood-water diffusional distance was significantly increased at the highest cadmium concentration.

A number of other studies have used modifications of the techniques of Hughes and co-workers to evaluate the effects of various substances on the diffusional capability of gills. Most often, the blood-water diffusional distance has been measured using the Merz grid to randomize the measurement sites. By this technique, Leino et al. (1987) demonstrated that blood-water diffusion distances were greater in fish from an experimentally acidified lake compared to those from a control lake. In laboratory experiments, Evans et al.(1988) found a trend towards increased diffusion distance in rainbow trout exposed to pH 4.7 with 20 μM Al, although the difference from control values was not significant due to the high variance of the measurements. Pinkney et al. (1989) detected thickening of the blood-water diffusion distance with exposure of mummichogs (*Fundulus heteroclitus*) to tributyltin (17.2 $\mu g/l$). Wilson et al. (1994) reported a significant increase in the blood-water diffusion distance in gills of juvenile rainbow trout exposed to pH 5.2 and 38 μg Al/l.

In many cases, the diffusional capability of gills appears to decrease with exposure to metals and low pH. Of course, this would be expected to interfere with gas exchange, and lower the capapacity to extract oxygen from water. However, gills serve as important diffusional areas for ions as well as oxygen. In fresh water, fish are in ionic disequilibrium with their environment: their body fluids are much higher in ionic strength then the water they live in. Freshwater fish must cope with osmotic uptake of water, and diffusional losses of ions as a consequence of this disequilibrium. Normally, gills of freshwater fish are relatively impermeable to ions to reduce passive losses (Evans, 1987). A number of studies have whown that increases in passive ion losses, presumably due to increased permeability, accompany low pH and metal exposure (Wood, 1992).

Earlier in this chapter, the "strategies" an organism might employ to respond to pollutant effects were discussed. In the face of increased diffusional losses of physiologically essential ions, it is clearly best to attempt to reduce the losses. One strategy to accomplish this would be a quantitative one: increase the blood-water diffusion distance, and decrease the gill surface area. This would involve a trade-off of diminished respiratory capacity for decreasing passive ion losses, as recognized by Gonzalez and McDonald (1992). Viewed this way, the increase in epithelial thickness and decrease in surface density characteristic of exposure to many dissolved metals is not necessarily gill "damage," but a response to decrease the toxicant effect, increased efflux of physiologically essential ions. Of course, the nature of the epithelial thickening is important: thickening due to edema with the collection of cellular debris implies damage to the tissue, whereas epithelial hyperplasia implies more of a compensatory response.

Gill ion fluxes also involve an active component. In fresh water, fish actively pump physiologically important ions such as calcium, chloride, and sodium from their environment by specialized cells (chloride cells) in the gills (Laurent and Perry, 1990; Perry et al., 1992). The mechanisms of ionic uptake in chloride cells of freshwater fishes are still not completely understood. In seawater, the mechanism of this active efflux is better understood; the same cell type pumps chloride out, with sodium following down the electrochemical gradient (Silva et al., 1977; Foskett and Scheffly, 1982). Active excretion of salts is necessary because saltwater teleosts have the opposite problem of freshwater dwellers. Their body fluids contain less salt than their environment, so they must get rid of the ions entering at the gills down their electrochemical gradient. Changes in these chloride cells represent another possible strategy of adapting to altered ionic flux patterns with pollutant exposure.

Changing in chloride cell number and structure have been reported in response to a number of pollutants. Leino and McCormick (1984) were among the first to quantitatively demonstrate that chloride cell number increased with exposure to low pH waters. They counted chloride cells along measured lengths of gill filaments in fish exposed to pH levels from 7.5 to 5.0. They also observed changes in chloride cell distribution: normally, these cells are pre-dominantly in the primary lamellar epithelium, but with exposure to low pH water, more are observed on the secondary lamellae of gills. Later studies (e.g., Leino et al., 1987) confirmed these observations in wild populations. Jagoe and Haines (1990) exposed Atlantic salmon to water from low- and high-pH rivers, and to limed water from a low-pH river, and counted and measured gill chloride cells. They observed increased chloride cell numbers with low pH exposure, although numbers of chloride cells on the secondary lamellae did not increase as pH decreased in this experiment.

Although effects of aluminum exposure were not evaluated in these earlier studies, later work showed that alterations in chloride cell numbers were caused by aluminum exposure. Evans et al. (1988) observed that chloride cell numbers decreased with acid exposure, presumably because the cells were accumulating aluminum and being damaged. Tietge et al. (1988), using a point-count method to determine volume density, showed that lamellar chloride cell densities increased as pH declined. Addition of aluminum did not cause an increase of morphologically normal chloride cells, but did cause an increase in "dense" cells, which may be a degenerate or damaged type of chloride cell. Mueller et al. (1991) also used a point-count method to demonstrate that the volume density of lamellar chloride cells decreased with exposure of juvenile brook trout of pH 5.2 with 75 µg Al/l. Recently, Leino and McCormick (1993) exposed young largemouth bass (*Micropterus salmoides*) to low pH (4.5) and slightly elevated aluminum (30 µg Al/l) for 113 d in simulated overwintering conditions. Fish held at low pH with aluminum had significantly lower num-bers of gill chloride cells. Interestingly, temperature affected chloride cell number as well: fish held at lower temperatures generally had more chloride cells than those at higher temperature.

A number of other metals have also been found to affect chloride cell number in freshwater fish. Many qualitative studies have reported increased chloride cell number with exposure to metals including cadmium and zinc (see Mallat, 1985; Evans, 1987). There are, however, relatively few quantitative studies. As an example, Oronsaye and Brafield (1984) counted the numbers of chloride cells of gills of stickleback (*Gasterosteus aculeatus*) exposed to up to 6μg Cd/l. Chloride cell numbers increased in all cadmium treatments, but later declined over time at the highest cadmium concentration.

Increased chloride cell number is another strategy for coping with increased ionic efflux due to low pH and metal exposure. This represents a quantitative response, increasing the number of ion-pumping cells to offset increased diffusional losses. Because ion transport via chloride cells is an energy-requiring process involving ATPase enzymes, this strategy will entail a metabolic cost. As Calow (1991) has pointed out, this will affect the budgets of other energy-requiring processes, such as growth and reproduction.

In addition to increasing chloride cell number, it may also be possible to increase chloride cell activity. Increased activity may be reflected in increased cell size. Chloride cell hypertrophy has been reported in response to a number of pollutants in qualitative studies (see Mallat, 1985; Evans, 1987). It is important to note that certain quantitative techniques cannot distinguish between hyperplasia, hypertrophy, or some combination of the two. For example, point-counting methods applied to entire tissue, as in the gills examined by Tietge et al. (1988), measure the fraction of the tissue occupied by a particular cell type. An increase in the volume fraction can by due to an increase in cell size, number, or both, but the relative contributions of hyperplasia or hypertrophy cannot be distinguished without counting individual cells, or measuring cell diameter. Jagoe and Haines (1990) measured profile areas of chloride cells in gills of Atlantic salmon exposed to low pH and limed water, and found that chloride cells were significantly larger at low pH.

Changes have been reported in the ultrastructural components of chloride cells upon low pH or metal exposure consistent with altered activity. A number of qualitative studies have reported altered chloride cell morphology with low pH and metal exposure (Chevalier et al., 1985; Karlsson-Norrgren et al., 1985; Holm et al., 1991). To my knowledge, none have quantified changes associated with energy production, such as increasing volume density of mitochondria, or with ATPase enzyme activity, such as volume or surface density of the tubular reticulum in response to pollutant exposure. However, several studies have reported increased frequencies of apical pits or crypts in chloride cells with low pH and metal exposure (Leino and McCormick, 1984; Leino et al., 1987; Jagoe et al., 1993. These crypts are typically found in seawater-acclimated fish, and are thought to reflect increased ion transport activity when present in freshwater fish. Chloride cell surface area is probably also related to transport activity, and a number of studies have quantitatively demonstrated increased apical areas due to stressors affecting ion regulation or acid-base balance (Laurent and Hebibi, 1989; Goss et al., 1994). Ultrastructural studies to measure features of the epithelium associated with passive diffusional fluxes may also be useful. Freda et al. (1991) measured the depth of the tight junctions between epithelial cells which normally seal the gill surface, and demonstrated that these junctions shortened as pH was lowered and sodium efflux increased.

In addition to chloride and pavement cells, mucous cells are an important cell type in gill epithelia. The mucous coating of the gill has been implicated in osmoregulatory function (Marshall, 1978), and the sequestration and removal of irritants, including metals (Playle and Wood, 1991). Mallat (1985) summarized a number of qualitative studies reporting increased mucous secretion in response to metals, including cadmium, copper, mercury, and zinc. Zuchelkowski et al. (1981) showed that the volume density of mucous cells increased in the skin of acid-stress catfish. Interestingly, this study also noted sexual dimorphism in the response; in females, the increase in volume density was primarily due to hypertrophy, while in males, the increase was due to both hyperplasia and hypertrophy. Mueller et al. (1991) observed that mucous cell volume density nearly doubled when juvenile brook trout were exposed to pH 5.2 with added aluminum (75 µg Al/l). Jagoe and Haines (1990) observed increases in both mucous cell number and size when Atlantic salmon were exposed to acidic river water. However, Leino and McCormick (1993) and Leino et al. (1987) did not observe changes in mucous cell numbers with acid or aluminum exposure.

V. CONCLUSION

Although qualitative approaches to studies of pathologies caused by environmental pollutants are useful in many circumstances, the utility of quantitative methods is generally underappreciated. Quantitative results can often provide a much clearer picture of the ways in which organisms respond to stress. A number of relatively simple techniques are available to extract quantitative information from histological preparations. These allow the detection of more subtle effects, and better estimation of the relation between dose and response than purely descriptive approaches. Whereas other disciplines studying effects at the suborganismal level such as biochemistry and physiology have strong quantitative traditions, this mindset is often lacking in morphologists. By this measure, histopathology is not as well developed as a predictive science, or as capable of rigorous hypothesis testing as these other disciplines. However, the utility of histopathology as a tool to detect and understand pollutant effects is unquestionable. The further refinement and application of quantitative techniques can only help improve our comprehension of responses to pollutants at the cell and tissue levels.

ACKNOWLEDGMENTS

I am most grateful to an anonymous reviewer for suggestions that improved this manuscript. This work was supported by contract DE-AC09-76SR00-819 between the U.S. Department of Energy and the University of Georgia's Savannah River Ecology Laboratory.

REFERENCES

Adams, S.M. and M.S. Greeley, 1991. Assessment and evaluation of the ecological health of fish populations exposed to PCBs in Hartwell reservoir. Tennesee Valley Authority Report TVA/WA/AB 91–14. Tennessee Valley Authority, Chattanooga, TN.

Adams, S.M., A.M., Brown and R.W. Geode, 1993. A quantitative health assessment index for rapid evaluation of fish condition in the field. Trans. Am. Fish. Soc. 122, 63–73.

Adams, S.M. (Ed.), 1990. Biological Indicators of Stress in Fish, Symposium 8. American Fisheries Society, Bethesda, MD>

Alberts, B., D. Bray, J. Lewis, M. Raff, K. Roberts and J.D. Watson, 1983. Molecular Biology of the Cell Garland, New York.

Baker, J.T.P., 1969. Histological and electron microscopical observations on copper poisoning in the winter flounder (*Pseudopleuronectes americanus*). J. Fish. Res. Board Can. 26, 2785–2793.

Baker, J.T.P. and C.L. Schofield, 1985. Acidification impacts on fish populations: a review. In: Acid Deposition: Environmental, Economic and Policy Issues, edited by D.P. Adams and W.P. Page, Plenum Press, New York, pp. 183–221.

Baumann, P.C., W.D. Smith and W.K. Parland, 1987. Tumor frequencies and contaminant concentrations in brown bullheads from an industrialized river and a recreational lake. Trans. Am. Fish. Soc. 116, 79–86.

Baumann, P.C., M.J. Mac, S.B. Smith and J.C. Harshbarger, 1991. Tumor frequencies in walleye (*Stizostedion vitreum*) and brown bullhead (*Ictalurus nebulosus*) and sediment contaminants in tributaries of the Laurentian Great Lakes. Can. J. Fish. Aquat. Sci. 48, 1804–1810.

Bolender, R.P., 1979. Morphometric analysis in the assessment of the response of the liver to drugs. Pharmaco. Rev. 30, 429–443.

Bookstein, F., 1991. Morphometric Tools for Landmark Data, Geometry and Biology. Cambridge University Press, Cambridge.

Booth, C.E., D.G., McDonald, B.P. Simons and C.M. Wood, 1988. Effects of aluminum and low pH on net ion fluxes and ion balance in the brook trout (*Salvelinus fontinalis*). Can. J. Fish. Aquat. Sci. 45, 1563–1574.

Calow, P., 1991. Physiological costs of combating chemical toxicants: ecological implications. Comp. Biochem. Physiol. 100C 3–6.

Chalkey, H.W., J. Cornfield and H. Park, 1949. A method for estimating volume-surface ratios. Science 110, 295–297.

Chevalier, G., L. Gauthier and G. Moreau, 1985. Histopathological and electron microscopic studies of gills of brook trout, *Salvelinus fontinalis,* from acidified lakes. Can. J. Zool. 63, 2062–2070.

Conti, E., 1987. Acute toxicity of two detergents and three insecticides to the lugworm *Arenicola marina* (L.): a histological and scanning electron microscopic study. Aquat. Toxicol. 10, 324–334.

Couillard, C.M., R.A. Berman and J.C. Panisset, 1988. Histopathology of rainbow trout exposed to a bleached kraft pulp mill effluent. Arch. Environ. Contam. Toxicol. 17, 319–323.

Cronan, C.S. and C.L. Schofield, 1979. Aluminum leaching response to acid precipitation: effects on high elevation watersheds in the northeast. Science 204, 304–306.

Daoust, P.Y., G. Wobeser and J.D. Newstead, 1984. Acute pathological effects of inorganic mercuty and copper in gills of rainbow trout. Vet. Pathol. 21, 93–101.

Daye, P.G. and E.T. Garside, 1976. Histopathologic changes in surficial tissues of brook trouyt *Salvelinus fontinalis* (Mitchill) exposed to acute and chronic levels of pH. Can. J. Zool. 54, 2140–2155.

DeHoff, R.T. and F.N. Rhines, 1968. Quantitative Microscopy. McGraw-Hill, New York.

DiMichele, L. and M.H. Taylor, 1978. Histopathological and physiological responses of *Fundulus heteroclitus* to naphthalene exposure. J. Fish. Res. Board Can. 35, 1060–1066.

Driscoll, C.T., J.P. Baker, J.J. Bisogni and C.L. Schofield, 1980. Effect of aluminum speciation on fish in dilute acidified waters. Nature 284, 161–164.

Egginton, S., 1988. Effect of inter-animal variation on a nested sampling design for stereological analysis of skeletal muscle. Acta Stereol. 7, 81–89.

Elias, H. and D.M. Hyde, 1983. A Guide to Practical Stereology. S. Karger, New York,

Evans, D.H., 1987. The fish gill: site of action and model for toxic effects of environmental pollutants. Environ. Health Perspect. 71, 47–58.

Evans, R.E., S.B. Brown and I.J. Hara, 1988. The effects of aluminum and acid on the gill morphology in rainbow trout, *Salmo gairdneri*. Environ. Biol. Fishes 22, 299–311.

Ferguson, H.W., 1989. Systematic Pathology of Fish. Iowa State University Press, Ames, IA.

Forbus, W.D., 1952. Reaction to Injury: Pathology for Students of Disease. Williams & Wilkins, Baltimore, MD.

Foskett, J.K. and C. Scheffly, 1982. The chloride cell: definitive identification as the salt secretory cell in teleosts. Science 215, 164–166.

Freda, J., D.A. Sanchez and H.L. Bergman, 1991. Shortening of branchial tight junctions in acid-exposed rainbow trout (*Oncorhynchus myukiss*). Can. J. Fish. Aquat. Sci. 48, 2028–2033.

Gardner, G.R. and P.P. Yevich, 1970. Histological and hematological responses of an estuarine teleost to cadmium. J. Fish. Res. Board Can. 27, 2185–2196.

Gardner, G.R., 1975. Chemically induced lesions in estuarine or marine teleosts. In: The Pathology of Fishes, edited by W.C. Ribelin and G. Migaki, University of Wisconsin Press, Madison, WI, pp. 657–694.

Gonzalez, R.J. and D.G. McDonald, 1992. The relationship between oxygen consumption and ion loss in a freshwater fish. J. Exp. Biol. 163, 317–332.

Goss, G.G., C.M. Wood, P. Laurent and S.F. Perry, 1994. Morphological responses of the rainbow trout (*Onchorynchus mykiss*) gill to hyperoxia, base ($NaHCO_3$) and acid (CHl) infusions. Fish Physiol. Biochem. 12, 465–477.

Grizzle, J.M. and W.A. Rogers, 1976. Anatomy and Histology of the Channel Catfish. Auburn Press, Auburn, AL.

Gundersen, H.J.G. and R. Osterby, 1981. Optimizing sampling efficiency of stereological structures in biology: or 'Do more less well'. J. Microsc. 121, 65–73.

Gundersen, H.J.G., P. Bagger, T.F. Bendtsen, S.M. Evans, L. Korbo, N. Marcussen, A. Moller, K. Nielsen, J.R. Nyengaard, B. Pakkenberg, F.B. Sorensen, A. Besterby and M.J. West, 1988a. The new stereological tools: disector, fractor nucleator and point sampled intercepts and their use in pathological research and diagnosis. Apmis 96, 857–881.

Gundersen, H.J.G., T.F. Bendtsen, L. Korbo, N. Marcussen, A. Moller, K. Nielsen, J.R. Nyengaard, B. Pakkenberg, F.B. Sorensen, A. Vesterby and M.J. West, 1988b. Some new, simple and efficient sterological methods and their use in pathological research and diagnosis. Apmis 96, 379–394.

Gundersen, H.J.G., O. Gotzsche, E.B. Jensen and R. Osterby, 1988c. Designing efficient sampling schemes: automatic, semiautomatic or manual image analysis for
stereological studies in biology? In: Stereology and Morphometry in Electron
Microscopy, Problems and Solutions, edited by A. Reith and T.M. Mayhew,
Hemisphere, New York. pp. 15–21.

Gupta, M., T.M. Mayhew, K.S. Bedi, A.K. Sharma and F.H. White, 1983. Inter-animal
vairation and its influence on the overall precision of morphometric estimates
based on nested sampling designs. J. Microsc. 131, 147–154.

Haines, T.A., 1981. Acidic preciptiation and its consequences for aquatic ecosystems:
a review. Trans. Am. Fish. Soc. 110, 669–707.

Hayes, M.A., I.R. Smith, T.H. Rushmore, T.L. CRane, C. Thorn, T.E. Kocal and H.W.
Ferguson, 1990. Pathogenesis of skin and liver neoplasms in white suckers from
industrially polluted areas in Lake Ontario. Sci. Total Environ. 94, 105–123.

Hinton, D.E., J.E. Klauni and M.M. Lipsky, 1978. PCB-induced alterations in trout
liver: a model for environmental disease in fish. Mar. Fish. Rev. 40, 47–50.

Hinton, D.E., R.C. Lantz and J.A. Hampton, 1984. Effect of age and exposure to a
carcinogen on the structure of medaka liver: a morphometric study. Natl. Cancer
Inst. Monogr. 65, 239–249.

Hinton, D.E., R.C. Lantz, J.A. Hampton, P.R. McCuskey, and R.S. McCuskey, 1987.
Normal versus abnormal structure: considerations in morphologic responses of
teleosts to pollutants. Environ. Health Perspect. 71, 139–146.

Hinton, D.E., J.A. Couch, S.J. Teh and L.A. Courtney, 1988a. Cytological changes
during progression of neoplasia in selected fish species. Aquat. Toxicol. 11, 77–
112.

Hinton, D.E., D.J. Lauren, S.J. Teh and C.S. Giam, 1988b. Cellular composition and
ultrastructure of hepatic neoplasms induced by diethylnitrosamine in Oryzias
latipes. Mar. Environ. Res. 24, 307–310.

Hinton, D.E. and D.J. Lauren, 1990. Liver Structural Alterations Accompanying Chronic
Toxicity in Fishes: Potential Biomarkers of Exposure. Lewis Publishers, Boca
Raton, FL.

Hinton, D.E. and D.J. Lauren, 1990. Integrative histopathological approaches to detecting effects of environmental stressors on fish. In: Biological Indicators of Stress
in Fish, Symposium 8, edited by S.M. Adams, American Fisheries Society, Bethesda,
MD, pp. 51–66.

Hinton, D.E., P.C. Baumann, G.R. Gardner, W.E. Hawkins, J.D. Hendricks, R.A.
Murchelano and M.S. Okihiro, 1992. Histopathological biomarkers. In: Biomarkers:
Biochemical, Physiological and Histological Makres of Anthropogenic Stress,
edited by R.J. Huggett, R.A. Kimerle, P.M. Mehrle and H.L. Bergman, Lewis
Publishers, Boca Raton, FL, pp. 155–209.

Hinton, D.E., 1993. Toxicologic histopathology of fishes: a systematic approach and
overview. In: Pathobiology of Marine and Estuarine Organisms, edited by J.A.
Couch and J.A. Fournie, CRC Press, Boca Raton, FL, pp. 177–215.

Hochachka, P.W. and G.N. Somero, 1973. Strategies of Biochemical Adaptation. W.B.
Saunders, Philadelphia.

Holm, G., L. Norrgren and O. Linden, 1991. Reproductive and histopathological
effects of long-term experimental exposure to bis(tributyl)tin oxide (TBTO) on
the three spined stickleback Gasterosteus aculeatus (Linnaeus). J. Fish Biol. 38,
373–386.

Hughes, G.M., 1972. Morphometrics of fish gills. Respir. Physiol. 14, 1–25.

Hughes, G.M. and S.F. Perry, 1976. Morphometric study of trout gills: a light-microscopic method suitable for the evaluation of pollutant action. J. Exp. Biol. 64, 447–460.

Hughes, G.M., S.M. Perry and V.M. Brown, 1979. A morphometric study of the effects of nickel, chromium and cadmium on the secondary lamellae of rainbow trout gills. Water Res. 13, 665–679.

Jagoe, C.H. and T.A. Haines, 1983. Alterations in gill epithelial morphology of yearling Sunapee trout exposed to acute acid stress. Trans. Am. Fish Soc. 112, 689–695.

Jagoe, C.H. and T.A. Haines, 1990. Morphometric effects of low pH and limed water on the gills of Atlantic salmon (*Salmo salar*). Can. J. Fish. Aquat. Sci. 47, 2451–2460.

Jagoe, C.H., V.E. Matey, T.A. Haines and V.T. Komov, 1993. Effect of beryllium on fish in acid water is analogous to aluminum toxicity. Aquat. Toxicol. 24, 241–256.

Johnson, L.L., E. Castilias, T.K. Collier, B.B. McCain and U. Varanasi, 1988. Contaminant effects on ovarian development in English sole (*Parophrys vetulus*) from Puget Sound, Washington. Can. J. Fish. Aquat. Sci. 45, 2133–2146.

Karlsson-Norrgren, L., P. Runn, C. Haux and L. Forlin, 1985. Cadmium-indiced changes in gill morphometry of zebrafish *Brachydanio rerio* (Hamilton-Buchanan) and rainbow trout *Salmo gairdneri* Richardson. J. Fish Biol. 27, 81–95.

Khan, R.A., D.E. Barker, R. Hooper, E.M. Lee, K. Ryan and K. Nag, 1994. Histopathology in winter flounder *Pleuronectes americanus* living adjacent to a pulp and paper mill. Arch. Environ. Contam. Toxicol. 26, 95–102.

Klaunig, J.E., M.M. Lipsky, B. Trump and D.E. Hinton, 1979. Biochemical and ultrastructural changes in teleost liver following subacute exposure to PCB. J. Exp. Pathol. Toxicol. 2, 953–963.

Lanno, R.P. B. Hicks and J.W. Hilton, 1987. Histological observations on interhepatocytic copper-containing granules in rainbow trout reared on diets containing elevated amounts of copper. Aquat. Toxicol. 10, 251–263.

Laurent, P. and N. Hebibi, 1989. Gill morphometry and fish osmoregulation. Can. J. Zool. 67, 3055–3063.

Laurent, P. and S.F. Perry, 1990. Effects of cortisol on gill chloride cell morphology and ionic uptake in the freshwater trout, *Salmo gairdneri*. Cell Tissue Res. 259, 429–442.

Leino, R.L. and J.H. McCormick, 1994. Morphological and morphometrical changes in chloride cells of the gills of *Pimephales promelas* after chronic exposure to acid water. Cell Tissue Res. 236, 121–128.

Leino, R.L., P. Wilkinson and J.G. Anderson, 1987. Histopathological changes in the gills of pearl dace *Semotilus margarita* and fathead minnows *Pimephales promelas* from experimentally acidified Canadian lakes. Can. J. Fish. Aquat. Sci. 44, 126–143.

Leino, R.L., J.H. McCormick and K.M. Jensen, 1990. Multiple effects of acid and aluminum on brood stick and progeny of fathead minnows, with emphasis on histopathology. Can. J. Zool. 68, 234–244.

Leino, R.L. and J.H. McCormick, 1993. Responses of juvenile largemouth bass to different pH and aluminum levels at overwintering temperatures: effects on gill morphology, electrolyte balance, scale calcium, liver glycogen and depot fat. Can. J. Zool. 71, 531–543.

Mailins, D.C., B.B. McCain, D.W. Brown, S.L. Chan, M.S. Myers, J.T. Landahl, P.G. Prohaska, A. J. Friedman, L.D. Rhodes, D.G. Burrows, W.D. Gronlund and H.O. Hodgins, 1984. Chemical pollutants in sediments and diseases in bottom dwelling fish in Puget Sound, Washington. Environ. Sci. Technol. 18, 705–713.

Mallat, J., 1985. Fish gill structural changes induced by toxicants and other irritants: a statistical review. Can. J. Fish. Aquat. Sci. 42, 630–648.

Marshall, W.S., 1978. On the involvement of mucus secretion in teleost osmoregulation. Can. J. Zool. 56, 1088–1091.

Matey, V.Y., A.D. Kharazova and G.A. Vinogradov, 1981. Response of Gasterosteus aculeatus L. gill epithelium chloride cells to changes in pH and salinity in the environment. Tsitologiya 23, 159–165.

Mathieu, O., L.M. Cruz-Orive, H. Hoppeler and E.R. Weibel, 1981. Measuring error and sampling variation in stereology: comparison of the efficiency of various methods for planar image analysis. J. Microsc. 121, 75–88.

Mayhew, T.M., 1982. Towards economy of effort in quantitative ultrastructural pathology: efficient sampling schemes for studying experimental carcinogenesis. J. Pathol. 138, 179–191.

McCarthy, J.F. and L.R. Shugart, 1990. Biological markers of environmental contamination. In: Biomarkers of Environmental Contamination, edited by J.F. McCarthy and L.R. Shugart, Lewis Publishers, Boca Raton, FL pp. 3–14.

McCormick, J.H., G.N. Stokes and R.O. Hermanatz, 1989. Oocyte atresia and reproductive success in fathead minnows (Pimephales promelas) exposed to acidified hardwater environments. Arch. Environ. Contam. Toxicol. 18, 207–214.

McDonald, D.G., 1983. The effects of H^+ upon the gills of freshwater fish. Can. J. Zool. 61, 691–703.

Meyers, T.R. and J.D. Hendricks, 1984. Histopathology. In: Fundamentals of Aquatic Toxicology, edited by G.M. Rand and S.R. Petrocelli, Hemisphere, Washington, D.C.

Mierle, G., 1990. Aqueous inputs of mercury to precambrian shield lakes in Ontario. Environ. Toxicol. Chem. 9, 843-851.

Mills, K.H., S.M. Chalanchuk, L.C. Mohr and I.J. Davies, 1987. Responses of fish populations in Lake 223 to 8 years of experimental acidification. Can. J. Fish. Aquat. Sci. 44, 114–125.

Mueller, M.E., D.A. Sanchez, H.L. Bergman, D.G. McDonald, R.G. Rhem and C.M. Wood, 1991. Nature and time course of acclimation to aluminum in juvenile brook trout (Salvelinus fontinalis). II. Gill histology. Can. J. Fish. Aquat. Sci. 48, 2016-2027.

Muniz, I.P. and H. Leivestad, 1980. Acidification-effects on freshwater fish. In: Ecological Impact of Acid Precipitation, edited by D. Drablos and A. Tollan, SNSF Project, Aas, Norway, pp. 84–92.

Myers, M.S., C.M. Stehr, O.P. Olson, L.L. Johnson, B.B. McCain, S.L. Chan and U. Varanasi, 1994. Relationships between toxicopathic hepatic lesions and exposure to chemical contaminants in English sole (Pleuronectes vetulus), starry flounder (Platichthys stellatus), and white croaker (Genyonemus lineatus) from selected marine sites on the Pacific coast, USA. Environ. Health Perspect. 102, 200-215.

Myking, A.O., 1988. Studies on the volumetric composition of lymph nodes: problems of efficient sampling and the use of point counting versus digitizer tablets. In: Stereology and Morphometry in Electron Microscopy: Problems and Solutions, edited by A. Reith and T.M. Mayhew, Hemisphere, New York, pp. 47-62.

Oronsaye, J.A.O. and A.E. Brafield, 1984. The effects of dissolved cadmium on the chloride cells of the gills of the stickleback, Gasterosteus aculeatus L. J. Fish Biol. 25, 253-258.

Overstreet, R.M., 1988. Aquatic pollution problems, southeastern U.S. coasts: histopathological indicators. Aquat. Toxicol. 11, 213-239.

Perry, S.F., G.G. Goss and P. Laurent, 1992. The interrelationships between gill chloride cell morphology and ionic uptake in four freshwater teleosts. Can. J. Zool. 70, 1775-1786.

Pinkney, A.E., D.A. Wright and G.M. Hughes, 1989. A morphometric study of the effects of tributyltin compounds on the gills of the mummichog (*Fundulus heteroclitus*). J. Fish Biol. 34, 665-677.

Playle, R.C., G.G. Goss and C.M. Wood, 1989. Physiological disturbances in rainbow trout (*Salmo gairdneri*) during acid and aluminum exposures in soft water of two calcium concentrations. Can. J. Zool. 67, 314-324.

Playle, R.C. and C.M. Wood, 1991. Mechanisms of aluminium extraction and accumulation at the gills of rainbow trout, *Oncorhynchus mykiss* (Walbaum), in acidic soft water. J. Fish Biol. 38, 791-805.

Reith, A. and T.M. Mayhew, 1988. Stereology and Morphometry in Electron Microscopy: Problems and Solutions. Hemisphere, New York.

Rigaut, J.P., 1988. Analyzing electron microscopic images by computer: a guided tour. In: Stereology and Morphometry in Electron Microscopy, Problems and Solutions, edited by A. Reith and T.M. Mayhew, Hemisphere, New York, pp. 161-192.

Rojik, I., J. Nemcsok and L. Boross, 1983. Morphological and biochemical studies on liver, kidney and gill of fishes affected by pesticides. Acta Biol. Hung. 34, 81-92.

SAS, 1990. SAS/Stat user's guide. SAS Institute, Cary, NC.

Shay, J., 1975. Economy of effort in electron microscope morphometry. Am. J. Pathol. 81, 503-511.

Silva, P., R. Solomon, K. Spokes and F.H. Epstein, 1977. Ouabain inhibition of gill Na, K ATP-ase: relationship to active chloride transport. J. Exp. Zool. 199, 419-426.

Sindermann, C.J., 1979. Pollution-associated diseases and abnormalities of fish and shellfish. Fish. Bull. 76, 717-749.

Sindermann, C.J., 1990. Principal diseases of marine fish and shellfish, second edition. Academic Press, New York.

Skidmore, J.F. and P.W.A. Tovell, 1972. Toxic effects of zinc sulphate on the gills of rainbow trout. Water Res. 6, 217-230.

Smith, C.E., T.H. Peck, R.J. Klauda and J.B. McLaren, 1979. Hepatomas in tomcod (*Microgadus tomcod*)(Walbaum) collected in the Hudson River estuary in New York. J. Fish Dis. 2, 313-319.

Sokal, R.R. and F.J. Rohlf, 1981. Biometry. W.H. Freeman, San Francisco.

Solangi, M.A. and R.M. Overstreet, 1982. Histopathological changes in two estuarine fishes, *Menidia beryllina* (Cope) and *Trinectes maculatus* (Bloch and Schneider), exposed to crude oil and its water-soluble fractions. J. Fish Dis. 5, 13-35.

Stennes, E., 1990. Lead, cadmium and other metals in Scandinavian surface waters, with emphasis on acidification and atmospheric deposition. Environ. Toxicol. Chem. 9, 825-831.

Stott, G.G., N.H. McArthur, R. Tarpley, V. Jacobs and R.S. Sis, 1981. Histopathologic survey of ovaries of fish from petroleum production and control sites in the Gulf of Mexico. J. Fish Biol. 18, 261-269.

Tietge, J.E., R.D. Johnson and H.L. Bergman, 1988. Morphometric changes in gill secondary lamellae of brook trout (*Salvelinus fontinalis*) after long-term exposure to acid and aluminum. Can. J. Fish. Aquat. Sci. 45, 1643-1648.

Underwood, E.E., 1970. Quantitative Stereology. Addison-Wesley, Reading, MA.

Vesely, J., P. Benes and K. Sevcik, 1989. Occurrence and speciation of beryllium in acidified freshwaters. Water Res. 23, 711-717.

Weibel, E.R. and B.W. Knight, 1964. A morphometric study on the thickness of the pulmonary air-blood barrier. J. Cell Biol. 21, 367-384.

Weibel, E.W., G.S. Kistler and W.F. Scherle, 1966. Practical methods for morphometric cytology. J. Cell Biol. 30, 23-38.

Weibel, E.R., 1979. Stereological Methods, Vol. 1, Practical Methods for Biological Morphometry. Academic Press, New York.

Wilber, C.G., 1980. Beryllium, a potential environmental contaminant. C.T. Thomas, Springfield, IL.

Wilson, R.W., H.L. Bergman and C.M. Wood, 1994. Metabolic costs and physiological consequences of acclimation to aluminum in juvenile rainbow trout (*Oncorhynchus mykiss*). II: Gill morphology, swimming performance and aerobic scope. Can. J. Fish. Aquat. Sci. 51, 536-544.

Winfrey, M.R. and J.W.M. Rudd, 1990. Environmental factors affecting the formation of methylmercury in low pH lakes. Environ. Toxicol. Chem. 9, 853-869.

Witters, H.E., 1986. Acute acid exposure of rainbow trout, *Salmo gairdneri* Richardson: effects of aluminium and calcium on ion balance and hematology. Aquat. Toxicol. 8, 197-210.

Wood, C.M., R.C. Playle, B.P. Simons, G.G. Goss and D.G. McDonald, 1988. Blood gases, acid base status, ions and hematology in adult brook trout (*Salvelinus fontinalis*) under acid/aluminum exposure. Can. J. Fish. Aquat. Sci. 45, 1575-1586.

Wood, C.M., 1992. Flux measurements as indices of H^+ and metal effects on freshwater fish. Aquat. Toxicol. 22, 239-264.

Yasutake, W.T. and J.H. Wales, 1983. Microscopic anatomy of salmonids: an atlas. Resource Publication 150. U.S. Fish and Wildlife Service, Washington, D.C.

Zuchelkowski, E.M., R.C. Lantz and D.E. Hinton, 1981. Effects of acid-stress on epidermal mucous cells of the brown bullhead *Ictalurus nebulosus* (LeSeur): a morphometric study. Anat. Rec. 200, 33-39.

CHAPTER 7

Effects of Pollutants on Individual Life Histories and Population Growth Rates

Richard M. Sibly

I. INTRODUCTION

This chapter examines the effects of pollutants on individual organisms and begins to consider the implications for populations and gene pools. It takes some account of the mechanisms and constraints operating at various biochemical and physiological levels, as described in the preceding chapters, and begins to explore the ramifications at higher levels, described in detail in succeeding chapters. The translation from a pollutant's effects on individuals to its effects on the population can be accomplished using life-history analysis to calculate the effect on the population's growth rate. This area has been the subject of some experimental interest since the early 1980s and is reviewed in detail in the next section. A full understanding of a pollutant's effects on a population can be gained using the methods of population ecology, and these are outlined in the third section. In the final section, I consider the effects of chronic pollution on the gene pool, identifying evolutionary outcomes again using life-history analysis.

The key concept in this chapter is that of population growth rate. It is used to analyze population dynamics in Section III, to identify evolutionary outcomes in Section IV, and to evaluate a pollutant's mortality and production effects in Section II. Although this chapter is perhaps unusual in its emphasis on the integrating role of population growth rate, the importance of population growth rate is widely recognized. Population growth rate is used, for instance, to compare spray efficacies of pesticides in the field: examples include Rieske and Raffa's (1990) study of the effects of lindane on weevil pests of Christmas tree farms, Oomen's (1982) study of the scarlet mite, which is a pest of tea in Indonesia, and Swezey and Cano-Vesquez's (1991) study of citrus blackfly in Nicaragua. Population growth rate has been used to assess the effects of

1-56670-1127-9/95/$0.00+$.50

pollution, for example, on sea birds along the Dutch coast (Nolet, 1988). Population growth rate is also widely used to evaluate pesticide effects in simulation models, for instance of budworm-forest management in Maine (Fleming and Shoemaker, 1992), in modeling the evolution of pesticide resistance (Taylor and Georghiou, 1979; Shaw, 1993), and in evaluating the effects of different regimes of pesticide application (Inoue and Ohgushi, 1976; Chi, 1990; Iwasa and Mochizuki, 1988).

The importance of invertebrate life tables in assessing pollution has been emphasized by Schindler (1987), who regarded them as the most sensitive early indicators of stress in ecosystems. He used them to show that primary production is reduced at an earlier stage of air pollution stress in terrestrial ecosystems than in aquatic systems, and Bakhetia (1986) saw them as a necessary ingredient in the management of pests of rapeseed and mustard crops in India.

This chapter overlaps Chapter 9 in linking the individual and population levels using life tables. Data and analytical results relevant to one chapter are, therefore, often relevant to the other. The chapters differ in the details of their methodologies and in emphasis and coverage. Chapter 9 uses projection matrix methods, the development and application of which in population biology are largely due to Caswell, expounded in his 1989 book. Here, continuous time methods are used.

II. EFFECTS OF POLLUTANTS ON INDIVIDUAL ORGANISMS, AND THE CONSEQUENCES FOR POPULATION GROWTH RATE

Pollutants may damage organisms with immediate lethal consequences, as discussed in Chapter 8. The effect on the life cycle is an increase in the per capita mortality rate of at least one age class. Alternatively there may be damage to, or effects on, the machinery of resource acquisition and uptake, with sublethal effect, as discussed in Chapters 3, 5, 6, and 11. In this case "production rate" is reduced, with consequent reduction of birth rate and/or somatic growth rate. However, production rate may also be affected when pollutants have no direct effects as a result of effective detoxification mechanisms, because energy used in detoxification is not available for production. The effectiveness of such mechanisms is discussed in the next section. It seems, then, that either damage or detoxification may result in reduced production in a polluted environment.

The effect of pollution on production is usually measured by its effect on "scope for growth," defined as the difference between energy intake and total metabolic losses (Warren and Davis, 1967; Widdows and Donkin, 1992; Figure 1). An example showing the effects of tributyltin (TBT) concentration on scope for growth in *Mytilus edulis* is given in Figure 2. Note that above a threshold of about 3 µg/g, scope for growth declines as TBT concentration increases, indicating a loss of production. In the field, this decline could

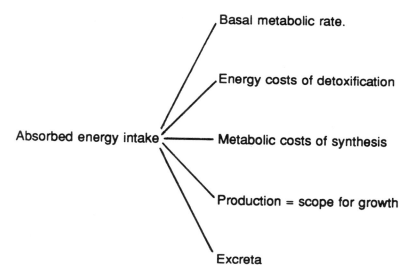

Figure 1   Energy/nutrient allocation diagram illustrating the definition of "scope for growth."

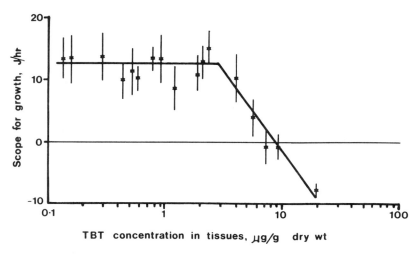

Figure 2   Effect of TBT (tributylin) on scope for growth in *Mytilus edulis*. (From Widdows J. and P. Donkin, 1992. In: The Mussel Mytilus: Ecology, Physiology, Genetics and Culture, edited by E. Gosling, Elsevier Amsterdam, pp. 383–424. With permission.)

translate into a lower abundance of animals (see next section). There is here no effect of TBT on scope for growth at low levels of TBT. However, in the case of essential nutrients (e.g., some metals), there is a decrease in scope for growth at very low levels of the nutrient. Scope for growth has been particularly useful in assessing the effects of pollution on aquatic animals (e.g., *Gammarus*, Maltby, et al., 1990; fish, Crossland, 1988).

The effects of pollutants on the mortality rates or production rates of individual organisms are readily measured, at least in some cases. Such measurements are usually made with population density fixed, generally at a low level. The production rates at different ages determine the organism's somatic growth rate and birth rate, and with mortality rates, provide a complete description of the organism's life history. If we know all the effects of a pollutant on the life history then we can calculate the effects on population growth rate, at the population density at which the measurements were made.

Specifically, we need to know how the pollutant alters somatic growth, birth, and death rates, since these together determine survival to the ages at which individual reproduction occurs, and the birth rates then achieved. These ages, t_i, birth rates, n_i, and survivorships, l_i, together determine population growth rate, r, from the Euler-Lotka equation

$$1 = \frac{1}{2} \sum_{i=1}^{\Omega} n_i l_i e^{-rt_i} \tag{1}$$

where i indexes breeding attempts, i = 1, 2, ... Ω, and Ω is the age at which the last breeding attempt occurs, if the organism lives that long. The ages, birth rates, and survivorships are traditionally tabulated in a life table, from which population growth rate is calculated using Equation 1. Several studies have measured the effects of pollutants on life table components, and then calculated the effects on population growth rates (Table 1).

Knowing how a pollutant alters the life history, it should be possible to discover from the life table whether the main effects of the pollutant on the population are caused by the pollutant's effects on death rate or on production (= somatic growth and birth rate). This is important to know because otherwise it is not possible to extrapolate reliably to the field from tests measuring effects on death rate or scope for growth in the laboratory. A general method of analysis has been devised by Caswell (1989) and is described in Chapter 9. What follows is a modification targeting the specific question of whether a pollutant's effects via death or production have the greater effect on the population, and using a 4-parameter continuous-time model instead of a projection matrix model.

At this point it is useful to consider some examples, and I have chosen two from the 1981 article of R.E. Daniels and J.D. Allan, who investigated the effects of the pesticide dieldrin on a copepod, *Eurytemora affinis*, and on a cladoceran, *Daphnia pulex*.

The experiments were carried out in the laboratory. Temperature and food supply were constant. The experiments consisted in recording the life histories of cohorts of newly hatched larvae subjected to different concentrations of dieldrin. The numbers of survivors and births were counted daily, and population growth rates were calculated from Equation 1. Comparing first the effects on population growth rate (Figure 3), it appears that in *E. affinis* the

Table 1 Studies of the Effects of Pollutants on Life Tables Retrieved by the Literature Search Described in Appendix 1

Species	Type	Pollutant	Authors	Year
Daphnia pulex	Cladoceran	Gamma radiation	Marshall	1962
Daphnia (4 spp.)	Cladoceran	Copper	Winner and Farrell	1976
Lepidodermella sp.	Gastrotrich	DDT	Hummon and Hummon	1976
Daphnia pulex	Cladoceran	Dieldrin	Daniels and Allan	1981
Eurytemora affinis	Copepod	Dieldrin	Daniels and Allan	1981
Eurytemora affinis	Copepod	Kepone	Allan and Daniels	1982
Daphnia pulex	Cladoceran	Simazine	Fitzmayer et al.	1982
Daphnia pulex	Cladoceran	pH	Walton et al.	1982
Tetranychus urticae	Spider mite	6 Pesticides	Boykin and Campbell	1982
Mysidopsis bahia	Mysid shrimp	Mercury, Nickel	Gentile et al.	1982
Chlorella pyrenoidosa	Alga	Aminocarb	Weinberger and Rea	1982
Sitophilus oryzae	Coleopteran	Pirimiphosmethyl/Deltamethrin	Longstaff and Desmarchelier	1983
Skeletonema costatum	Marine diatom	Hexachlorocyclopentadiene, EPN, Chlorpyrifos, Carbophenothion, Atrazine	Walsh	1983
Man		Radiation	Dunning et al.	1984
Daphnia magna	Cladoceran	Bromide	van Leeuwen et al.	1986
Daphnia pulex	Cladoceran	Cadmium/copper	Meyer et al.	1987
Echinisca triserialis	Cladoceran	Cadmium	Chandini	1988
Moina micrura	Cladoceran	Endosulfan/Carbaryl	Krishnan and Chockalingam	1989
Daphnia obtusa	Cladoceran	Chromium	Coniglio and Baudo	1989
Turbatrix aceti	Nematode	Microbial toxins	Meadows et al.	1990
Sitophilus zeamais	Coleopteran	Pirimiphos-methyl/Permethrin	Giga and Canhao	1993
Prostephanus truncata	Coleopteran	Pirimiphos-methyl/Permethrin	Giga and Canhao	1993
Daphnia magna	Cladoceran	Endosulfan	Fernandez-Casalderrey et al.	1993

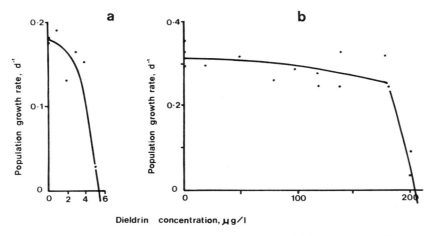

Figure 3   Effect of dieldrin on population growth rates of (a) *Eurytemora affinis* (b) *Daphnia pulex*. (Modified from Daniels and Allan, 1981.)

pollutant acts progressively up to a concentration of 4 to 5 µg/l, after which r → −∞ (not zero as erroneously stated in the article). In contrast, in *D. pulex* the pollutant has little if any effect up to 180 µg/l, but thereafter r → −∞ (but see Chapter 9 for further analysis of these data).

Are these effects on population growth rate of *E. affinis* primarily due to mortality? The survivorship curves for *E. affinis* are shown in Figure 4A. Although there is some variation it seems that at any given concentration up to 5 µg/l dieldrin, mortality rate is fairly constant until about day 30 (vital rates after day 30 had negligible effect on estimated population growth rates (Daniels and Allan, 1981) and are here ignored). Estimating those mortality rates allows us to compare the mortality effects of the pollutant as in Figure 5A. In this chapter, "mortality rate" means per capita mortality rate per unit time, measured in deaths per capita per unit time. These were estimated by eye, using the fact that constant mortality rate results in a straight line on a log survivorship curve (as plotted in Figure 4A), and giving a little more weight to data earlier in the life history, both because they have more effect on r and because they are more precise estimates, being derived from larger samples. This latter point can be taken into account using the methods of Chapter 8.

Examining the mortality effects of the pollutant (Figure 5A), it might seem that mortality effects alone could account for the effects on r shown in Figure 3. There is not much effect on mortality rate in the range 0 to 4 µg/l but then a jump between 4 and 5 µg/l. The question is how sensitive r is to the pollutant's effects on mortality, birth rate, and age at first reproduction.

Sensitivity has a natural mathematical formulation. Since mortality rate and the pattern of fecundity jointly determine r according to Equation 1, r can in principle be written as the mathematical function r = r(µ,t,b) where µ = mortality rate, t is age at first reproduction, and b is a parameter representing birth rate, considered in detail below. Because r is a mathematical function of

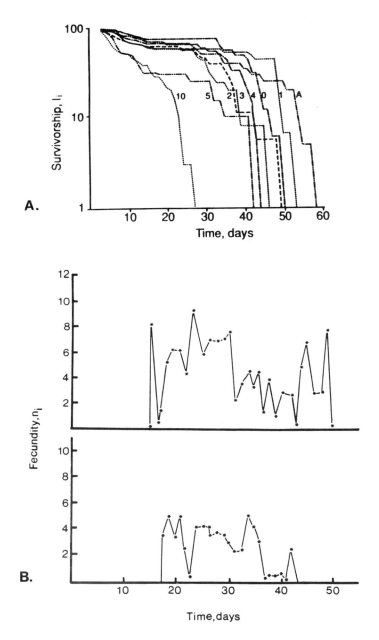

Figure 4 Effect of dieldrin on the life history of *Eurytemora affinis*. (A) Survivorship curves. Numbers indicate concentrations of dieldrin, in µg/l, A is an acetate control. (Modified from Daniels and Allan, 1981.) (B) Birth rate (eggs/day) was obtained at all concentrations, but only two representative concentrations, 2 and 4 µg/l, are shown here (upper and lower graph, respectively).

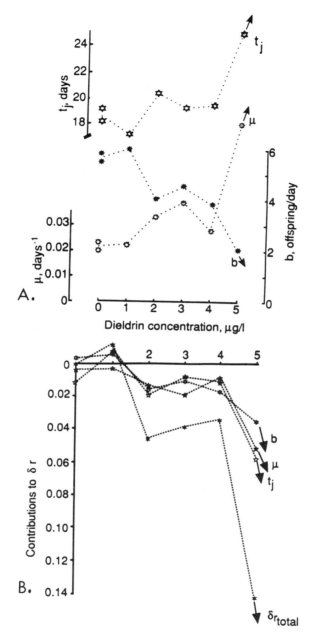

Figure 5 Effects of dieldrin concentration on *E. affinis*. (A) Effects on mortality rate, μ, and birth rate, b, were estimated as described in the text. Age at first reproduction, t_j, was obtained from Daniels and Allan (1981). (B) Contributions of μ, b, and t_j, to population growth rate change, δr_{total}. The contribution of μ is defined as $\frac{\partial r}{\partial \mu} \delta \mu$. See text for further details.

μ, t, and b, it can be rewritten to indicate the pollutant's effects. This is achieved by expanding the function in a Taylor series about its value when no pollutant is present:

$$\delta r \simeq \frac{\partial r}{\partial \mu}\delta\mu + \frac{\partial r}{\partial t}\delta t + \frac{\partial r}{\partial b}\delta b, \qquad (2)$$

where δr, $\delta\mu$, δt, and δb represent the pollutant's effects, assumed small, on r, μ, t and b, respectively. Here the partial derivative $\frac{\partial r}{\partial \mu}$ indicates the sensitivity of r to μ, i.e., the increase in r that results from a small increment in μ, t, and b being held constant.

These "mortality sensitivities" are readily calculated (see Appendix 2); they are –1 for both *E. affinis* and *D. pulex*. A sensitivity of –1 means that an increase of 0.01 in mortality rate μ produces a decrease of 0.01 in population growth rate r. Since up to 4 μg/l mortality rate in *E. affinis* only varies between 0.020 and 0.039 (Figure 5A), the effect of mortality on r is at most 0.019. The mortality contribution to δr, i.e., $\frac{\partial r}{\partial \mu}\delta\mu$, is plotted in Figure 5B. It transpires that, at each concentration of dieldrin, mortality-effects contribute a little under half of the total change in δr. This shows how r is affected by the pollutant's mortality effects.

A pollutant's effects on production may result in increased age at first reproduction, or lower birth rate. Age at first reproduction, t, does vary over the ranges of interest (Figure 5A) and so it is not surprising that it also contributes to r (Figure 5B). I modeled birth rate as age invariant, ignoring vital rates after 30 d as with mortality rate (Figure 4B). Birth rate decreased as the concentration of dieldrin increased (Figure 5A), and r was fairly sensitive to these fecundity changes (Figure 5B).

The overall effects of the pollutant on population growth rate, r, are shown in Figure 5B. Together with the total effect (δr_{total}), the contributions to δr_{total} due to the pollutant's effects on mortality rate, birth rate, and age at first reproduction are shown. These contributions are $\frac{\partial r}{\partial \mu}\delta\mu$, $\frac{\partial r}{\partial b}\delta b$, and $\frac{\partial r}{\partial t}\delta t$, and they sum to δr_{total} as in Equation 2. As can be seen from Figure 5B, mortality, birth, and age at first reproduction make roughly equal contributions to δr_{total} at each level of pollution. First there is no effect below 2 μg/l, then a step between 2 and 4 μg/l, and then rapidly increasing effects. In this case, therefore, the pollutant affects death and production rates roughly equally, as judged by their impact on population growth rate.

It may be that the case examined here is exceptional and pollutants generally act either on production or on mortality (DiGiulo et al., 1993). Contrary to this view, however, an indication of the possible effects of pollution on mortality rate and somatic growth rate can be obtained by considering how pollution affects an individual's "health status" and physiological condition (Lloyd, 1972; Depledge et al., 1993). As pollution increases, organisms are considered to be progressively unable to maintain homeostasis ("stressed"), unable to compensate ("diseased"), and eventually suffer irreversible noncompensation and death. Since efforts to maintain homeostasis and to compensate for environmental insult are energetically expensive, somatic growth is impaired, but at the same time the loss of health and onset of "disease" have mortality implications. According to this view, then, increasing pollution leads to a simultaneous loss of production and increase in mortality — just as in the example in Figure 5. It would be interesting to examine more widely this question of whether pollutants have greater effects on death or production rates, using the studies listed in Table 1. Some further, more detailed analyses are presented in Chapter 9.

III. POPULATION DENSITY AND POPULATION ECOLOGY

In the last section, we considered how a pollutant achieves its effect on population growth rate, and population density was implicitly assumed to be fixed. In this section, we allow population density to vary, and we assume that it also affects population growth rate. Analysis of density-dependence is at the heart of population ecology. In reality, there are, of course, many other factors that also affect population growth rate, both physical factors (temperature, salinity, and so forth) and biotic factors, such as the availability of food, and the distributions of predators and competitors. However, many of these biotic factors act through population density. I am going to focus here on life history methods for exploring the joint effect of two factors, pollution and density, on population dynamics, but it should be remembered that they are just single examples of physical and biotic factors, respectively. I begin by considering the effects of density in the absence of pollution, and add the effects of pollution later.

The effects of population density on mortality rate are central to population ecology, and have been reviewed by Sinclair (1989). Considering the importance of the topic and the attention it has received over the years, it is perhaps disappointing to record Sinclair's conclusion that "we still have a poor understanding of where density dependence occurs in the life cycle of almost every group of animals."

An example of the effects of population density on mortality rate is in the 25-year study of sea-trout by J.M. Elliott and collaborators (Elliott, 1993a). By electric fishing at fixed times of year, population density was established at various points in the life history, as shown in Figure 6. k values (measures of mortality) were calculated using the formula $k_i = \ln R_{i-1}/R_i$, defined as in Figure 6. The per capita mortality rate in each life history phase is obtained from k_i simply by dividing by the time period T_i over which the mortality operates. If the time periods are all unity, then k_i is the same as mortality rate. If they vary, as here, then normalizing for time, per capita mortality rate (designated μ_i) is given by

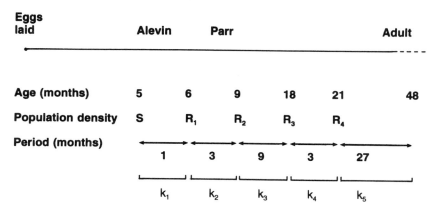

Figure 6 The life history of the sea trout at a stream in northwest England. The eggs hatch after about 5 months. The young trout are known as alevin from hatching until resorbtion of the yolk sac, after which they are known as parr. Population density at each age was measured by electric fishing and was designated S, R_1, R_2 ... as shown. The k-values are measures of mortality (see text).

$$\mu_i = \frac{k_i}{T_i} = \frac{1}{T_i} \ln \frac{R_{i-1}}{R_i} \tag{3}$$

Mortality rate is plotted against population density for five phases of the sea-trout life history in Figure 7. In the first two phases (alevin and young parr, Figure 7a and b), there is a clear positive relationship, but there is no relationship in the later phases of the life history. The effect of a positive relationship as shown in Figure 7a and b is to stabilize the population because higher mortality occurs at higher population density. This reduces population density when it is high. Conversely, at low population density, mortality rate is relatively low, and this allows the population to increase.

This example shows how population density affects mortality rate, and population density may also affect somatic growth rate and birth rate. And as always, mortality, somatic growth, and birth rates together determine population growth rate.

Many parameters are needed to fully describe a general life history, as mentioned before. However, in the case of the simplest life history, the organism only breeds once and only three parameters are needed: development period, fecundity, and juvenile mortality rate. Then the Euler-Lotka equation reduces to

$$\text{Population growth rate} = \frac{1}{\text{development period}} \ln\left(\frac{\text{fecundity}}{2}\right) - \text{mortality rate} \tag{4}$$

and, if we here define somatic growth rate as 1/(development period), Equation 4 can be rearranged as

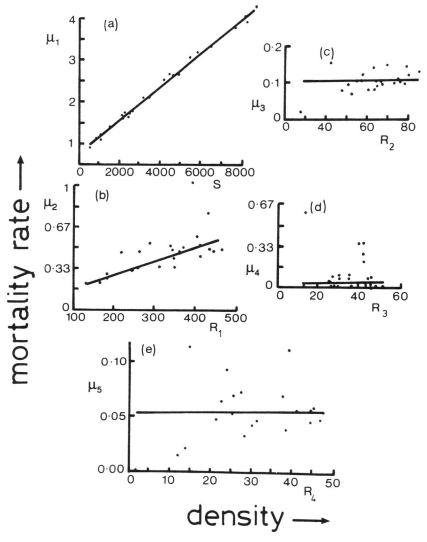

Figure 7 Mortality rates in relation to population density for sea-trout in each of the five periods depicted in Figure 6 (after Elliott, 1993a). Thus (a) refers to alevin, (b) to young parr, etc. Densities S, R_1 ... R_4 are defined in Figure 6 calculated μ_i from Equation 3 using the T_i values in Figure 6, but it is more accurate to use year-specific T_i values, as in Elliott (1993b).

$$\text{Mortality rate} = \text{somatic growth rate} * \ln\left(\frac{\text{fecundity}}{2}\right) - \text{population growth rate} \quad (5)$$

If fecundity is fixed, Equation 5 is a linear equation relating mortality, somatic and population growth rates. It can be used to connect mortality rates and somatic growth rates that result in the same population growth rate: the

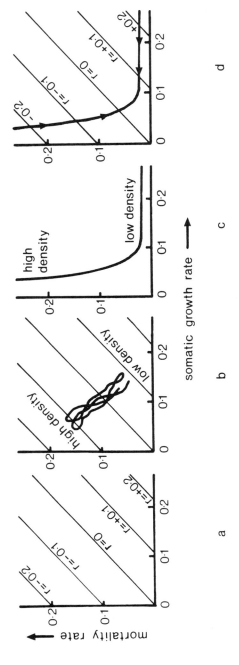

Figure 8   (a) Isoclines of population growth rate. Each line connects mortality and growth rates that result in a particular population growth rate, r. Fecundity is here taken to be 5. (b) Possible trajectories over many generations. At high densities the population declines (negative r), at low densities it increases (positive r). (c) Population trajectory calculated from the *Daphnia* model of Nisbet et al. (1989). Population growth rates can be calculated by superimposing (a) and (c) as in (d). Note that the population trajectory tends towards stability where it crosses the r = 0 isocline. (From Sibly, R.M., 1994. In: Water Quality and Stress Indicators in Marine and Freshwater Ecosystems: Linking Levels of Organisation (Individuals, Populations, Communities), edited by D.W. Sutcliffe, Freshwater Biological Association, Ambleside, pp. 63–74. With permission.)

resulting line is an "isocline" on which population growth rate is constant, as in Figure 8a.

Now we have the (graphical) means with which to analyze the joint effects of pollution and density: start with density. For a population to persist in its environment, it must increase when at low density and decline when at high density and, over many generations, the population will therefore trace out a trajectory, of which a possible example is shown in Figure 8b. An example from Nisbet et al.'s (1989) mathematical model of *Daphnia* is shown in Figure 8c, constructed as follows. Nisbet et al. assumed that ingestion rate was the result of dividing the food supply by the number of animals in the population; thus, if the food supply was constant, high population density resulted in low ingestion rates, and low density in high ingestion rates. During the juvenile phase, higher ingestion rates were assumed to result in higher somatic growth rates, according to their Equation (22). Low ingestion rates were assumed to result in increased mortality rates (their Equation (25)). Taking out ingestion rate between their Equations (22) and (25), we are left with the relationship between mortality rate and somatic growth rate depicted in Figure 8c. Note especially that, at low population density, somatic growth rate was high and mortality rate was low (right-hand end of curve) — this was because ingestion rates were then high. To calculate population growth rate in a simple worked example, I now assume the animals die after producing five offspring at first reproduction (for a review of brood size in relation to food availability, see McCauley et al., 1990), thus the population growth isoclines of Figure 8a apply. Calculating population growth rate for the right-hand end of the curve in Figure 8c, as in Figure 8d, shows the population increases (r is positive). Conversely high population density results in population decline (r is negative) at the left-hand end of the curve, and this occurs because high population density results in low ingestion rates, which results in low somatic growth rate and high mortality rate.

In a major early analysis of the effect of population density on the life history, Frank et al. (1957) measured *Daphnia pulex* life tables at seven densities with six replicates at each density. They found that population growth rate decreased linearly with density (as in the logistic model), being negative at the two highest densities. They concluded that this decrease in population growth rate was due to decreased production (birth rate and somatic growth rate), not death rate, which they considered also decreased with density. In my opinion, it would be worth repeating the death rate analysis using the more sophisticated methods now available (Chapters 8 and 9), since giving greater weight to observations earlier in the life history might alter the conclusion of Frank et al. (1957). Another example of the effects of population density on the life history is provided by the work of Pesch et al. (1987) on the semelparous nereid worm *Neanthes arenaceadentata*, undertaken with the effects of pollution specifically in mind. Worms were cultured in a laboratory experiment at densities of 40, 80, and 160 worms per 840 cm². Most, but not all, life history traits were measured. Density had an adverse effect on 70-d juvenile survivorship, time to spawning, percentage of females that reproduced, and number of eggs per reproducing female. Other things being equal, density would therefore have an adverse effect on population growth rate. It would be interesting to see in studies like these how the effects of density on death rate and production rate compare with the effects of pollution. Ideally, the contributions to population growth would be calculated and compared, extending the analysis of Figure 5 to include also the effects of density.

Figure 8 has shown how density alone may affect population growth rate, but what happens when the population is affected by pollution? Mortality effects push the population trajectory vertically upwards (vertical arrow in Figure 9a). If the trajectory no longer crosses the $r = 0$ isocline then r is always negative so the population goes extinct. On the other hand, a decrease in somatic growth rate (horizontal arrow in Figure 9a) pushes the population trajectory horizontally to the left, and again in the case shown in Figure 9a the result is extinction.

Of course, populations affected by pollution do not inevitably go extinct and it would be interesting to know, in terms of Figure 9, the mechanism by which extinction is avoided. The net result must be that although the population trajectory is displaced, it still reaches the $r = 0$ isocline. What mechanisms might bring this about?

One possibility is that, as a result of mortality stress, population densities become lower, leaving more food per head among the survivors so that somatic growth rates increase, shifting the entire population trajectory horizontally to the right (Figure 9b, arrow i). Similarly, production stress (e.g., a reduction in scope for growth) may reduce population density. Whether or how ecological compensation might then occur is unclear, but perhaps the increased availability of food among survivors might return somatic growth rates close to their former values (arrow ii in Figure 9b).

Although the analysis has been presented primarily with chronic pollution and a nonseasonal organism in mind, it could be developed more generally. Thus, short-term pollution may have its effect primarily through mortality or primarily through reduced production, or it may affect both (Figure 9a), and recovery of the afflicted population may be facilitated by density-dependent effects, as in Figure 9b. Density decreases for many temperate species during overwintering so they start the productive season at a low density (bottom right in Figure 8); as the season progresses, densities increase (move towards the top left in Figure 8).

So far, we have seen how pollution and density could jointly affect life history characters such as growth, reproduction, and mortality rates, and we have used life history theory to calculate the potential population consequences. This method of analysis allowed us to investigate directly the twin effects of pollution and density on population growth rate, as shown in Figure 9.

The joint effects of pesticides and density on population growth rate have also been analyzed using large simulation models to provide greater realism than is possible in the simple models discussed so far. Large-scale models of budworm-forest dynamics are used to evaluate the efficacy of pesticide spraying regimes, for example (see Fleming and Shoemaker, 1992 for a review).

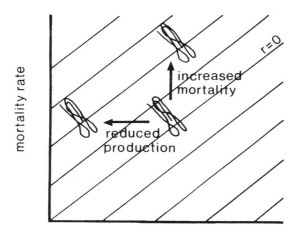

somatic growth rate

a

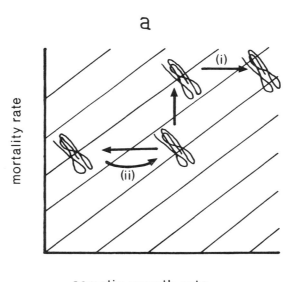

somatic growth rate

b

Figure 9   (a) Pollution may affect the population trajectory of Figure 8b via mortality (vertical arrow) or decreased production (horizontal arrow): if the trajectory no longer crosses the r = 0 isocline the population goes extinct. (b) Ecological compensation may prevent extinction via routes indicated by arrows (i) and (ii) discussed in the text. (From Sibly, R.M., 1994. In: Water Quality and Stress Indicators in Marine and Freshwater Ecosystems: Linking Levels of Organisation (Individuals, Populations, Communities), edited by D.W. Sutcliffe, Freshwater Biological Association, Ambleside, pp. 63–74. With permission.)

There have also been a number of studies investigating the joint effects of pollutants and predators or competitors on population growth rate. If pollution reduced the number of competitors, this could free resources for the focal species and increase its somatic growth rate. If the ratio of predators to prey declined, this could reduce the mortality rate. Both of these processes would affect the "ecological compensation" discussed in relation to Figure 9. That the outcome of competition can be affected by pollution has been demonstrated by Gilbert (1990), who showed in a community of cladocerans and rotifers, that the presence of the filamentous cyanobacterium, *Anabaena affinis*, prevented suppression of a rotifer by a large cladoceran and led to dominance by the rotifer.

Many studies have shown that removal of predators by pollutants leads to an increase in population densities of prey species (Dempster, 1975; Schindler, this volume). For example, the red spider mite appeared as a pest on outdoor fruit trees after the elimination of the slow-breeding predatory insects which previously controlled the mites. Fruit farmers used to spray pesticides on orchards in Britain as many as 20 times in a season, and it seems this upset the natural balance which kept the mite population under control (Mellanby, 1967). More recently Inoue et al. (1986), investigating the effects of spraying Kanzawa spider mites with six kinds of insecticides and three kinds of fungicide, showed that the population density of the Kanzawa spider mites increased with the application of certain insecticides and fungicides, probably because of their adverse effects on the natural predators (three species of phytoseiid mite, three species of insect, and a spider). This result led to the development of an integrated control program (Ashihara et al., 1992).

IV. EVOLUTIONARY ANALYSIS

A method of population analysis that potentially allows us to link the levels of individuals and populations was outlined above. An analogous method allows consideration of the evolutionary effects of stress, of which pollution is one example. The analogy occurs because at the heart of the evolutionary analysis is population growth rate, but this time the population is not the population of organisms, but the population of a certain type of allele. Evolutionary analysis considers whether alleles of a certain type increase or decline, and a measure of their spread or decline is the growth rate of the population of those alleles. This per copy growth rate of an allele will be referred to as its (Darwinian) fitness (Sibly and Antonovics, 1992; Sibly and Curnow, 1993). In this sense, the fitness of an allele is the direct analogue of the growth rate of a population.

Life history theory can again be useful, this time in linking fitness to the characteristics of individual copies of the allele. These characteristics depend on the characteristics of the organisms that carry them. It is usual to consider that new copies are created at the moment that syngamy (fertilization, conception) occurs (thus the period between meiosis and syngamy is ignored, and only gametes achieving syngamy are counted). The subsequent life history of the allele depends on its carrier. Ages of allele reproduction and survivorship are obtained directly from the carrier's characteristics. The number of copies made at the time of reproduction will, on average, be half the number of offspring produced by the carrier, if it is diploid, by Mendel's laws. In this way, the life history characteristics of a particular type of allele can be calculated from the life history properties of its carriers (for further details, see Sibly and Curnow, 1993).

Of course, different types of alleles may differ in their life history effects. Some may increase somatic growth rates; others may reduce mortality rates. Plotting the possible characteristics of different types of alleles, as in Figure 10a, we arrive at a set of "genetic options" for the study species, and the boundary of this set represents a "trade-off curve" which shows the form of trade-off to which the species is subject. Thus, in Figure 10a some alleles (top right) produce fast somatic growth rate, but at a cost of a high mortality rate, whereas others (bottom right) produce a lower mortality rate but at a cost of decreased somatic growth rate. In this way, mortality rate and somatic growth rate may be involved in a trade-off. Of course, genes only achieve such effects via physiology, and it is worth noting the trade-off could stem from energy allocation, because resources spent on defense (reducing mortality rate, e.g., to the starred level in Figure 10a) are not available for somatic growth rate.

Such a trade-off could come about in many ways (shells, spines, vigilance, etc.) (Sibly and Calow, 1989), but here we are especially concerned with defenses against toxicants, i.e., tolerance and resistance. Possible methods of defense include relatively impermeable exterior membranes (e.g., Little et al., 1989; Oppenoorth, 1985), more frequent molts (and consequent removal of toxicant in shed skin, e.g., Bengtsson et al., 1985), a more comprehensive immune system, and detoxification enzymes (Oppenoorth, 1985; Terriere, 1984). Many examples are given in this book. Although it is clear such defenses generally have energy costs (Hoffman and Parsons, 1991; Sibly and Calow, 1989), these could be small, for example in the case of inducible enzyme responses. Even with inducible responses, however, there must be a cost, since amino acids are required at every stage of the genetic response, and since all genetic mechanisms involve outlays (molecular checking, DNA turn-over, and disposal of waste).

There is strong circumstantial evidence that defense is costly and some-times has a genetic basis. From an extensive review, Hoffman and Parsons (1991) conclude that increased resistance to specific chemical stressors such as herbicides, pesticides, and heavy metals is often associated with lower fitness in optimal conditions. Similarly, it is generally the case in plants that tolerant strains, while more successful in polluted sites, are outcompeted in unpolluted sites (Baker, 1987). One suggestion is that the tolerant strains grow more slowly than susceptible strains in unpolluted sites because they allocate energy

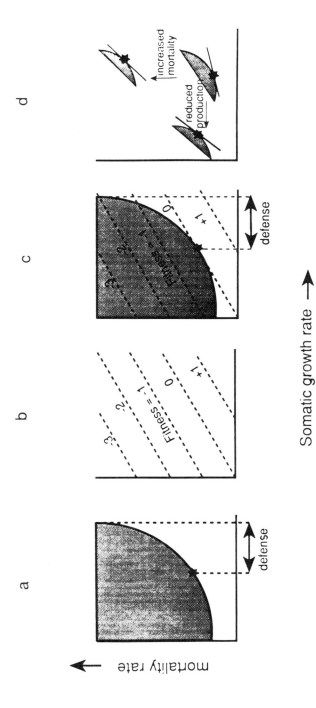

Figure 10   (a) Genetic options set (shaded) and trade-off curve. Note that defense compromises growth because of energy allocation but decreases mortality rate. (b) Fitness isoclines. Note that the zero-fitness isocline goes through the origin. (c) Superimposing (a) and (b) allows identification of the evolutionary outcome as the allele achieving highest fitness. In (c) and (d) evolutionary outcomes (optimal strategies) are represented by*; (d) shows the effects on evolutionary outcomes of long-term mortality and production stresses. The straight lines are zero-fitness isoclines. (From Sibly, R.M., 1994. In: Water Quality and Stress Indicators in Marine and Freshwater Ecosystems: Linking Levels of Organisation (Individuals, Populations, Communities), edited by D.W. Sutcliffe, Freshwater Biological Association, Ambleside, pp. 63–74. With permission.)

to defense which could otherwise, as in the susceptible strains, be used for somatic growth (Ernst, 1976).

The evolutionary implications of such trade-offs are a major theme of life history theory (Stearns, 1992; Roff, 1992; Sibly and Antonovics, 1992). Briefly, the fitness of each type of allele can be calculated as described above, and so, on a plot of mortality versus somatic growth rates, lines can be drawn connecting alleles that have the same fitness (Figure 10b). These "fitness isoclines" are exact analogues of the population growth isoclines plotted in Figure 8, and both types of isocline are given by essentially the same equation (Equation 4). Figure 10b shows that alleles conferring low mortality and high somatic growth rate are favored (and will increase) because their Darwinian fitnesses are relatively high, whereas those producing high mortality and low somatic growth rate are selected against. The likely outcome of the evolutionary process is the allele achieving highest fitness, and this can be identified by superimposing Figures 10a and b, as in Figure 10c. Note that density effects are here ignored and it is simply assumed that if an allele persists in a population, its rate of increase is 0 in the long run (Sibly and Calow, 1987). It turns out that the zero-fitness isocline passes through the origin of mortality rate/somatic growth rate space (Figure 10).

The evolutionary implications of stress acting on such a trade-off have been considered by Sibly and Calow (1989). Suppose for simplicity that stress affects the position but not the shape of the trade-off curve. Mortality stress shifts the trade-off curve vertically upwards, whereas a stress reducing somatic growth rate shifts the trade-off curve horizontally to the left (Figure 10d). Inspection of the evolutionary outcomes shows that less is spent on defense in both mortality- and production-stressed populations.

This prediction of less defense in polluted environments appears paradoxical, and it should be emphasized that it only applies if the shape of the trade-off curve remains unchanged when its position is shifted. If more defense evolves in polluted environments then the inference must be that the shape of the trade-off curve changes, as shown in Figure 11. In the simple case that the "allocation to defense" is genetically controlled, with the corollary as in Figure 10c that the somatic growth rate is genetically controlled, then genes for optimal defense are selected in the polluted environment, and genes for no defense are selected in the unpolluted environment, as shown by the stars in Figure 11.

Figures 10 and 11 have not allowed the possibility of inducible responses; defense is a "fixed response" here. If the response can be induced by the environment, however, as with inducible enzyme responses, then there is the possibility of producing a near optimal response regardless of the environment. Evolutionary biologists refer to such responses as "optimal reaction norms" or "optimal phenotypic plasticity," and they currently attract considerable experimental attention (Stearns, 1992). Since inducible responses must have a cost, if only a small one, in terms of the maintenance of, for instance, extra protein, it follows that inducible-response organisms would be outcompeted by fixed-response organisms in homogeneous environments. Hence inducible responses

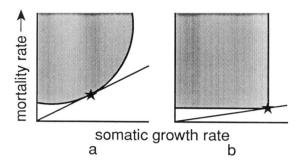

Figure 11   If the trade-off curves in (a) polluted and (b) unpolluted environments have different shapes, then the evolutionary outcomes (*) may involve more defense in polluted environments, as shown.

can only evolve in heterogeneous environments. Not surprisingly, the characteristics of the heterogeneity are important, and care is necessary in calculating optimal strategies in heterogeneous environments (Houston and McNamara, 1992; Kawecki and Stearns, 1993; Sibly and Atkinson, 1994; Sibly, 1995).

Because of the rigorous links between trade-off shapes and evolutionary outcomes, it is to be hoped that evolutionary analyses may play a useful part in understanding stress responses in the future.

V. CONCLUSIONS

Section II of this chapter considers the effects of pollutants on individual organisms, asking whether these are mainly on mortality or production. Since there is no generally accepted procedure for answering the question, I suggest a method that involves calculating the contributions, via mortality and via production, to population growth rate. A worked example, taken from Daniels and Allan (1981), is provided in which mortality and production contributions are about the same. In general, I imagine this will not be the case, and one will dominate the other. However, this remains to be tested. Section II, then, is concerned with the evaluation of pollutant effects on individual organisms, using as "common currency" their effects on population growth. This allows us to compare mortality effects with production effects, the comparison being made using the currency of the implications for population growth.

In the field, of course, population growth is affected by many other factors too. Section III considers the effects of population density in some detail. The treatment is necessarily simplistic, but it does allow analysis of the joint effects of pollution and density on population growth. Section IV uses an evolutionary analysis to identify likely evolutionary outcomes when an organism's genetic options are subject to trade-offs. The trade-off considered here is between somatic growth rate and mortality rate. Such a trade-off could arise because of an energy allocation trade-off if resources spent on defense (reducing mortality rate) are not available for somatic growth rate. The evolutionary implications of pollution acting on such a trade-off are outlined.

Both the population and the evolutionary analyses described here represented the state of a population in a "state space" with axes, mortality rate and somatic growth rate. Both analyses considered possible effects of pollutants on the population's state, the obvious simple possibilities being vertical and horizontal shifts increasing mortality or decreasing growth. Although the analysis has been presented for a semelparous life history in terms of three parameters (juvenile mortality rate, somatic growth rate, and birth rate), it could be developed for the general life history if density and pollution only affect juveniles. What is then necessary is to replace birth rate in Equation 5 with reproductive value at maturity, as can be shown from the Euler-Lotka equation (Equation 1). The effect of this on the population growth isoclines of Figure 8 is to lower the slope of the higher r isoclines. If adults as well as juveniles are affected by pollution and density then a multidimensional approach is inevitable (see Caswell, 1989 and Brown et al., 1993 for examples).

ACKNOWLEDGMENTS

I am very grateful to Dr. H. Caswell for some useful suggestions and to Drs. R. Laskowski, G.F. Warner, and three referees for their comments on the manuscript.

APPENDIX 1

A Biosis search was carried out on April 27, 1994 using search terms (Pollutant# or Pesticide# or Toxin#) and {(Life table#) or (Population growth) or (Intrinsic or natural) (S) Rate# (S) Increase}, where # indicates that s may be present, and (S) means that the terms on either side must both be present. Publications were retrieved numbering 112, of which 53 were irrelevant.

APPENDIX 2   SENSITIVITY OF r TO μ,t AND b

If an organism starts breeding at age t and finishes at age t_Ω, producing female offspring at rate b in between, and if the mortality rate is μ throughout its life, then the Euler-Lotka Equation 1 takes the form:

$$1 = \frac{b}{r+\mu}\left[e^{-(r+\mu)t} - e^{-(r+\mu)t_\Omega}\right] \tag{6}$$

from which the following partial derivatives may be obtained:

$$\frac{\partial r}{\partial b} = \frac{r + \mu}{b^2 \left[\frac{1}{b} + e^{-(r+\mu)t} \, t - e^{-(r+\mu)t_\Omega} \, t_\Omega \right]} \tag{7}$$

$$\frac{\partial r}{\partial t} = - \frac{(r + \mu)e^{-(r+\mu)t}}{\frac{1}{b} + e^{-(r+\mu)t} \, t - e^{-(r+\mu)t_\Omega} \, t_\Omega} \tag{8}$$

$$\frac{\partial r}{\partial t_\Omega} = \frac{(r + \mu)e^{-(r+\mu)t_\Omega}}{\frac{1}{b} + e^{-(r+\mu)t} \, t - e^{-(r+\mu)t_\Omega} \, t_\Omega} \tag{9}$$

$$\frac{\partial r}{\partial \mu} = -1 \tag{10}$$

In the calculations used in drawing Figure 5b, these sensitivities were evaluated at zero concentration of dieldrin.

REFERENCES

Allan, J.D. and R.E. Daniels, 1982. Life table evaluation of chronic exposure of *Eurytemora affinis* (Copepoda) to kepone. Mar. Biol. 66, 179–184.

Ashihara, W., K. Inoue, M. Osakabe and T. Hamamura, 1992. Effectiveness of *Phytoseiulus persimilis* Athias-Henriot Acarina: Phytoseiidae as a control agent for the Kanzawa spider mite *Tetranychus kanzawai* Kishida Acarina: Tetranychidae and occurrence of native natural enemies of the spider mite on grapevine in glasshouse. Bull. Fruit Tree Res. Stn. Ser. E, 109–130.

Baker, A.J.M., 1987. Metal tolerance. New Phytol. 106 (Suppl.), 93–111.

Bakhetia, D.R., 1986. Pest management in rapeseed and mustard. Pesticides (Bombay) 20, 32–38.

Bengtsson, G., T. Gunnarsson and S. Rundgren, 1985. Influence of metals on reproduction, mortality and population growth in *Onychiurus armatus* (Collembola). J. Appl. Ecol. 22, 967–978

Boykin, L.S. and W.V. Campbell, 1982. Rate of population increase of the two-spotted spider mite *Tetranychus urticae* Acari Tetranychidae on peanut *Arachis hypogaea* leaves treated with pesticides. J. Econ. Entomol. 75, 966–971.

Brown, D., N.D.E. Alexander, R.W. Marrs and S. Albon, 1993. Structured accounting of the variance of demographic change. J. Anim. Ecol. 62, 490–502.

Caswell, H., 1989. Matrix Population Models. Sinauer Associates., Sunderland, MA. 328 pp.

Chandini, T., 1988. Effects of different food chlorella concentrations on the chronic toxicity of cadmium to survivorship growth and reproduction of *Echinisca triserialis* Crustacea: Cladocera. Environ. Pollut. 54, 139–154.

Chi, H., 1990. Timing of control based on the stage structure of pest populations: a simulation approach. J. Econ. Entomol. 83, 1143–1150.

Coniglio, L. and R. Baudo, 1989. Life-table of *Daphnia obtusa* (Kurz) survivorship of exposure to toxic concentrations of chromium. Hydrobiologia 188/189, 407–410.

Crossland, N.O., 1988. A method for evaluating effects of toxic chemicals on fish growth rates. In: Aquatic Toxicology and Hazard Assessment; Vol. 10. edited by W.J. Adams, G.A. Chapman and W.G. Landis, American Society for Testing and Materials, Philadelphia, PA, pp. 463–467.

Daniels, R.E. and J.D. Allan, 1981. Life table evaluation of chronic exposure to pesticide. Can. J. Fish. Aquat. Sci. 38, 485–494.

Dempster, J. P., 1975. Effects of organochlorine insecticides on animal populations. In: Organochlorine Pesticides: Persistent Organic Pollutants, edited by F. Moriarty, Academic Press, London pp. 231–248.

Depledge, M.H., J.J. Amaral-Mendes, B. Daniel, R.S. Halbrook, P. Koepper-Sams, M.N. Moore and D.B. Peakall, 1993. The conceptual basis of the biomarker approach. In: Biomarkers, edited by D.B. Peakall and L.R. Shugart, Springer-Verlag, Berlin pp. 15–30.

DiGiulo, R. T., J.K. Chipman, M. Feeley, W.E. Hawkins, K. Smith, G. Suter and G. Winston, 1993. Interpretation of biomarker responses. In: Biomarkers, edited by D.B. Peakall and L.R. Shugart, Springer-Verlag, Berlin pp. 49–62.

Dunning, D.E., Jr, R.W. Leggett and R.E. Sullivan, 1984. An assessment of health risk from radiation exposures. Health Phys. 46, 1035–1052.

Elliott, J.M., 1993a. A 25-year study of production of juvenile sea-trout, *Salmo trutta*, in an English Lake District stream. Can. Spec. Publ. Fish. Aquat. Sci. 118, 109–122.

Elliott, J.M., 1993b. The pattern of natural mortality throughout the life cycle in contrasting populations of brown trout, *Salmo trutta* L. Fish. Res. 17, 123–136.

Ernst, W., 1976. Physiological and biochemical aspects of metal tolerance. In: Effects of Air Pollutants on Plants, edited by T.A. Mansfield, Cambridge University Press, Cambridge, pp. 115–133.

Fernandez-Casalderrey, A., M.D. Ferrando and E. Andreu-Moliner, 1993. Effects of endosulfan on survival, growth and reproduction of *Daphnia magna*. Comp. Biochem. Physiol. C Comp. Pharmacol. Toxicol. 106, 437–441.

Fitzmayer, K.M., J.G. Geiger and M.J. Zan Den Avyl, 1982. Effects of chronic exposure to semazine on the cladoceran, *Daphnia pulex*. Arch. Environ. Contam. Toxicol. 11, 603–609.

Fleming, R.A. and C.A. Shoemaker, 1992. Evaluating models for spruce budworm-forest management comparing output with regional field data. Ecol. Appl. 2, 460–477.

Frank, P.W., C.D. Boll and R.W. Kelly, 1957. Vital statistics of laboratory cultures of *Daphnia pulex* DeGeer as related to density. Physiol. Zool. 30, 287–305.

Gentile, J.H., S.M. Gentile, N.G. Hairston and B.K. Sullivan, 1982. The use of life tables for evaluating the chronic toxicity of pollutants to Mysidopsis bahia. Hydrobiologia 93, 179–187.

Giga, D.P. and J. Canhao, 1993. Effects of sublethal doses of insecticides on population growth of mixed populations of *Prostephanus truncatus* Horn Coleoptera: Bostrichidae and *Sitophilus zeamais* Motschulsky Coleoptera: Curculionidae at two temperatures. Afr. Entomol. 1, 93–99.

Gilbert, J.J., 1990. Differential effects of *Anabaena affinis* on Cladocerans and Rotifers: mechanisms and implications. Ecology 71, 1727–1740.

Hoffmann, A.A. and P.A. Parsons, 1991. Evolutionary Genetics and Environmental Stress. Oxford University Press, Oxford.

Houston, A.I. and J.M. McNamara, 1992. Phenotypic plasticity as a state dependent life-history decision. Evol. Ecol. 6, 243–253.

Hummon, W.D. and Hummon, M.R., 1976. Use of life table data in tolerance experiments. Can. Biol. Mar. 16, 743–749.

Inoue, K., M. Osakabe, W. Ashihard, and T. Hamamura, 1986. Factors affecting abundance of the Kanzawa spider mite *Tetranychus kanzawai* on grapevine in a glasshouse influence of pesticidal application on occurrence of the Kanzawa spider mite and its predators. Bull. Fruit Tree Res. Stn. Ser. E, 103–116.

Inoue, T. and R-I. Ohgushi, 1976. A simulation model of the arrowhead scale population on a citrus tree in relation to control programs, I. Res. Popul. Ecol. 18, 89–104.

Iwasa, Y. and H. Mochizuki, 1988. Probability of population extinction accompanying a temporary decrease of population size. Res. Popul. Ecol. 30, 145–164.

Kawecki, T.J. and S.C. Stearns, 1993. The evolution of life histories in spatially heterogeneous environments: optimal reaction norms revisited. Evol. Ecol. 7, 155–174.

Krishnan, M. and S. Chockalingam, 1989. Toxic and sublethal effects of endosulfan and carbaryl on growth and egg production of *Moina micrura* Kurz Cladocera: Moinidae. Environ. Pollut. 56, 319–326.

Little, E. J., A.R. McCaffery, C.H. Walker and T. Parker, T., 1989. Evidence for an enhanced metabolism of cypermethrin by a monooxygenase in a pyrethroid-resistent strain of the tobacco budworm (*Heliothis virescens* F.). Pest. Biochem. Physiol. 34, 58–68.

Lloyd, R., 1972. Problems in determining water quality criteria for freshwater fisheries. Proc. R. Soc. London Ser. B 180, 439–449.

Longstaff, B.C. and J.M. Desmarchelier, 1983. Effects of the temperature toxicity relationships of certain pesticides upon the population growth of *Sitophilus oryzae* Coleoptera: Curculionidae. J. Stored Prod. Res. 19, 25–30.

Maltby, L., C. Naylor and P. Calow, 1990. Effect of stress on a freshwater Benthic Detritivore: scope for growth in *Gammarus pulex*. Ecotoxicol. Environ. Saf. 19, 285–291.

Marshall, J.S., 1962. The effects of continuous gamma radiation on the intrinsic rate of natural increase on *Daphnia pulex*. Ecology 43, 598–607.

McCauley, E., W.W. Murdoch and R.M. Nisbet, 1990. Growth, reproduction, and mortality of *Daphnia pulex* Leydig: life at low food. Funct. Ecol. 4, 505–514.

Meadows, J., S.S. Gill and L.W. Bone, 1990. *Bacillus thuringiensis* strains affect population growth of the free-living nematode *Turbatrix aceti*. Invertebr. Reprod. Dev. 17, 73–76.

Mellanby, K., 1967. Pesticides and Pollution. Collins, London.

Meyer, J.S., C.G. Ingersol and L.L. McDonald, 1987. Sensitivity analysis of population growth rates estimated from Cladoceran chronic toxicity tests. Environ. Toxicol. Chem. 6, 115–126.

Nisbet, R.M., W.S.C. Gurney, W.W. Murdoch and E. McCauley, 1989. Structured population models: a tool for linking effects at individual and population level. Biol. J. Linn. Soc. 37, 79–99.

Nolet, B.A., 1988. Breeding success of some coastal birds in a herring gull *Larus argentatus* colony. Limosa 61, 79–84.

Oomen, P.A., 1982. Studies on population dynamics of the scarlet mite *Brevipalpus phoenicis*, a pest of tea in Indonesia. Meded. Landbouwhogesch. Wageningen 82, 1–88.

Oppenoorth, F. J. 1985. Biochemistry and genetics of insecticide resistance. In: Comprehensive Insect Physiology Biochemistry and Pharmacology, Vol. 12, edited by G. A. Kerkut and L. I. Gilbert, Pergamon Press, Oxford.

Pesch, C.E., R.N. Zajac, R.B. Whitlatch and M.A. Balboni, 1987. Effect of intraspecific density on life history traits and population growth rate of *Neanthes arenaceodentata* (Polychaeta: nereidae) in the laboratory. Mar. Biol. 96, 545–554.

Rieske, L.K. and K.F. Raffa, 1990. Use of a monitoring system to evaluate pesticide efficacy and residual activity against two pine root weevils *Hylobius pales* and *Pachylobius picivorus*. Coleoptera: Curculionidae in Christmas tree farms. Gt. Lakes Entomol. 23, 189–194.

Roff, D.A., 1992. The Evolution of Life Histories: Theory and Analysis, Chapman and Hall, London.

Schindler, D.W., 1987. Detecting ecosystem responses to anthropogenic stress. Can. J. Fish. Aquat. Sci. 44, 6–25.

Shaw, M.W., 1993. Theoretical analysis of the effect of interacting activities on the rate of selection for combined resistance to fungicide mixtures. Crop Prot. 12, 120–126.

Sibly, R.M., 1994. From organism to population: linking levels of organization. In: Water Quality and Stress Indicators in Marine and Freshwater Ecosystems: Linking Levels of Organisation (Individuals, Populations, Communities), edited by D.W. Sutcliffe, Freshwater Biological Association, Ambleside, England, pp. 63–74.

Sibly, R.M., 1995. Life-history evolution in spatially-heterogeneous environments with and without phenotypic plasticity. Evol. Ecol. 9, 242–257.

Sibly, R.M. and J. Antonovics, 1992. Life history evolution. In: Genes in Ecology, edited by R.J. Berry, T.J. Crawford and G.M. Hewitt, Blackwell Scientific Publications, Oxford, pp. 87–122.

Sibly, R.M. and D. Atkinson, 1994. How rearing temperature affects optimal adult size in ectotherms. Funct. Ecol. 8, 486–493.

Sibly, R.M. and P. Calow, 1986. Physiological Ecology of Animals, Blackwell Scientific Publications, Oxford.

Sibly, R.M. and P. Calow, 1987. Ecological compensation — a complication for testing life-history theory. J. Theor. Biol. 125, 177–186.

Sibly, R.M. and P. Calow, 1989. A life-cycle theory of responses to stress. Biol. J. Linn. Soc. 37, 101–116.

Sibly, R.M. and R.N. Curnow, 1993. An allelocentric view of life-history evolution. J. Theor. Biol. 160, 533–546.

Sinclair, A.R.E., 1989. Population regulation in animals. In: Ecological Concepts, edited by J.M. Cherrett, Blackwell Scientific Publications, Oxford pp. 197–241.

Stearns, S.C., 1992. The Evolution of Life Histories. Oxford University Press, Oxford.

Swezey, S.L. and E. Cano-Vesquez, 1991. Biological control of citrus blackfly *Homoptera aleyrodidae* in Nicaragua. Environ. Entomol. 20, 1691–1698.

Taylor, C.E. and G.P. Georghiou, 1979. Suppression of insecticide resistance by alteration of gene dominance and migration. J. Econ. Entomol. 72, 105–109.

Terriere, L. C., 1984. Induction of detoxification enzymes in insects. Annu. Rev. Entomol. 29, 71–88.

Van Leeuwen, C.J., M. Rijkeboer and G. Niebeek, 1986. Population dynamics of *Daphnia magna* as modified by chronic bromide stress. Hydrobiologia 133, 277–285.

Walsh, G.E., 1983. Cell death and inhibition of population growth of marine unicellular algae by pesticide. Aquat. Toxicol. 3, 209–214.

Walton, W.E., S.M. Compton, J.D. Allan and J.D. Daniels, 1982. The effect of acid stress on survivorship and reproduction of *Daphnia pulex* (Crustacea: Cladocera). Can. J. Zool. 60, 573–579.

Warren, C.E. and G.L. Davis, 1967. Laboratory studies on the feeding of fishes. In: The Biological Basis of Freshwater Fish Production, edited by S.D. Gerking, Blackwell Scientific Publications, Oxford, pp.175–214.

Weinberger, P. and M.S. Rea, 1982. Effects of aminocarb and its formulation adjuncts on the growth of *Chlorella pyrenoidosa* Chick. Environ. Exp. Bot. 22, 491–496.

Widdows, J. and P. Donkin, 1992. Mussels and environmental contaminants: bioaccumulation and physiological aspects, In: The Mussel Mytilus: Ecology, Physiology, Genetics and Culture, edited by E. Gosling, Elsevier, Amsterdam, pp. 383–424.

Winner, R.W. and M.P. Farrell, 1976. Acute and chronic toxicity of copper to four species of *Daphnia*. J. Fish. Res. Board Can. 33, 1685–1691.

CHAPTER 8

Ecologically Meaningful Estimates of Lethal Effect in Individuals

Michael C. Newman and Philip M. Dixon

I. OVERVIEW

An individual organism's fitness is diminished in the presence of sufficiently high concentrations of a toxicant. As survival is the most easily measured fitness component, most of the early efforts to protect aquatic biota focused on lethality. Methods, borrowed from mammalian toxicology, were first applied to acute or intense toxicant exposures and later extended to chronic exposures. For decades, the foundation of methods and assessment procedures (e.g., Sprague, 1969, 1971; EPA, 1975; Stephanz, 1977; Buikema et al., 1982; ASTM, 1989) was classic work in mammalian toxicology (e.g., Bliss, 1935; Litchfield and Wilcoxon, 1949; Armitage and Allen, 1950; Gaddum, 1953; Finney, 1964).

The predominant approach adopted in the U.S. (dose- or concentration-response approach) quantified mortality at a predetermined time endpoint to estimate the toxicant concentration producing a certain level of mortality, e.g., 96-h LC50 (Sprague, 1969). The predicted concentration at which 50% of exposed individuals survived exposure of a specified duration (LC50) was selected as the primary measure of toxic effect as it was the most statistically reliable estimate, i.e., the estimate with the narrowest 95% confidence interval (Trevan, 1927). It was not selected because it was the most ecologically meaningful estimate. An extensive regulatory structure was built with this approach as the cornerstone.

Despite their early utility, direct application of dose-response or time endpoint techniques became increasingly awkward as attention shifted from acute exposure toward chronic exposure scenarios. Temporal dynamics of toxicant effects, less pertinent in early mammalian toxicity testing, grew in importance during assessment of long-term toxicant impacts on aquatic

1-56670-1127-9/95/$0.00+$.50

populations. Modification of time endpoint methods became necessary to incorporate variation of exposure duration into effects models. The incipient LC50 (the predicted concentration below which at least 50% of exposed individuals would live indefinitely, relative to the lethal effects of the toxicant) was formulated from plots of LC50 versus exposure time. This estimate of toxic incipiency generated from LC50 values also has ambiguous ecological significance (Newman, 1995). The long-term consequences of 50% mortality on population viability is impossible to assess from such studies. Empirical models continue to be developed in attempts to encompass temporal dynamics with the time endpoint approach, e.g., Wang and Hanson (1985) and Mayer et al. (1994).

Sprague (1969) argued unsuccessfully for equal consideration of the alternate, time-response approach. Instead of noting deaths only at one specific time, the times-to-death for individuals exposed to various treatments were to be noted. He reemphasized Finney's (1964) point that ignoring mortality information prior to a time endpoint seriously reduces statistical power. He quoted Gaddum (1953), "... theoretically it may be expected that about half the information is lost, so that twice as many observations will be needed for any given degree of accuracy." Sprague followed Gaddum's estimate with Burdick's (1960) more extreme estimate that ten times more replicates would be needed if time endpoint methods were used instead of time-to-death methods. Despite these sound arguments, use of the endpoint methods was deemed "good enough" and expanded to dominate ecotoxicology while the survival time approach did not progress much beyond the Litchfield (1949) method for estimating median time of survival (LT50). Today, endpoint methods are used without critical comparison to alternative approaches, such as the survival time approach, and without much understanding of underlying assumptions.

Consequently, it is the purpose of this chapter to provide a brief treatment of the often-neglected basis for time endpoint methods and then to detail the implementation of the heretofore neglected survival time methods. Discussion of survival time methods will pull together material scattered throughout several of our past publications, including Diamond et al. (1989, 1991), Dixon and Newman (1991), Heagler et al. (1993), Keklak et al. (1994), Newman et al. (1989, 1994), Newman and Aplin (1992), and Newman (1995). Data sets from Diamond et al. (1989) and Newman and Aplin (1992) will be used throughout to illustrate application of survival time methods. Time endpoint methods will not be described in as much detail because specifics are provided in many other sources (e.g., Hamilton et al., 1977; Stephan, 1977; Buikema et al., 1982; Newman, 1995).

II. THE DOSE-RESPONSE (TIME ENDPOINT) APPROACH

A wide range of time endpoint methods are applied in ecotoxicology (Figure 1). Most are used to estimate the LC50 and its 95% confidence

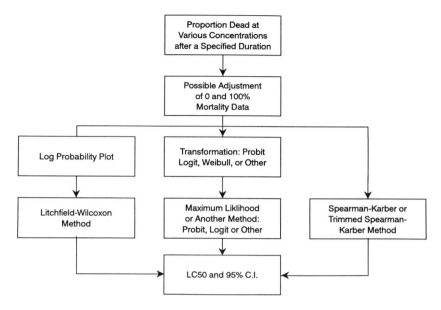

Figure 1 Dose-response (time endpoint) methods for analyzing mortality data.

interval.* One such method, the semigraphical Litchfield-Wilcoxon approach (Litchfield and Wilcoxon, 1949), is subject to error because a line is fit by eye in one step. The Litchfield-Wilcoxon method can be improved by using statistically fit data, but methods such as maximum likelihood estimation fit data assuming a specific underlying distribution and provide more reliable estimates than the Litchfield-Wilcoxon approach. Maximum likelihood methods have good precision yet generate slightly biased estimators.** But this bias is small relative to those of alternative methods except when there are very few observations (Finney, 1971). A range of underlying distributions can be assumed, including a log normal (probit transformations), log logistic (logit transformations), or Weibull. Finney (1964) also describes the less common Wilson-Worcester, Cauchy-Urban, and "linear" or rectangular sigmoid models. The arcsine square root transformation may also be used, although its use is based less on conformity to any specific underlying model than on the desire to produce transformed data with constant variance over the range of exposure concentrations. These techniques are implemented conveniently with a wide range of software including CT-TOX (CT Dept. of Environmental Protection, 1990), PROBIT (EPA, 1988), SAS (SAS Institute, 1988), and TOXSTAT (WEST, Inc. and Gulley, 1994).

* Although concentrations killing other proportions of exposed individuals, e.g., LC10, can also be estimated, the precision of the estimate is generally poorer than that of the LC50.

** Alternatively, the Spearman-Karber method is applicable if a symmetrical but unspecified distribution is assumed.

Fitting a probit model with a maximum likelihood method carries with it the assumption of a log normal distribution for the mortality response. The probit* of the proportion dead (P) plotted against the log of exposure concentration should generate a straight line. Bliss (1935) explains the process underlying this distribution of mortalities within a population using the concept of individual lethal dose, i.e., there exists a characteristic minimal amount of toxicant needed to kill any particular individual. Gaddum (1953) also explains the basis for the log normal model in an identical manner but refers to the individual lethal dose as the individual effective dose or I.E.D. He gives the example of slow intravenous infusion of digitalis into individual cats until just enough is present to stop the heart. The distribution of I.E.D.s among individuals in a population often displays a log normal distribution because "... in biological material the variation often shows a geometrical rather than an arithmetic distribution, an observation which has been confirmed by several investigators in respect to toxicological characteristics" (Bliss, 1935). This explanation is forwarded to support the present-day use of probits in ecotoxicology although I.E.D.s are never measured and the validity of extrapolating from poisoned "biological material" to whole organism exposure from an aqueous media is questionable.

The use of logits implies a log logistic model. A plot of the logit (logit = $\ln[P/(1 - P)]$) against the log of exposure concentration may produce a straight line.** Use of the logit has been defended by linkage to either the Hill equation for enzyme kinetics or Langmuir adsorption models (Gaddum, 1953). Also, the differential form of the logistic model describes many diverse and pertinent processes including autocatalysis, bimolecular reactions, and enzyme-mediated hydrolysis (Berkson, 1951). Regardless of similarities of curves and underlying mechanisms, the rate of change in the logistic model is proportional to an "amount" to be acted upon (e.g., number of individuals alive at any moment) and is also proportional to some additional factor that increases as this "amount" decreases (Berkson, 1951). In support of the log logistic model, Berkson (1951) points out that the log normal model assumes a static distribution of tolerances but the log logistic model is based on dynamic mechanics. He discusses a study of human tolerance (I.E.D.) to high altitude conditions that, upon repetition, did not support the assumption that the I.E.D. was an inherent quality of an individual. Because the I.E.D. was the usual explanation of the log normal (probit) model, Berkson argued that this study illustrated the inadequacy of the log normal model. Gaddum (1953) countered this criticism by suggesting that the log normal model reflects a process in which several "hits" are needed at a target site to result in death. Finney (1964) provides more

* The probit is the normal equivalent deviate + 5. The normal equivalent deviate is the distance from the mean of a normal distribution expressed in units of standard deviations. The normal equivalent deviate is also called the z, standard normal deviate, normal equivalent deviation, and standard score in various statistical textbooks.

** Often a transformed logit (transformed logit = logit/2 + 5) is used because the resulting transformed values are very similar to probit values.

detail but no definitive resolution regarding the theoretical foundations for the probit or logit transformations.

More recently the Weibull model, a flexible generalization of the exponential model, was advocated for fitting survival data (Pinder et al., 1978), including time endpoint data (Christensen, 1984; Christensen and Nyholm, 1984; Newman, 1995). A straight line is produced by plotting the Weibull transform $(\ln(-\ln(1-P)))$ against the log of exposure concentration. This model has been applied successfully to algal (Christensen, 1984; Christensen and Nyholm, 1984) and fish (Christensen, 1984; Newman, 1995) toxicity data. Speculation about the underlying mechanism for a Weibull model has included its conformity to one hit or multiple hit models of carcinogenesis (Christensen, 1984). Christensen and Nyholm (1984) link the Weibull model to Teisser's equation also. They assume that the slope of the Weibull transform curve is a measure of the number of toxicant molecules interacting with each receptor site. No evidence has been presented to date to support or refute this insightful yet speculative assertion.

There are two important points to be made regarding the application of these methods to time endpoint data. First, these transforms produce very similar estimates near the LC50 but estimates become increasingly divergent toward the extremes, e.g., LC10. Also, confidence intervals at the extremes become larger, regardless of the model. Consequently, effective model selection* becomes more important as ecotoxicology shifts its focus downward away from the LC50. Second, these well-established methods have no theoretical superiority to alternative approaches and, as applied today, are simply empirical or redescription models (see Chapter 1).

III. THE TIME-RESPONSE (SURVIVAL TIME) APPROACH

A. INTRODUCTION

The time-response (survival time, failure time, time-to-death, resistance time, or waiting time) approach is a powerful alternative to dose-response methods. These methods make more effective use of mortality data than do dose-response methods (see Figure 2). In Figure 2, mortality is noted through time at five concentrations. Only four data points (marked by ▲'s) would be used in applying a dose-response method to estimate an LC50. Complete mortality before 96 h at the highest concentration would render this treatment useless or marginally useful in calculations. In contrast, almost an order of magnitude more data would be available from the same experiment if mortality were noted every 12 h. This larger data set would include data from the highest concentration. This approach entails some extra effort; however, it is customary with toxicity testing to periodically remove dead individuals and to note

* Newman (1995) provides a detailed discussion and example of such comparisons of probit, logit, and Weibull transformations.

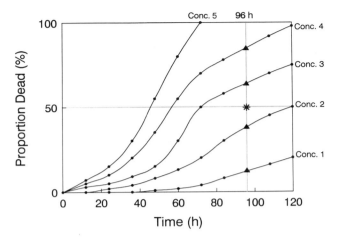

Figure 2   More data are collected from a toxicity test if times-to-death are noted. Only four data points are available if proportion dead by 96 h is the measurement noted.

mortality, e.g., the sample data sheet for toxicity tests provided in Parrish (1985).

Litchfield's method (Litchfield, 1949) for estimating median lethal time (LT50) and its 95% confidence interval is the only time-response method routinely presented to aquatic toxiciologists by instructors, textbooks, and methods manuals. Although this graphical method is useful, the time-response approach embraces a much richer suite of methods routinely used in clinical sciences, economics, engineering, and epidemiology (Figure 3). Nonparametric methods, including product-limit (Kaplan-Meier) and life table methods, are available. Neither requires that a specific distribution be assigned to the survival curve. Product-limit methods have the advantage of allowing observation intervals to vary in length and carry fewer assumptions than life table methods (Newman, 1995). Life (actuarial) tables have the advantage of linkage to many ecological concepts and parameters. (The reader is referred to Chapter 9 in this volume and Miller (1981) for details regarding life table methods.) Survival curves for different groups or classes of exposed individuals can be tested for equality using log-rank, Wilcoxon, or other methods.

The remaining methods in Figure 3 compare covariate effects to a reference survival curve, e.g., survival of smokers relative to that of nonsmokers. A Cox proportional hazard model can be used if no underlying model can or need be assumed for the reference survival curve. The hazards (proneness to die) of the various classes are assumed to be proportional. If a specific underlying distribution is assumed for the survival curves, data can be fit using one of two parametric formulations, proportional hazard or accelerated failure time. The hazards of various classes are assumed proportional in the parametric proportional hazard model as with the semiparametric Cox proportional hazard model. With the accelerated failure time model, covariates act to modify the ln of time-to-death (TTD). A covariate may shorten the ln TTD of an individual

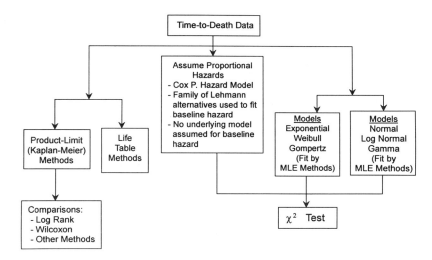

Figure 3   Time-response (survival time) methods for analyzing mortality data.

or group of similar individuals, that is "accelerate" the process resulting in death. Hypothesis tests are used to test for differences among classes or effects of continuous variables with these semiparametric and parametric models.

Nonparametric, semiparametric, and fully parametric methods are easily implemented with a wide range of popular software packages, including BMDP (Dixon, 1985), GLIM 4 (Francis et al., 1993), SAS (SAS Institute, 1988), S-Plus (Statistical Sciences, 1993), Survcalc (Lachenbruch, 1986), and SYSTAT (Steinberg and Colla, 1988). Code for the widely used SAS package (SAS Institute, 1988) will be used to illustrate implementation here.

Before detailing each method, it is necessary to define several unfamiliar terms. F(t) is the cumulative mortality distribution function and f(t) is the associated probability density function. The F(t) is 0 at t = 0 and slowly increases toward 1 as the cumulative number of deaths increases through time. The survival function (S(t)) begins at 1 and slowly decreases toward 0 through time. S(t) is equal to $1 - F(t)$. The hazard function, h(t), describes the probability of dying in a time interval. H(t) is the cumulative hazard function. Dixon and Newman (1991) provide further details and examples of hazard and survival curves.

Another conspicuous feature of time-to-death data is the common occurrence of censoring. Often, the exact time-to-death is not known for some individuals. For example, the only information available from a censored individual may be that it survived at least a certain number of hours of exposure. Censoring often involves termination of the exposure before all individuals are dead. The time-to-death for a survivor is some undetermined time greater than the duration of the experiment. This is referred to as right censoring because the missing times-to-death are from the right-hand side of the survival curve. Alternatively, individuals may have been removed at set times during the course of the exposure to assay some physiological or biochemical change during exposure. A series of censoring times would be generated for sets of fish removed during the trial. This is referred to as multiple, right censoring. Finally, time-to-death may be measured within a sufficiently long interval that time-to-death can only be noted to have occurred between the beginning and end of the interval, e.g., between 24 and 30 h of exposure. This interval censoring can bias results if intervals are large relative to the total duration of the exposure. Newman et al. (1994) and Newman (1995) provide a detailed example of interval censoring in survival model development.

B. NONPARAMETRIC (PRODUCT-LIMIT) METHODS

S(t) or the cumulative survival density function can be described by a nonparametric product-limit estimator (Kaplan and Meier, 1958) as detailed in Miller (1981), Blackstone (1986), Dixon and Newman (1991), and Newman (1995). It is initially 1 and declines over time as individuals die. It remains undefined beyond the end of the exposure period (T) if right censoring is present, i.e., there are survivors.

$$\hat{S}(t_i) = \prod_{j=1}^{i}\left(1 - \frac{d_j}{n_j}\right) \tag{1}$$

where n_j = the number of individuals alive just before t_j, and d_j = the number of individuals dying at t_j. Greenwood's formula (Equation 2) estimates the variance for the product-limit $\hat{S}(t_i)$ which becomes Equation 3 for all times prior to the end of the exposure if no individuals were censored before the end of the experiment.

$$\hat{\sigma}^2 = \sum_{j=1}^{i} \frac{d_j}{n_j s_j} \tag{2}$$

where $s_j = n_j - d_j$.

$$\hat{\sigma}^2(t_i) = \frac{\hat{S}(t_i)[1 - \hat{S}(t_i)]}{N} \tag{3}$$

where N = the total number of individuals exposed.

The standard error for $\hat{S}(t_i)$ is the square root of the variance calculated above. This standard error and $z_{1-\alpha/2}$ can be used to produce confidence intervals as described in SAS Institute (1988), Dixon and Newman (1991), and Newman (1995).

Dixon and Newman (1991) provide a detailed application of these methods to survival data, including SAS code to produce survival curves with confidence intervals. They also illustrate the use of log-rank and Wilcoxon tests for equivalence of survival curves for treatment classes of tuberculosis-infected mice, trout exposed to different dissolved oxygen concentrations, and mosquitofish of different sexes and genotypes exposed to arsenate. With these methods, Newman (1995) tested and accepted the null hypothesis that survival time data for duplicate tanks in a salt toxicity test (Newman and Aplin, 1992) came from the same distribution. Appendix 1 provides SAS program code to implement such a comparison. This code also generates product-limit estimates and associated standard errors. More detailed description of these results is given in Newman (1995). Rejection of the null hypothesis using the calculated χ^2 statistic implies that the duplicates were not acceptably similar, much as Fisher's exact test might be used to test whether duplicate treatments in a dose-response test were homogeneous. The Wilcoxon test tends to be more sensitive than the log-rank test to deviations at the onset of exposure (Dixon and Newman, 1991).

C. SEMIPARAMETRIC COX PROPORTIONAL HAZARD METHODS

This approach is based on the assumption that the effect of a covariate such as toxicant concentration is to shift the baseline hazard (probability of dying during an interval). The amount by which the baseline hazard is shifted remains constant with time: the effect of each covariate does not change over the duration of exposure (Equation 4).

$$h(t, x_i) = e^{f(x_i)} h_0(t) \qquad (4)$$

where $h(t,x_i)$ = the hazard at t given the covariate value x_i, $h_0(t)$ = the baseline hazard, and $e^{f(xi)}$ = the function of the covariate x acting on the baseline hazard.

More precisely, the Cox proportional hazard model assumes "linearity and additivity of the predictors with respect to log hazard" (Harrell, 1988). No specific distribution is assumed for the baseline hazard to which hazards of other types are scaled. Instead an empirical function (from a family of Lehmann alternatives) is fit for the baseline hazard ($h_0(t)$) using the rank order of the TTDs. (See Miller, 1981, or Cox and Oakes, 1984 for further discussion of Lehmann alternatives.) The neglect of the specific form of the baseline hazard may be by design or necessity. In many clinical trials, the exact form of the underlying hazard is much less important than the effect of various treatments on diminishing the likelihood of dying. Alternatively, Heagler et al. (1993) exposed mosquitofish to inorganic mercury and found no acceptable model for the hazard function: a Cox proportional hazard model was necessary. Steadman et al. (1991) also used such a model to describe the effect of trout pre-exposure to No. 2 fuel oil on TTD during a second exposure.

The term, $e^{f(x_i)}$ scales hazards to the baseline hazard. The significance of a covariate is tested with a χ^2 statistic. The mercury-mosquitofish data from

Table 1   Summary of Cox Proportional Hazard Model for Diamond et al. (1989) data

Variable	df	Estimate (S.E.[a])	χ^2	P of obtaining this χ^2 value by chance alone
NTANK	1	0.1207 (0.0861)	1.965	0.1610
NSEX	1	-0.8033 (0.0894)	80.830	0.0001
LWT	1	-2.7551 (0.2470)	124.474	0.0001

[a] Standard error of the estimate.

Diamond et al. (1989) are used here to illustrate this approach (Table 1). The SAS code to perform the associated analyses is provided in Appendix 2. The cumulative mortality curves for the two exposure tanks and one control tank are provided in Figure 4. The effects of exposure tank (NTANK), fish sex (NSEX), and fish weight (ln of wet weight, LWT) are tested with the SAS PHGLM procedure. Note that two of the covariates (NTANK and NSEX) are class variables and one (LWT) is a continuous variable. The tank effect is not significant at $\alpha = 0.05$; however, mosquitofish sex and wet weight are highly significant. The maximum likelihood estimate of $f(x_i)$ for ln of wet weight was negative, indicating that the hazard increased as size decreased. Smaller fish were more sensitive than larger fish. Sex also has a negative estimate. Note from the code in Appendix 2 that females are arbitrarily designated as 1 and males are designated as 0. Therefore, a decrease in NSEX from 1 to 0 (risk associated with being female is changed to that of a male) increased the hazard. Males were more sensitive than females.

These differences can be expressed as relative risks. (See Dixon and Newman, 1991 for more details.) For a class variable such as sex, the relative risk is $e^{\hat{T}_i}$ where $\hat{T}_i$ is the estimated effect of the ith type within the class. The relative risk of females and males in this Cox proportional hazard model are $e^{-0.8033 \times 1}$ (=0.447) and $e^{-0.8033 \times 0}$ (=1), respectively. Males are more than twice (1/0.447 = 2.23) as likely to die as females. For continuous variables such as ln of wet weight, relative risk is $e^{\hat{\beta}_i \Delta x}$ where $\hat{\beta}_i$ is the estimated effect of the continuous variable and Δ_x is the change in the variable x. For example, the relative risk associated with the difference between a 0.1 g (ln 0.1 = -2.30258) and 0.5 g (ln 0.5 = -0.69315) fish is $e^{-2.7751 \times (-2.30258 - (-0.69315))}$ or approximately 87. A 0.1-g fish is 87 times more likely than a 0.5-g fish to die at any time during exposure. Under the assumption of additivity of covariate effects on the ln hazard, the risk of a 0.1-g male fish relative to a 0.5-g female is approximately 87*2 or 174 times higher.

D. PARAMETRIC METHODS

i. General

In other cases, it is possible and advantageous to define an underlying distribution for the baseline mortality. There are two general forms of such parametric models. A proportional hazard model with the same general form as Equation 4 can be used; however, $h_0(t)$ is now fit to a specific distribution. The exponential and Weibull distributions are those that result in a proportional

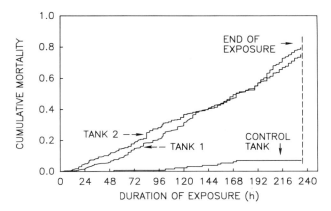

Figure 4 Cumulative time-to-death of mercury-exposed (tanks 1 and 2) and control fish. (Fom Diamond, S.A. et al., 1989. Environ. Toxicol. Chem. 8, 613–622. Reproduced by permission of Pergamon Press.)

hazard model. The second form is the accelerated failure time model, which has the form given in Equation 5.

$$\ln t_i = f(x_i) + \varepsilon_i \tag{5}$$

where t_i = time-to-death, $f(x_i)$ = a function of the covariate effecting $\ln t_i$, and ε_i = an error term. The distributions most often applied to an accelerated failure model are the log normal, log logistic, and gamma. Cox and Oakes (1984) describe less commonly used distributions in their Table 2.1.

Why do the different distributions result in proportional hazard or accelerated failure time models? Cox and Oakes (1984), and Dixon and Newman (1991) point out that the accelerated failure time model (Equation 5) can be expressed also as a model of hazard (Equation 6). Under the assumption of an exponential or Weibull distribution, this general hazard model is a proportional hazard model. Hazard is either constant (exponential distribution) or changing monotonically (Weibull distribution) with time for these two distributions. With the other distributions mentioned above, hazards do not remain proportional through time and an accelerated failure model (Equation 5) is appropriate.

$$h(t, x_i) = e^{f(x_i)} h_0(t, e^{f(x_i)}) \tag{6}$$

The $f(x_i)$ can take on a wide range of forms in both the proportional hazard and accelerated failure time models. Dixon and Newman (1991), and Newman et al. (1994) provide several examples. The function may simply describe the mean response of each class if the variable in question is a class variable, e.g., the mean response for males and females. Alternatively, it may be some function of a continuous variable, e.g., $f(\text{Weight}_i) = \alpha + \beta \ln (\text{Weight}_i)$.

Inclusion of several covariates in the model is easily accomplished. For example, Newman et al. (1994) describe size-dependent mortality during

Table 2   Linearizing Transformations for Selecting Among Candidate Underlying Distributions in Formulating Survival Time Models

Distribution	X	Y
Exponential	ln $\hat{S}(t)$	t
Weibull	ln($-$ln $\hat{S}(t)$)	ln t
Normal	Probit ($\hat{F}(t)$)	t
Log normal	Probit($\hat{F}(t)$)	ln t
Log logistic	ln($\hat{S}(t)/\hat{F}(t)$)	ln t

sodium chloride exposure with the model, $f([NaCl_i], Weight_j) = \alpha + \beta_s \log[NaCl_i] + \beta_w \log Weight_j$. A vector of $\hat{\beta}$ values and $\hat{\alpha}$ are estimated and multiplied by the data matrix to predict TTD. Maximum quasi-likelihood estimates ($\hat{\beta}$s and $\hat{\alpha}$) are generated assuming a distribution for ε.

ii. Application of Parametric Models

Several tools are available for selecting the best model for survival data. A series of linearizing transformations can be done on the data and their effectiveness examined visually. The most common linearizing transformations are provided in Table 2. No linearizing transformation is available for the gamma distribution. (Appendix 3 contains SAS program code for plotting the various linearizing transformations for data from Diamond et al., 1989 shown in Figure 4.) For example, Figure 5 is a plot of the data from Newman and Aplin (1992) assessing the Weibull model. $\hat{S}(t)$ is estimated up to the end of the exposure as one minus the cumulative mortality at each time, i.e., one minus the total number of fish dying by time t divided by the total number of exposed fish. (Similarly, $\hat{F}(t)$ used in other transformations in Table 2 is estimated as the total number of fish dying by time t divided by the total number of exposed fish.) The relatively straight and parallel lines for the five salt treatments suggest that the Weibull model is appropriate for these data.

The log likelihood statistic may also be used to compare candidate models much as sums of squares are used to compare model fit with least-squares methods. The model with the largest (least negative) log likelihood statistic has the best fit to the data. The SAS code provided in Appendix 4 generates the log likelihood statistics required to assess the relative goodness of fit for various underlying distributions (exponential, Weibull, log normal, log logistic, and gamma) in combination with various covariate (salt concentration and fish wet weight) transformations. The log likelihood statistics cannot be used directly in this situation because the number of parameters varies among the candidate models. Instead Akaike's information criterion (AIC), which adjusts the log likelihood value to account for the varying number of parameters, can be used for model comparison (Atkinson, 1980; Harrell, 1988).

$$AIC = -2(\log \text{ likelihood}) + 2P \qquad (7)$$

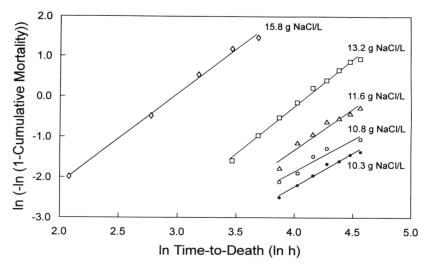

Figure 5 Transformed sodium chloride TTD data. (From Newman, M.C. and M.S. Aplin, 1992. Aquat. Toxicol. (Amst.) 23, 85–96. Reproduced by permission of Elsevier Science Ltd.)

where P = the number of parameters fit with the model. In the exponential model, three parameters are estimated, the two $\hat{\beta}$s for the effects of size and salt concentration plus the exponential parameter ($\hat{\mu}$). In the Weibull model, the two $\hat{\beta}$s for the effects of size and salt concentration plus the two Weibull parameters (the scale and shape parameters) are estimated. Similarly, the log normal, log logistic, and gamma models involve estimation of the two $\hat{\beta}$s plus the parameters required to define the log normal (two parameters), log logistic (two parameters), or gamma (three parameters) distributions. As the log logistic model using ln transformed wet weight and ln transformed salt concentration has the smallest AIC, it fits these data best (Table 3). The parameter estimates generated for these data using the log logistic model with log transformed covariates are provided in Table 4. Notice that two additional parameters are provided for the parametric, log logistic model (Table 4) relative to the Cox proportional hazard model (e.g., Table 1). The values of $\hat{\mu}$ and $\hat{\sigma}$ are estimates of the central tendency and logistic scale parameter. The log logistic model fit to these data is the following.

$$\text{Ln TTD} = \hat{\mu} + \hat{\beta}_s \ln[\text{NaCl}] + \hat{\beta}_w \ln \text{ Weight} + \hat{\varepsilon} \qquad (8)$$

or

$$\text{TTD} = e^{\hat{\mu}} e^{\hat{\beta}_s \ln[\text{NaCl}]} e^{\hat{\beta}_w \ln \text{ Weight}} e^{\hat{\varepsilon}} \qquad (9)$$

The scale parameter and the value for 50% from a standardized error distribution assuming a logistic distribution are used for estimation of the

Table 3 Comparison of Candidate Models for Describing Survival Time of Mosquitofish Exposed to a Series of Sodium Chloride Concentrations

Transformations[a] of weight/salt concentration	Distribution	Log likelihood	No. of parameters	AIC
A/A	Exponential	−385	3	776
A/A	Weibull	−194	4	396
A/A	Log normal	−202	4	412
A/A	Log logistic	−198	4	404
A/A	Gamma	−193	5	396
A/L	Exponential	−383	3	772
A/L	Weibull	−193	4	394
A/L	Log normal	−193	4	394
A/L	Log logistic	−190	4	388
A/L	Gamma	−189	5	388
L/A	Exponential	−377	3	760
L/A	Weibull	−196	4	400
L/A	Log normal	−195	4	398
L/A	Log logistic	−190	4	388
L/A	Gamma	−191	5	392
L/L	Exponential	−376	3	758
L/L	Weibull	−194	4	396
L/L	Log normal	−186	4	380
L/L	Log logistic	−182	4	372
L/L	Gamma	−185	5	380

[a] A, arithmetic; L, ln of covariate. A/L would indicate that wet weight was used in the arithmetic and salt concentration was transformed to its natural logarithm in the model.

From Newman, M.C. and M.S. Aplin, 1992. Aquat. Toxicol. (Amst.) 23, 85–96. With permission of Elsevier Science Ltd.

Table 4 Summary of the Log Logistic Model for the Mosquitofish-Sodium Chloride Toxicity Data of Newman and Aplin (1992)

Variable	df	Estimate (S.E.[a])	χ^2	Probability of obtaining this χ^2 value by chance alone
Intercept (μ)	1	15.2860 (0.2563)	3555.98	<0.0001
Ln [NaCl] (β_s)	1	−4.2129 (0.0830)	2575.80	<0.0001
Ln Weight (β_w)	1	0.2545 (0.0386)	43.50	0.0001
Scale (σ)[b]	1	0.2081 (0.0104)		

[a] Standard error of the estimate.

[b] The scale parameter for the logistic distribution.

median TTD. The value ($L_{0.5}$) corresponding to 50% for a logistic distribution (0) is taken from a table such as Appendix 7 in Newman (1995). The median TTD (MTTD) for this log logistic model is predicted with Equation 10. Equation 10 reduces to Equation 12. (Appendix 5 also provides SAS program code to calculate MTTD and its associated confidence interval.)

$$MTTD = e^{\hat{\mu}} e^{\hat{\beta}_s \ln[NaCl]} e^{\hat{\beta}_w \ln \text{Weight}} e^{\hat{\sigma} L_{0.5}} \qquad (10)$$

$$MTTD = e^{\hat{\mu}} e^{\hat{\beta}_s \ln[NaCl]} e^{\hat{\beta}_w \ln \text{Weight}} e^{\hat{\sigma}*0} \qquad (11)$$

$$MTTD = e^{\hat{\mu}} e^{\hat{\beta}_s \ln[NaCl]} e^{\hat{\beta}_w \ln \text{Weight}} \qquad (12)$$

These MTTD can be estimated for a wide range of salt concentrations (Figure 6). If the sodium chloride data had been used to estimate 96-h LC50 with a conventional dose-response method such as the trimmed Spearman-Karber method, only one point and its confidence interval would have been generated in Figure 6. However, a series of estimates of toxic effect, including confidence intervals, can be generated using the methods just described for these same data. Even more useful predictions can be generated from these models. For example, Figure 7 displays predicted MTTD at various sodium chloride concentrations for various sized fish.

Other TTD for other percentages of mortality can be calculated using the corresponding $L_{0.X}$ values from the standardized error distribution for the logistic distribution. In the SAS program code in Appendix 5, the requested quantile (Q = 0.5 for median) can be easily changed to facilitate these calculations. Further, the parameter estimates for other models, such as the Weibull model discussed immediately below, can be used to estimate TTD for any percentile using the appropriate standardized error distribution. Respectively, $W_{0.5} = -0.36651$ and $N_{0.5} = 0.00000$ would be used instead of $L_{0.5}$ in equations such as those above for estimation of the MTTD for the Weibull and log normal models.

The Weibull model for these data (Table 5) can be used to illustrate the estimation of relative risk with a parametric proportional hazard model. The Weibull model employing arithmetic covariates (salt concentration and fish wet weight) suffices as it has an AIC value only slightly larger than the log logistic (transformed covariates) model and it generates similar predictions in the pertinent toxicant concentration range (Figure 6). The approach is similar to that described previously for the Cox proportional hazard model, except the scale parameter estimate is incorporated. The relative risk is $e^{-T_i/s}$ for a class variable such as fish sex or $e^{-\beta_i \Delta/s}$ for a continuous variable such as salt concentration or fish wet weight. Using this Weibull model with untransformed covariates as an example, the risk of a 0.1-g fish relative to a 1.0-g fish is $e^{(-1.0602*-0.9)/0.3046}$ or approximately 23 times higher during salt exposure. Combined risks associated with differences in two or more covariates may be estimated as done with the Cox proportional hazard model results.

Conventional estimates of toxicity such as the 96-h LC50 can be generated with these parametric models (Newman and Aplin, 1992). Equation 13 predicts LC50 for the Weibull model with untransformed covariates. LC50 could be estimated for any time by changing ln 96 to the ln of the particular time of interest. The LC50 for fish of various weights could be calculated by specifying the particular weight of interest (Weight).

$$96 - h\ LC50 = \frac{\ln 96 - \hat{\mu} - \hat{\beta}_w \text{Weight} - \hat{\sigma} W_{0.5}}{\hat{\beta}_s} \qquad (13)$$

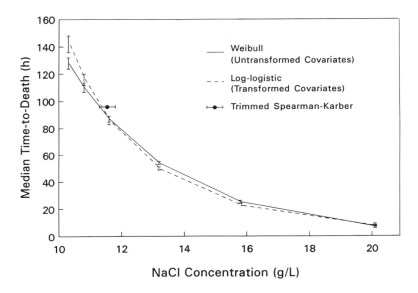

Figure 6   Prediction of median times-to-death during sodium chloride exposure using a
Weibull model with untransformed covariates and a log logistic model with
transformed covariates. (From Newman M. C. and M.S. Aplin, 1992. Aquat.
Toxicol. (Amst.) 23, 85–96. Reproduced by permission of Elsevier Science Ltd.)

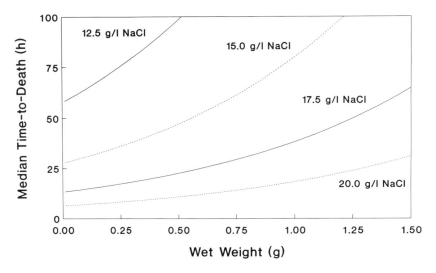

Figure 7   Prediction of median times-to-death for different concentrations of sodium
chloride and mosquitofish sizes. (From Newman M.C. and M.S. Aplin, 1992.
Aquat. Toxicol. (Amst.) 23, 85–96. Reproduced by permission of Elsevier
Science Ltd.)

Table 5 Summary of the Weibull Model for the Mosquitofish-Sodium Chloride Toxicity Data of Newman and Aplin (1992)

Variable	df	Estimate (S.E.[a])	χ^2	Probability of obtaining this χ^2 value by chance alone
Intercept (μ)	1	7.8579 (0.0853)	8487.27	<0.0001
[NaCl] (β_s)	1	−0.2953 (0.0052)	3257.53	<0.0001
Weight (β_w)	1	1.0602 (0.2566)	17.07	0.0001
Scale (σ)[b]	1	0.3046 (0.0137)		

[a] Standard error of the estimate.

[b] The scale parameter for the Weibull distribution.

iii. Multiple Comparisons

The class variables considered to this point have been those involving only two types: tank 1 versus tank 2, or male versus female. Means of comparing among several types within a class, such as several genotypes at a locus, are presented in this section using data generated initially by Diamond et al. (1989). In this study, mosquitofish in duplicate tanks were exposed for 240 h to 0.964 µg/l of inorganic mercury. The TTD for each fish was noted within 3-h intervals. The TTD, fish weight, fish sex, and genotype at eight loci were noted. Data for fish from the duplicate tanks were pooled, as preliminary analyses indicated no significant tank effect on TTD (Table 1). The effect of genotype (six genotypes) at the glucosephosphate isomerase-2 locus (PGI-2) on TTD is analyzed here in addition to the effects of fish sex and size noted as significant in Table 1. The $\log_{10}$ of wet weight is used in contrast to the original analysis of Diamond et al. (1989) to improve goodness-of-fit. The code detailed in Appendix 6 applied to these data produced the model estimates in Table 6.

Note that both sex and GPI-2 genotype are treated as class variables. For sex classes, males were arbitrarily selected as the reference sex and the effect of being female estimated based on the differences between the means of the two sexes. In contrast with continuous variables for which the value of the variable is used, an indicator variable is used for class variables. One is used for all types within a class, except the reference type, which is given an indicator value of zero. The effects of all GPI-2 genotypes are estimated relative to the 38/38 genotype: the 38/38 genotype has an indicator variable of zero and the remaining genotypes have indicator variables of one. For example, the MTTD for a male fish of 0.15 g weight ($\log_{10}$ of 0.15 = −0.8239) and genotype of 100/100 would be 124 h as calculated with Equation 14. That for a female fish with the same weight and genotype would be 182 h as calculated with Equation 15.

$$\text{MTTD} = e^{5.7191} e^{(0*0)} e^{(1.2971* -0.8239)} e^{(0.3571*1)} e^{(0.5082* -0.3665)} \tag{14}$$

$$\text{MTTD} = e^{5.7191} e^{(0.3806*1)} e^{(1.2971* -0.8239)} e^{(0.3571*1)} e^{(0.5082* -0.3665)} \tag{15}$$

Table 6   Summary of the Weibull Model for the Mosquitofish-Mercury Toxicity Data of Diamond et al. (1989)

Variable	df	Estimate (S.E.[a])	χ^2	Probability of obtaining this χ^2 value by chance alone
Intercept (μ)	1	5.7191 (0.1478)	1496.93	<0.0001
Sex (β_s)	1		70.00	0.0001
Female	1	0.3806 (0.0455)	70.00	0.0001
Male	0	0(0)	—	—
Log$_{10}$ Weight (β_w)	1	1.2971 (0.1254)	107.03	0.0001
PGI-2 genotype	5		16.86	0.0048
100/100	1	0.3571 (0.1162)	9.44	0.0021
66/100	1	0.4544 (0.1152)	15.55	0.0001
38/100	1	0.3551 (0.1216)	8.53	0.0035
66/66	1	0.3737 (0.1236)	9.14	0.0025
38/66	1	0.3235 (0.1393)	5.39	0.0202
38/38	1	0(0)		
Scale (σ)[b]	1	0.5082 (0.0189)		

[a] Standard error of the estimate.
[b] The scale parameter for the Weibull distribution.

The effect of each class variable is indicated by an overall χ^2, e.g., 16.86 with df = 5 for an overall effect of GPI-2 genotype on TTD (Table 6). There was clearly a significant effect of GPI-2 genotype on TTD with the 38/38 being the most sensitive genotype. Additional χ^2 values are provided for pairwise comparisons of the reference genotype (38/38) to each of the other five genotypes. Adjusting the experimentwise α level to account for five comparisons, differences between the reference genotype and each of the other genotypes can be tested. However, additional information is required to compare all $k(k-1)/2$ possible genotype pairs. (k is the number of types in the class, e.g., six GPI-2 genotypes will require 15 pairwise comparisons.) The convenient SAS code provided by Fox (1993) is modified for this purpose in Appendix 7. Values from the parameter covariance matrix output from the SAS program are used in Equation 16 (Fox, 1993) to estimate a z score.

$$Z = \frac{\left|\hat{\beta}_i - \hat{\beta}_j\right|}{\sqrt{V_{ii} + V_{jj} - 2(V_{ij})}} \tag{16}$$

where $\hat{\beta}_i$ and $\hat{\beta}_j$ = the estimates for the ith and jth type, and V_{ii}, V_{jj}, V_{ij} = the iith, jjth, and ijth element of the covariance matrix.

The associated probabilities (*P*) are calculated or taken from a table using these z scores. These probabilities are then compared to an experimentwise α. Fox (1993) uses the following adjustment (Dunn-Šidák adjustment with a one-sided interval) to ensure that the experimentwise α remains constant (e.g., 0.05).

$$\alpha' = 1 - \left(\frac{1}{2}\right)^{1/p} (2 - \alpha)^{1/p} \tag{17}$$

or

$$\alpha' = 1 - \left(1 - \frac{\alpha}{2}\right)^{1/p} \qquad (18)$$

where p = the number of comparisons. Fox's use of the one-sided interval is not appropriate and, in our modified SAS program code in Appendix 7, Dunn-Šidák adjustment for a two-sided interval is made according to Equation 19 instead. Other adjustments such as the Bonferroni adjustment may also be used, but they are generally more conservative than the Dunn-Šidák adjustment.

$$\alpha' = 1 - (1 - \alpha)^{1/p} \qquad (19)$$

The results suggest that the 38/38 genotype is significantly different (experimentwise $\alpha = 0.05$) from all genotypes except the 38/66 genotype (Table 7). There are no other significant differences between genotypes.

iv. Summary of Parametric Methods

The mechanisms underpinning the most common models remain as speculative as those discussed earlier for the dose-response models. For example, the exponential model suggests a simple process with a constant hazard over time. The constant hazard implies no system memory, i.e., the hazard at any time interval is unrelated to what occurred in previous intervals (Cox and Oakes, 1984). The best example of this type of process is radioactive decay, in which the probability of a radionuclide atom decaying is independent of any processes that took place in past time intervals. Obviously, this type of model would be inappropriate if a slow accumulation of damage or "wear" over time was suspected to result in death (Nelson, 1969). Cox and Oakes (1984) give an example of a distribution of "loads" described by a Poisson process: there are many "loads" or toxicant damages that the organism could experience during exposure. Failure occurs the first time that an extreme "load" is experienced. This explanation is similar to that given earlier for a dose-response model based on the I.E.D. Use of the more general Weibull model suggests a "weak link" failure mode (Dixon and Newman, 1991). The Weibull shape parameter is speculated to reflect the number of "components," "parts," or "links in a chain" available to fail and cause death. "... if a unit can be regarded as a system of many parts each with a failure time from the same distribution and if the unit fails with the first part failure, then a Weibull distribution may be appropriate" (Nelson, 1969). It follows mathematically and conceptually that the Weibull model with a shape parameter of one (one component available to fail) reduces to an exponential model.

Regardless of the predominantly nonmechanistic motivations behind their application, a wide spectrum of powerful parametric models for survival time can be used to predict survival as a function of exposure time and covariates. In addition to the methods described here, means of incorporating covariates

Table 7   Summary of Comparisons Among GPI-2 Genotypes

Genotype	100/100	66/100	38/100	66/66	38/66	38/38
100/100	—	N	N	N	N	Y
66/100		—	N	N	N	Y
38/100			—	N	N	Y
66/66				—	N	Y
38/66					—	N
38/38						–

An experimentwise $\alpha = 0.05$ was maintained by adjusting the pairwise αs using the Dunn-Šidàk method ($\alpha' = 0.003$).

Note: Comparisons were based on the modified code of Fox (1993) as provided in Appendix 7. See Table 6 for estimates of effect for each genotype.

that change over the duration of the exposure, e.g., toxicant accumulating in a target organ, are available. Chapter 8 in Cox and Oakes (1984) and Chapter 6 in Miller (1981) provide general discussions of time-dependent covariates. Several statistical computer programs, e.g., SAS (SAS Institute Inc., 1988) allow inclusion of time-dependent covariates. Such models have promise in linking bioaccumulation to toxicant effect. Concentration present in a critical tissue or cumulative amount of exposure in the tissue during exposure, i.e., the area under the time-tissue concentration curve, could be explored as the time-dependent covariate depending on the assumptions made regarding the mode of failure. These methods can also be extended to estimation of other ecologically relevant qualities, such as genetic fitness (Manly, 1985; Newman, 1995), seedling emergence and flowering (Fox, 1990a, b, 1993), and foraging behavior (Muenchow, 1986).

IV. CONCLUSION

Relative to dose-response methods, those based on time-response permit more meaningful inclusion of time, increased precision of estimates, and, because of their enhanced power, more effective incorporation of covariates. Enhancing the precision of estimates becomes increasingly important as our focus moves downward from 50% mortality to lower percentages. Resulting models may be linked to demographic and population genetics models. For these reasons, use of the time-response methods described herein will enhance our ability to predict (or test the significance of) covariate effects on toxicity. They provide more ecologically meaningful estimates of lethal effect than dose-response methods.

ACKNOWLEDGMENTS

This work was supported by contract DE-AC09-76SROO-819 between the U.S. Department of Energy and the University of Georgia. We appreciated the thoughtful reviews by Drs. G. Fox, T. La Point, P. Landrum, F. Mayer, M. Mulvey, and C. Strojan of an earlier version of the manuscript.

Appendix 1 Statistical Analysis System (SAS) Code for Nonparametric Tests of Equality Between Classes (Duplicate Tanks of Fish Exposed to Various Salt Concentrations)

```
DATA TOXICITY;
 INFILE "B:TOXICITY.DAT";
 INPUT TTD 1-2 TANK $ 4-5 PPT 6-10 WETWGT 12-16 STDLGTH 18-20;
 IF TTD>96 THEN FLAG=1;
  ELSE FLAG=2;
RUN;
PROC SORT;
 BY PPT TANK TTD;
RUN;
PROC LIFETEST;
 TIME TTD*FLAG(1);
 STRATA TANK;
 BY PPT;
RUN;
```

The data in file "TOXICITY.DAT" are identified and their format defined. The time-to-death (TTD), replicate tank (TANK), sodium chloride concentration in g/l (PPT), fish wet weight (WETWGT), and fish standard length (STDLGTH) are defined. If the TTD is greater than 96 h (longer than the duration of the experiment), the fish survived to the end of the exposure and it is identified as censored (FLAG = 1). After these data are sorted, the LIFETEST procedure is used to test for significant differences between duplicates at each salt concentration. Product-limit survival estimates (including standard errors) are tabulated in the resulting output. A χ^2 statistic and an associated P is calculated to facilitate testing of differences based on the Log-rank, Wilcoxon, and −2 log (likelihood ratio) methods.

Note: These data come from Newman and Aplin (1992). A detailed description of this example is contained in Newman (1995).

Appendix 2 Statistical Analysis System (SAS) Code for Using a Cox Proportional Hazard to Analyze the Mercury-Exposed Mosquitofish Data (Diamond et al., 1989)

Fish sex (NSEX as a numerical variable), fish weight (LWT as the log of wet weight), and duplicate tank (NTANK as a numerical variable) are tested for significant effect.

```
DATA FISH;
 INFILE "B:FISH.DAT";
 INPUT TTD 5-9 TANK $ 11 SEX $ 13 SIZE 15-18;
 IF TANK NE "C";
 IF SEX = "F" THEN NSEX=1;
  ELSE NSEX=0;
 IF TANK="1" THEN NTANK=1;
  ELSE NTANK=0;
 LWT=LOG10(SIZE);
 IF TTD>234 THEN FLAG=1;
  ELSE FLAG=2;
RUN;
PROC SORT;
 BY NTANK NSEX;
RUN;
PROC PHREG;
 CLASS NTANK NSEX;
 MODEL TTD*FLAG(1)=NTANK NSEX LWT;
RUN;
```

No underlying distribution is assumed for the baseline hazard in this model. However, hazards are assumed proportional among classes (duplicate tanks and sexes).

**Appendix 3 Statistical Analysis System (SAS) Code for Plotting
Various Linear Transformations of Mercury Survival
Data from Diamond et al. (1989)**

Cumulative mortality (CDF_TTD) in two, replicate tanks containing 367 (TANK 1) and 375 (TANK 2) mosquitofish are plotted to visually assess the appropriateness of the exponential, Weibull, log normal, and log logistic distributions.

```
DATA FISH;
 INFILE "B:FISH.DAT";
 INPUT ID 1-3 TTD 5-9 TANK $ 11 SEX $ 13 SIZE 15-18;
 IF TREAT NE "C";
 IF TTD>234 THEN FLAG=1;
  ELSE FLAG=2;
RUN;
PROC SORT;
 BY TANK TTD;
RUN;
DATA FIGURE;
 KEEP TANK TTD CUM_TTD CDF_TTD ST FT PRIT E H ODD LODD LOG_H
 LOG_T;
 SET FISH;
 BY TANK TTD;
 RETAIN CUM_TTD 0;
  IF FIRST.TANK THEN CUM_TTD=0;
  CUM_TTD=CUM_TTD+1;
  IF LAST.TTD THEN DO;
   IF TANK="1" THEN CDF_TTD=CUM_TTD/367;
    ELSE CDF_TTD=CUM_TTD/375;
  ST=1-CDF_TTD;
  FT=1-ST;
  PRIT=PROBIT(FT);    /* Five may be added to Prit (normal equivalent deviate or */
  E=LOG(ST);          /* inverse normal distribution function) to get the probit   */
  H=-LOG(ST);
  ODD=ST/FT;
  LODD=LOG(ODD);
  LOG_H=LOG(H);
  LOG_T=LOG(TTD);
  OUTPUT;
 END;
RUN;
PROC PLOT;
 PLOT E*TTD=TANK;           /* EXPONENTIAL   */
 PLOT LOG_H*LOG_T=TANK;     /* WEIBULL       */
 PLOT PRIT*LOG_T=TANK;      /* LOG NORMAL    */
 PLOT LODD*LOG_T=TANK;      /* LOG LOGISTIC  */
RUN;
```

A series of potentially linearizing plots are generated with this code. Gross assessment of goodness-of-fit of the data to the underlying model can be made with these plots.

**Appendix 4 Statistical Analysis System (SAS) Code for Selection Among
Various Model Formulations by Comparing Underlying Distributions
and Variable Transformations of the Newman and Aplin (1992) data**

```
DATA TOXICITY;
 INFILE "B:TOXICITY.DAT";
 INPUT TTD 1-2 TANK $ 4-5 PPT 6-10 WETWGT 12-16 STDLGTH 18-20;
 IF PPT>0;
 IF TTD>96 THEN FLAG=1;
  ELSE FLAG=2;
 LWETWGT=LOG(WETWGT);
```

```
    LPPT=LOG(PPT);
RUN;
PROC SORT;
  BY PPT WETWGT TTD;
RUN;
PROC LIFEREG;
    MODEL TTD*FLAG(1)=PPT WETWGT/DISTRIBUTION=EXPONENTIAL;
    MODEL TTD*FLAG(1)=PPT WETWGT/DISTRIBUTION=WEIBULL;
    MODEL TTD*FLAG(1)=PPT WETWGT/DISTRIBUTION=LNORMAL;
    MODEL TTD*FLAG(1)=PPT WETWGT/DISTRIBUTION=LLOGISTIC;
    MODEL TTD*FLAG(1)=PPT WETWGT/DISTRIBUTION=GAMMA;
    MODEL TTD*FLAG(1)=PPT LWETWGT/DISTRIBUTION=EXPONENTIAL;
    MODEL TTD*FLAG(1)=PPT LWETWGT/DISTRIBUTION=WEIBULL;
    MODEL TTD*FLAG(1)=PPT LWETWGT/DISTRIBUTION=LNORMAL;
    MODEL TTD*FLAG(1)=PPT LWETWGT/DISTRIBUTION=LLOGISTIC;
    MODEL TTD*FLAG(1)=PPT LWETWGT/DISTRIBUTION=GAMMA;
    MODEL TTD*FLAG(1)=LPPT WETWGT/DISTRIBUTION=EXPONENTIAL;
    MODEL TTD*FLAG(1)=LPPT WETWGT/DISTRIBUTION=WEIBULL;
    MODEL TTD*FLAG(1)=LPPT WETWGT/DISTRIBUTION=LNORMAL;
    MODEL TTD*FLAG(1)=LPPT WETWGT/DISTRIBUTION=LLOGISTIC;
    MODEL TTD*FLAG(1)=LPPT WETWGT/DISTRIBUTION=GAMMA;
    MODEL TTD*FLAG(1)=LPPT LWETWGT/DISTRIBUTION=EXPONENTIAL;
    MODEL TTD*FLAG(1)=LPPT LWETWGT/DISTRIBUTION=WEIBULL;
    MODEL TTD*FLAG(1)=LPPT LWETWGT/DISTRIBUTION=LNORMAL;
    MODEL TTD*FLAG(1)=LPPT LWETWGT/DISTRIBUTION=LLOGISTIC;
    MODEL TTD*FLAG(1)=LPPT LWETWGT/DISTRIBUTION=GAMMA;
RUN;
```

The data in file "TOXICITY.DAT" are identified and their format defined as detailed in Appendix 1. Two additional variables are created (LPPT = natural log of salt concentration and LWETWGT = natural log of wet weight). After sorting, the LIFEREG procedure is used to generate models using the various distributions and variable transformations. The log likelihood statistic is used indirectly to compare relative goodness of fit for the candidate models.

Appendix 5 Statistical Analysis System (SAS) Code for Generating Predicted Times-to-Death and Associated Standard Errors for the Newman and Aplin (1992) Data Set

```
DATA TOXICITY;
  INFILE "B:TOXICITY.DAT";
  INPUT TTD 1-2 TANK $ 4-5 PPT 6-10 WETWGT 12-16 STDLGTH 18-20;
  IF PPT>0;
  IF TTD>96 THEN FLAG=1;
    ELSE FLAG=2;
  LWETWGT=LOG(WETWGT);
  LPPT=LOG(PPT);
RUN;
PROC SORT;
  BY LPPT LWETWGT TTD;
RUN;
PROC LIFEREG;
  MODEL TTD*FLAG(1)=LPPT LWETWGT/DISTRIBUTION=LLOGISTIC;
  OUTPUT OUT=TIMEL Q=0.5 P=LLPRED CDF=LLALIVE STD=LLSTD;
RUN;
PROC PRINT;
  VAR PPT WETWGT TTD LLPRED LLSTD.
```

The data in file "TOXICITY.DAT" are identified and their format defined as detailed in Appendix 1. Two additional variables are created (LPPT = natural log of salt concentration and LWETWGT = natural log of wet weight). After sorting, the LIFEREG procedure is used to generate a log logistic model. The predicted TTD for each record (fish) and its associated standard deviation is calculated, output to a file (TIMEL), and printed. Predictions can also be generated with SAS for fish with other qualities, e.g., predicted median TTD and associated standard deviations for 0.1-g fish held at various salt concentrations. Lines of data with a negative ID, an average standard weight, missing TTD, and different exposure concentrations of interest (e.g., PPT of 10, 12.5, 15.0, and 15.5) are added to the original data set. Associated predicted TTD will be those of "average fish" identical except for the concentration to which they were exposed. Because the TTDs are missing for these fabricated records, these records will not be used for the calculations although predicted values will be generated for each. To conveniently output the predictions from these records only, the code is modified so that only those records with negative IDs are printed.

Appendix 6 Statistical Analysis System (SAS) Code for Generating a Proportional Hazard Model (Weibull Distribution) to Incorporate the Effects of Mosquitofish Sex (SEX), Wet Weight (log of wet weight in g or SIZE), and Genotype at the Glucosephosphate Isomerase-2 Locus (PGI-2) on Time-to-Death During Exposure to Inorganic Mercury

The covariance matrix is requested at the end of the model statement for use later in estimating the significance of differences between all genotypes.

```
DATA HG;
 INFILE "B:HG.DAT";
 INPUT ID 1-3 PGI2 $ 38-39;
RUN;
PROC SORT;
 BY ID;
DATA FISH;
 INFILE "B:FISH.DAT";
 INPUT ID 1-3 TTD 5-9 TANK $ 11 SEX $ 13 SIZE 15-18;
 LSIZE=LOG10(SIZE);
 IF TREAT NE "CONTROL";
 IF TTD>234 THEN FLAG=1;
  ELSE FLAG=2;
RUN;
PROC SORT;
 BY ID;
RUN;
DATA ALL;
 MERGE HG FISH;
 BY ID;
RUN;
PROC SORT DATA=ALL;
 BY SEX PGI2;
RUN;
PROC LIFEREG;
 CLASS SEX PGI2;
 MODEL TTD*FLAG(1)=SEX LSIZE PGI2/COVB;
RUN;
```

A model is generated that estimates β's for the effects of fish sex, log of wet weight, and GPI-2 genotype on TTD. Tests of significant differences (χ^2) are performed. The addition of /COVB to the model statement results in an output containing the covariance matrix. This matrix can then be used for making multiple comparisons, e.g., pairwise comparisons among all genotypes (see Appendix 7).

Appendix 7 Statistical Analysis System (SAS) Code for Doing Multiple Comparisons Between GPI-2 Genotypes in the Mercury-Exposed Mosquitofish Data

The code detailed in Appendix 6 is used to generate estimates of β for each genotype and a covariance matrix. Three files are then made with the output from the code: names of the genotypes (NAMES.DAT), the covariance matrix with 0s in the table for the reference genotype (COVAR.DAT), and the β estimates for each genotype (STATS.DAT). This code is modified from that of Fox (1993). The results are placed into a file (MERCURY.OUT).

```
DATA TOT;
 ARRAY NAMES[6] $;
 ARRAY STATS[6];
 ARRAY COVAR[6,6];
 ARRAY ZSCORE[6,6];
 ARRAY PROB[6,6];
 INFILE "B:NAMES.DAT";
  INPUT NAMES [*] $;
 INFILE "B:STATS.DAT";
  INPUT STATS[*];
 INFILE "B:COVAR.DAT";
  INPUT COVAR[*];
FILE "B:MERCURY.OUT";
PUT 'MULTIPLE COMPARISONS OF PGI-2 GENOTYPES';
PUT 'BASED ON LIFEREG BETA STATISTICS';
PUT;
DO I=1 TO 6;
 DO J=(I+1) TO 6;
    ZSCORE[I,J]=ABS(STATS[I]-STATS[J])/
          SQRT(COVAR[I,I]+COVAR[J,J]-2*COVAR[I,J]);
    PROB[I,J]=1-PROBNORM(ZSCORE[I,J]);
    PUT 'COMPARISON:
    'NAMES[I] '&' NAMES [J]':'
       'Z = ' ZSCORE[I,J]
       'PR{FROM SAME SAMPLE} ='
       PROB[I,J];
     PUT;
   END;
  END;
  PUT;
  PUT 'THESE GIVE PROBABILITIES FOR SINGLE COMPARISONS ONLY.';
  PUT 'FOR MULTIPLE COMPARISONS, USE THE FOLLOWING:';
ALPHA=0.05;
NPOP=6;
NCOMPARE=NPOP*(NPOP-1)/2;
INVCOMP=1/NCOMPARE;
ADJUST=1-(1-ALPHA)**INVCOMP; /* DUNN-SIDAK ADJUSTMENT */
ZADJUST=PROBIT(ADJUST);
PUT 'TO ACCEPT A DIFFERENCE AS SIGNIFICANT AT THE ' ALPHA
    'LEVEL,';
PUT 'USE ONLY Z-VALUES WITH P < '
    ADJUST;
RUN;
```

The SAS data set NAMES.DAT is the following: 100100 10066 10038 6666 6638 3838.

These values are the six genotypes for this three-allele locus, e.g., the 100/100 genotype is designated 100100.

The SAS data set STATS.DAT containing the estimates of β is the following: 0.35713182 0.45437342 0.35505211 0.37369819 0.3235346 0.00000.

These estimates could also have been statistics generated with the nonparametric product-moment method. Indeed, this code was modified slightly from that of Fox (1993), which used the Wilcoxon scores as generated by the SAS procedure LIFETEST. In Fox's analysis, the covariance matrix in COVAR below would contain the covariance matrix for the Wilcoxon scores.

The SAS data set COVAR.DAT is the following:

```
0.013513 0.011806 0.011805 0.011779 0.011768 0.000000
0.011806 0.013278 0.011823 0.011793 0.011779 0.000000
0.011805 0.011823 0.014787 0.011776 0.011799 0.000000
0.011779 0.011793 0.011776 0.015270 0.011762 0.000000
0.011768 0.011779 0.011799 0.011762 0.019413 0.000000
0.000000 0.000000 0.000000 0.000000 0.000000 0.000000
```

The columns containing 0.000000 are those associated with the reference genotype (3838). The remaining values were extracted from the appropriate columns in the covariance matrix generated by SAS. For example, 0.013513 is that associated value with the 100100, 100100 element in the matrix. Below it in the matrix is the value 0.011806 which is associated with the 100100, 10066 element of the matrix.

REFERENCES

Armitage, P. and I. Allen, 1950. Methods of estimating the LD50 in quantal response data. J. Hyg. 48, 298–322.

ASTM, 1989. Standard Guide for Conducting Acute Toxicity Tests with Fishes, Macroinvertebrates, and Amphibians. ASTM E 729–88. American Society for Testing and Materials, Philadelphia, PA, pp. 480–499.

Atkinson, A.C., 1980. A note on the generalized information criterion for choice of a model. Biometrika 67, 413–418.

Berkson, J., 1951. Why I prefer logits to probits. Biometrics 7, 327–339.

Blackstone, E.H., 1986. Analysis of death (survival analysis) and other time-related events. In: Current Status of Clinical Cardiology, edited by F.J. Macartney, MTP Press, Boston, MA, pp. 55–101.

Bliss, C.I., 1935. The calculation of the dosage-mortality curve. Ann. Appl. Biol. 22, 134–167.

Buikema, Jr., A.L., B.R. Niederlehner and J. Cairns, 1982. Biological monitoring. IV. Toxicity testing. Water Res. 16, 239–262.

Burdick, G.E., 1960. The use of bioassays by the water pollution control agency. Biol. Problems Water Pollution Transactions of Second Seminar, 1959, U.S. Public Health Service, R.A. Taft Sanit. Eng. Center Tech. Report W 60–3, U.S. Department of Health, Education and Welfare, Atlanta, 145–148.

Christensen, E.R., 1984. Dose-response functions in aquatic toxicity testing and the Weibull model. Water Res. 18, 213–221.

Christensen, E.R. and N. Nyholm, 1984. Ecotoxicological assays with algae: Weibull dose-response curves. Environ. Sci. Technol. 18, 713–718.

Cox, D.R. and D. Oakes, 1984. Analysis of Survival Data. Chapman and Hall, New York.

CT Department of Environmental Protection, 1990. CT-TOX MULTI-METHOD PROGRAM. Bureau of Water Management, CT Dept. Environ. Protect., Water Toxics Lab., 122 Washington St., Hartford, CT 06106.

Diamond, S.A., M.C. Newman, M. Mulvey, P.M. Dixon and D. Martinson, 1989. Allozyme genotype and time to death of mosquitofish, *Gambusia affinis* (Baird and Girard), during acute exposure to inorganic mercury. Environ. Toxicol. Chem. 8, 613–622.

Diamond, S.A., M.C. Newman, M. Mulvey and S.I. Guttman, 1991. Allozyme genotype and time-to-death of mosquitofish, *Gambusia holbrooki*, during acute inorganic mercury exposure: a comparison of populations. Aquat. Toxicol. (Amst.) 21, 119–134.

Dixon, W.J., 1985. BMDP statistical software 1985. University of California Press, Berkley, CA.

Dixon, P.M. and M.C. Newman, 1991. Analyzing toxicity data using statistical models for time-to-death: an introduction. In: Metal Ecotoxicology. Concepts and Applications, edited by M.C. Newman and A.W. McIntosh, Lewis Publishers, Chelsea, MI, pp. 207–242.

EPA, 1975. Methods for Acute Toxicity Tests with Fish, Macroinvertebrates, and Amphibians. Ecological Research Series No. EPA-660/3-75-013. U.S. Environmental Protection Agency, Washington, D.C.

EPA, 1988. Users guide for a computer program for probit's analysis of data from acute and short-term chronic toxicity tests with aquatic organisms. Biological Methods Branch, Environ. Mon. and Support Lab., Office of Res. and Devel., U.S. Environmental Protection Agency, Cincinnati, OH 45268.

Finney, D.J., 1964. Statistical Method in Biological Assay. Hafner, New York.

Finney, D.J., 1971. Probit Analysis, Cambridge University Press, London.

Fox, G.A., 1990a. Components of flowering time variation in a desert annual. Evolution 44, 1404–1423.

Fox, G.A., 1990b. Perennation and the persistence of annual life histories. Am. Nat. 135, 829–840.

Fox, G.A., 1993. Failure-time analysis: emergence, flowering, survivorship, and other waiting times. In: Design and Analysis of Ecological Experiments, edited by S.M. Scheiner and J. Gurevitch, Chapman and Hall, New York, pp. 253–425.

Francis, B., M. Green and C. Payne, 1993. The GLIM system, release 4 manual. Clarendon Press, Oxford.

Gaddum, J.H., 1953. Bioassays and mathematics. Pharmacol. Rev. 5, 87–134.

Hamilton, M.A., R.C. Russo and R.V. Thurston, 1977. Trimmed Spearmen-Karber method for estimating median lethal concentrations in toxicity bioassays. Environ. Sci. Technol. 11, 714–719.

Harrell, F.E., Jr., 1988. Survival and Risk Analysis. Duke University Medical Center, Durham, NC.

Heagler, M.G., M.C. Newman, M. Mulvey and P.M. Dixon, 1993. Allozyme genotype in mosquitofish, *Gambusia holbrooki*, during mercury exposure: temporal stability, concentration effects and field verification. Environ. Toxicol. Chem. 12, 385–395.

Kaplan, E.L. and P. Meier, 1958. Nonparametric estimation from incomplete observations. J. Am. Stat. Assoc. 53, 457–481.

Keklak, M.M., M.C. Newman and M. Mulvey, 1994. Enhanced uranium tolerance of an exposed population of the Eastern mosquitofish (*Gambusia holbrooki* Girard 1859). Arch. Environ. Contam. Toxicol. 27, 20–24.

Lachenbruch, P.A., 1986. SurvCalc. Survival Analysis on the IBM PC. John Wiley & Sons, New York.

Litchfield, Jr., J.T., 1949. A method for rapid graphic solution of time-per cent effects curves. J. Pharmacol. Exp. Ther. 97, 399–408.

Litchfield, Jr., J.T. and F. Wilcoxon, 1949. A simplified method of evaluating dose-effect experiments. J. Pharmacol. Exp. Ther. 96, 99–113.

Manly, B.F.J., 1985. The statistics of natural selection on animal populations. Chapman and Hall, London.

Mayer, F.L., G.F. Krause, D.R. Buckler, M.R. Ellersieck and G. Lee, 1994. Predicting chronic lethality of chemicals from acute toxicity test data: concepts and linear regression analysis. Environ. Toxicol. Chem. 13, 671–678.

Miller, R.G., Jr., 1981. Survival Analysis. John Wiley & Sons, New York.

Muenchow, G., 1986. Ecological use of failure time analysis. Ecology 67, 246–250.

Nelson, W., 1969. Hazard plotting for incomplete failure data. J. Qual. Technol. 1, 27–52.

Newman, M.C. 1995. Quantitative Methods in Aquatic Ecotoxicology. Lewis Publishers, Chelsea, MI.

Newman, M.C. and M.S. Aplin, 1992. Enhancing toxicity data interpretation and prediction of ecological risk with survival time modeling: an illustration using sodium chloride toxicity to mosquitofish (Gambusia holbrooki). Aquat. Toxicol. (Amst.) 23, 85–96.

Newman, M.C., S.A. Diamond, M. Mulvey and P. Dixon, 1989. Allozyme genotype and time to death of mosquitofish, Gambusia affinis (Baird and Girard) during acute toxicant exposure: a comparison of arsenate and inorganic mercury. Aquat. Toxicol. (Amst.) 15, 141–156.

Newman, M.C., M.M. Keklak and M.S. Doggett, 1994. Quantifying animal size effects on toxicity: a general approach. Aquat. Toxicol. (Amst.) 28, 1–12.

Parrish, P.R., 1985. Acute toxicity tests. In: Fundamentals of Aquatic Toxicology, edited by G.M. Rand and S.R. Petrocelli, Hemisphere Publishing, Washington, D.C., pp. 31–57.

Pinder, J.E., III, J.G. Wiener and M.H. Smith, 1978. The Weibull distribution: a new method of summarizing survivorship data. Ecology 59, 175–179.

SAS Institute Inc., 1988. SAS/STAT User's Guide, Release 6.03 edition. SAS Institute Inc., Cary, NC.

Sprague, J.B., 1969. Measurement of pollutant toxicity to fish. I. Bioassay methods for acute toxicity. Water Res. 3, 793–821.

Sprague, J.B., 1971. Measurement of pollutant toxicity to fish. III. Sublethal effects and "safe" concentrations. Water Res. 5, 245–266.

Statistical Sciences, 1993. S-PLUS Guide to Statistical and Mathematical Analysis, Version 3.2. Mathsoft, Inc., Seattle, WA, pp. 17-1–17-36.

Steadman, B.L., W.A. Stubblefield, T.W. LaPoint and H.L. Bergman, 1991. Decreased survival of rainbow trout exposed to No. 2 fuel oil caused by sublethal preexposure. Environ. Toxicol. Chem. 10, 355–363.

Steinberg, D. and P. Colla, 1988. SURVIVAL: A supplementary module for SYSTAT. SYSTAT, Inc., Evanston, IL.

Stephan, C.E., 1977. Methods for calculating an LC50. In: Aquatic Toxicology and Hazard Evaluation, ASTM STP 634, edited by F.L. Mayer and J.L. Hamelink, American Society for Testing and Materials, Philadelphia, PA, pp. 65–84.

Trevan, J.W., 1927. The error of determination of toxicity. Proc. R. Soc. London Ser. B Biol. Sci. 101, 483–514.

Wang, M.P. and S.A. Hanson, 1985. The acute toxicity of chlorine on freshwater organisms: time-concentration relationships of constant and intermittent exposures. In: Aquatic Toxicology and Hazard Evaluation, ASTM STP 891, edited by R.C. Bahner and D.J. Hansen, American Society for Testing and Materials, Philadelphia, PA, pp. 213–232.

WEST, Inc. and D.D. Gulley, 1994. TOXSTAT 3.4. Western Ecosystems Technology, Inc., Cheyenne, WY.

Demography Meets Ecotoxicology: Untangling the Population Level Effects of Toxic Substances

Hal Caswell

I. INTRODUCTION

Toxic substances have effects at the level of cellular biochemistry, but their ecological consequences are at the levels of the population, community, and ecosystem. Thus, there is a translation problem at the core of ecotoxicology: how to translate mechanisms at one level into effects at another. This problem is not unique to ecotoxicology, but arises in studies of any hierarchical system. In such systems, processes at one level take their mechanisms from the level below and find their consequences at the level above. The organismal physiology that is "mechanistic detail" to a population biologist is the pattern that the physiologist wants to explain and a higher-level integration in the eyes of a biochemist. Recognizing this principle makes it clear that there are no truly "fundamental" explanations, and makes it possible to move smoothly up and down the levels of a hierarchical system without falling into the traps of naive reductionism or pseudo-scientific holism. The chapters in this volume are nice examples.

In this chapter, I am concerned with the translation from individuals to populations using demographic population models as a link. Individual organisms are born, grow, reproduce, and die, and exposure to toxicants alters the risks of these occurrences. The dynamics of populations are determined by the rates of birth, growth, fertility, and mortality that are produced by these individual events. (Collectively, these rates are conveniently referred to as the "vital rates.") By incorporating individual rates into population models, the population effects of toxicant-induced changes in those rates can be calculated.

1-56670-1127-9/95/$0.00+$.50

The focus in this chapter is specifically on the use of population models in demographic bioassay studies; that is, experiments that manipulate toxicant exposure and measure the individual vital rates as endpoints. A recently developed analytical approach that greatly increases the power of such experiments is presented. This includes formulae for several important kinds of experimental designs; each accompanied by one or more examples to make clear the interpretation and value of the results.

A. DIVERSITY, STAGE-SPECIFICITY, AND POPULATION MODELS

Predicting the population-level effects of a toxic substance is challenging because the individual-level effects are *diverse* and *stage specific*. Any toxicant studied in any detail seems to have diverse effects on more than one process (survival, growth, development, maturation, reproduction, etc.). The effects on different processes differ in magnitude and, in some cases, in sign.

In addition, on close examination toxicant effects are often found to be stage specific. Juveniles may be more sensitive than adults, or large individuals more sensitive than small ones, or aquatic larvae more sensitive than terrestrial adults, but it is unlikely that any interesting toxicant will have effects that are the same for all stages of the life cycle. Examples of extreme stage specificity are not hard to find: the 96-h LC50 for copper varies 40-fold between instars of the copepod *Tigriopus californicus* (O'Brien et al., 1988); LC50 values for cadmium vary by as much as 1000-fold between instars of *Chironomus riparius* larvae (Williams et al., 1986), although sometimes the effects are smaller (Collyard et al., 1994).

Translating toxicant effects from the individual to the population thus requires a framework that explicitly includes the vital rates (so that diverse effects can be included) and the structure of the life cycle (so that the effects can be stage specific). Population models can provide such a framework. The simplest population models are written in terms of total abundance and of phenomenological parameters that are defined directly at the population level. A classic example of such a *phenomenological* population model is the logistic equation,

$$\frac{dN}{dt} = r_0 N\left(\frac{K-N}{K}\right) \tag{1}$$

where N is population size, r_0 is the population growth rate at low densities, and K is the carrying capacity of the environment. This model cannot incorporate stage specificity, because it is written in terms of N, and does not differentiate between individuals of different stages. Neither of its parameters (r_0 and K) can be written explicitly in terms of individual rates, so they incorporate no individual-level mechanisms. This makes them difficult to measure and, by extension, difficult to express as functions of toxicant exposure.

In spite of its lack of mechanistic detail, the logistic equation is a simple model that produces a sigmoid growth trajectory that adequately describes the result of simple density dependence. To apply such a model to ecotoxicology, the parameters must be expressed as functions of toxicant exposure. Hallam et al. (1983) analyze such a model, in which r_0 is written as a linear function of the internal concentration of the toxicant. The population equation is coupled to a simple pair of equations for internal and external toxicant concentration, and the resulting system analyzed for conditions that permit persistence or generate extinction.

More detailed are *demographic* models that classify individuals by stage (including, as a special case, age classification) and that describe the dynamics in terms of the stage-specific vital rates. These models (described in more detail in the next section) can be applied to toxicology by making the vital rates functions of toxicant exposure. The great advantage of these models is that the vital rates are well-defined properties of individual organisms, and there are powerful statistical methods for expressing them as functions of exposure (e.g., Newman and Dixon, Chapter 8 in this volume).

At a still finer level of resolution are what might be called *mechanistic* physiological models, which explicitly describe the metabolism, growth, energy allocation, reproduction, etc. of individuals. Hallam et al. (1990a), for example, developed a model for an individual *Daphnia* that includes compartments for labile and structural protein, for lipid storage, and for egg and carapace structures. The movement of material among these compartments is described in terms of ingestion, assimilation, and the losses of energy to work, maintenance, and reproduction. Hallam et al. (1990b) incorporated this individual model into a population model and explored the population-level consequences of toxicant effects at the individual level. Several other groups have also developed detailed models of individuals (e.g., McCauley et al., 1990; Kooijman, 1993). Extending such models from the individual to the population level is an area of active research (e.g., DeAngelis and Gross, 1992). The translation is always possible in principle, although in practice it may be difficult and the resulting models complicated.

Phenomenological, demographic, and mechanistic population models are clearly points along a continuum. I focus here on demographic models, which form a middle ground between phenomenological models that exclude population structure and mechanistic models that include the physiological processes determining the vital rates. Demographic models are most often used as a kind of demographic bioassay. In traditional bioassay experiments, some individual endpoint (death, say) is recorded as a function of toxicant exposure. In demographic bioassay, the endpoint is a set of vital rates, which together suffice to determine the dynamics of the population. A demographic model is used to calculate summary statistics, the most important being the rate of increase, for each exposure level. These statistics describe the integrated population-level consequences of the diverse and stage specific toxicant effects on the individual vital rates.

II. DEMOGRAPHIC MODELS

There are several approaches to modeling populations structured by age or stage. The most familiar to ecologists is the Euler-Lotka equation for the population growth rate r,

$$1 = \int_0^\infty m(a)l(a)e^{-ra}da \tag{2}$$

where the survivorship function $l(a)$ gives the probability of surviving from birth to age a, and $m(a)$ gives the expected number of offspring, per individual aged a, per unit time. The solution r to this equation gives the eventual per-capita rate of increase of a population characterized by $l(a)$ and $m(a)$, which are assumed not to vary with time. This equation, or equations similar to it, can be derived from three different types of models that actually describe the dynamics of the population structure over time. The types differ in whether time and stage are continuous or discrete variables.

When the life cycle is described in terms of a continuous variable like age or size, and time is continuous, the dynamics are described using partial differential equations. Let $n(a,t)$ denote the number of individuals of age a at time t, and let $\mu(a,t)$ and $m(a,t)$ denote the age-specific mortality and reproductive rates, respectively, at time t. The dynamics of the age distribution are given by

$$\frac{\partial n(a,t)}{\partial t} + \frac{\partial n(a,t)}{\partial a} = -\mu(a,t)n(a,t) \tag{3}$$

with a boundary condition describing reproduction (i.e., the production of new, age-0 individuals),

$$n(0,t) = \int_0^\infty n(a,t)m(a,t)da \tag{4}$$

This equation was introduced by McKendrick (1926); the complete story of this type of model is given by Metz and Diekmann (1986; see Metz et al., 1988 for a brief introduction and Kooijman and Metz, 1994 for a toxicological example).

A second approach to demographic modeling describes the dynamics of discrete stages (e.g., age or size *classes*, instars, developmental stages) in continuous time. A differential equation is written for the abundance of each stage; for stage i, for example, we might write

$$\frac{dN_i(t)}{dt} = R_i(t) - R_i(t - \tau_i)P_i(t) - \mu_i N_i(t) \tag{5}$$

Here μ_i is the mortality rate of stage i, R_i is the rate of recruitment from stage $i - 1$ into stage i, τ_i is the duration of stage i, and P_i the probability of survival through stage i. Recruitment into one stage depends on the time required to develop through the previous stage. Thus, the model includes time lags representing the durations of the different stages, and probabilities of surviving through those lags. These models have been used extensively by Nisbet and Gurney and their collaborators (e.g., Nisbet et al., 1989; Nisbet and Gurney, 1986; Crowley et al., 1987).

The third approach, more frequently encountered than the other two, uses matrix population models. These are described at length in Caswell (1989a), and very briefly below. Their main advantage is that they are easy to formulate and to analyze, and are flexible enough to accommodate classification by age, size, or more general developmental stage criteria.

A. MATRIX POPULATION MODELS

In a matrix population model, the life cycle is divided into s stages (age classes, size classes, instars, spatial locations, etc.). The state of the population at time t is given by the vector $\mathbf{n}(t)$, whose entries $n_i(t)$, $i = 1, ..., s$ give the abundance of each stage. The dynamics are specified by a $s \times s$ population projection matrix $\mathbf{A}_t$, where

$$\mathbf{n}(t+1) = \mathbf{A}_t \mathbf{n}(t) \tag{6}$$

The (i,j) entry of $\mathbf{A}_t$ is denoted by $a_{ij}(t)$; it gives the number of stage i individuals at $t + 1$ per stage j individual at time t. Thus, the a_{ij} describe the rates at which individuals move among life cycle stages by survival, growth, maturation, movement, reproduction, etc.; in short, the a_{ij} describe the vital rates. This formulation conceals one additional biological parameter: the *projection interval*, i.e., the interval between t and $t + 1$. Obviously, the transitions open to an individual, and thus the dynamics of the population, will appear different on a time scale of days, months, years, or decades.

Equation 6 is written purposely to indicate that the vital rates may vary through time, which includes the possibility that they vary in response to population density. However, much of what we can say analytically is restricted to the time-invariant model

$$\mathbf{n}(t+1) = \mathbf{A}\mathbf{n}(t) \tag{7}$$

For this model, it is known that (given some simple and almost always satisfied conditions on $\mathbf{A}$) the population will, from any nonzero initial condition, eventually converge to a structure proportional to $\mathbf{w}$ (which is therefore called the *stable stage distribution*) and grow exponentially at a rate λ (which is therefore called the *population growth rate* or *rate of increase*). The corresponding continuous-time rate of increase can be calculated as $r = \log \lambda$.

Mathematically, the growth rate λ is given by the largest eigenvalue of $\mathbf{A}$, and the stable stage distribution $\mathbf{w}$ by the corresponding right eigenvector, which satisfy

$$\mathbf{Aw} = \lambda\mathbf{w} \tag{8}$$

The corresponding left eigenvector $\mathbf{v}$, which satisfies

$$\mathbf{v}^* \mathbf{A} = \lambda\mathbf{v}^* \tag{9}$$

where $\mathbf{v}^*$ is the complex conjugate transpose of $\mathbf{v}$, gives the reproductive value of each stage. (For a complete treatment of these models, see Caswell, 1989a.)

The eigenvalue λ is obtained as the solution of the so-called characteristic equation of $\mathbf{A}$. Solving the characteristic equation can be a tricky numerical procedure, but microcomputer software is now widely available that makes the calculations so simple that there is no excuse for avoiding them. MATLAB® is recommended (The MathWorks, 24 Prime Park Way, Natick, MA 01760), but packages like MathCad®, Gauss®, and Mathematica® can also do many of the same tricks.

i. Projection and Prediction

It is important, especially in an ecotoxicological context, to pause and consider just what this analysis provides, and what it does not. No one seriously believes that the environment is constant, or that populations can continue to grow exponentially forever without encountering density-dependent effects. Thus, Equation 7 cannot be counted on to *predict* the future of a population.

However, it can *project* what the population *would do if the environment were to remain constant at its present state* (the distinction between projection and prediction is due to Keyfitz, 1968). This conditional, hypothetical, subjunctive usage is not mere verbal quibbling. It means that the results of Equation 7 reveal something about the current state of the environment, not the future state of the population. Think of the speedometer in your car. Interpreted as a prediction, a reading of 60 mph tells you that in 1 h you will be 60 miles away in a straight line. This is unlikely to be accurate as a prediction, but is extremely useful as a projection that tells you something important about your present state by telling you what would happen if it were to remain constant.

In the same way, the results of demographic analysis characterize the population consequences of the current environment. The population growth rate λ, in particular, integrates all of the vital rates into a single statistic that describes population growth. The critical value $\lambda = 1$ (equivalently, $r = 0$) separates environments that can support a population from those that cannot. An environment that produces vital rates yielding $\lambda = 2$ is a better environment than one that yields $\lambda = 1$, in the sense that the net effect of all the diverse and stage-specific differences between the two environments is that in the first the population is able to grow more successfully than in the second.

For this reason, demographic analysis is the main tool available for summarizing the link between the individual organism (which is born, lives, grows, and reproduces) and the population (which will increase or decrease depending on the propensities for those individual events). It is a particularly powerful tool in studies that compare more than one environment: this is precisely its utility in toxicology.

ii. Sensitivity Analysis: The Other Piece of the Puzzle

The population growth rate λ depends on the values of the vital rates. It is important in many contexts to understand how λ would change if the vital rates were to change. This is crucial to ecotoxicology, where we are interested in the consequences of changes in the vital rates caused by toxicants.

There is a remarkably simple formula for the sensitivity of λ to any of the matrix entries:

$$\frac{\partial \lambda}{\partial a_{ij}} = \frac{v_i w_j}{\mathbf{v} * \mathbf{w}} \tag{10}$$

where * denotes the complex conjugate transpose (Caswell, 1978; 1989a). This expression gives the slope of the relation between a_{ij} and λ. This simple formula means that the consequences of changes in each of the vital rates can be assessed directly from a single matrix, without having to actually change any of the entries. One simply calculates $\mathbf{w}$ and $\mathbf{v}$, and uses Equation 10 to get the sensitivities.

The sensitivity of λ varies widely from one vital rate to another, often by many orders of magnitude. This means that enormous changes in some vital rates may be less important, when measured in terms of population growth, than much smaller changes in other vital rates.

III. LIFE TABLE RESPONSE EXPERIMENTS

A life table response experiment (LTRE) is an experiment in which the response variable is the entire set of vital rates (Caswell, 1989b). The word "experiment" should be understood broadly, including not only designed manipulative experiments, but also comparative observations. In a toxicological context, such an experiment could equally well be called demographic bioassay. The basic structure of an LTRE is shown in Figure 1. A set of treatments T_i is imposed by the investigator, or occurs naturally. Each treatment produces a corresponding set of environmental conditions E_i. Each of these environments in turn has a diverse and stage specific set of effects on the vital rates.

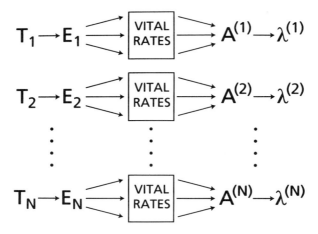

Figure 1   The basic structure of a life table response experiment. A series of treatments T_i — natural or imposed by an investigator — generate a series of environments E_i. Those environmental conditions have diverse and stage-specific effects on the vital rates. These rates are collected into a demographic model, represented here by a population projection matrix $\mathbf{A}^{(i)}$. From the demographic model, the population growth rate $\lambda^{(i)}$ is calculated. The population-level consequences of the treatments, mediated by all the vital rates, are summarized by the patterns of the $\lambda^{(i)}$.

The vital rates are integrated into a demographic model, expressed as a population projection matrix $\mathbf{A}^{(i)}$. From this matrix, one can calculate the whole spectrum of demographic indices, in particular the population growth rate $\lambda^{(i)}$.

This framework translates the individual-level effects of the environment into a set of population-level effects on λ. This approach was introduced in a landmark paper by the Australian ecologist L.C. Birch (1953), entitled "Experimental background to the study of the distribution and abundance of insects." Birch grew two species of flour beetles at various temperature and moisture conditions. He measured age-specific life tables for each species under each temperature-moisture combination, and calculated the population growth rate λ from this information. He expressed his results in a contour plot of λ as a function of temperature and moisture, which clearly revealed the differences in the population-level response of these species to these two environmental variables.

Birch's study was the first really complete LTRE on an animal. Earlier, Raymond Pearl had pioneered the use of the survivorship curve as a dependent variable in an extensive series of studies of the effects of density, food, temperature, and toxicants on *Drosophila*, but, as far as I know, he never combined this information with reproductive data to compute rates of increase or other demographic statistics (e.g., Pearl and Parker, 1922, 1924; Pearl et al., 1927; see Pearl, 1928 for a review). Lotka (1936) had compared values of r for human populations in different states in the U.S., and related them to social variables, but he never extended the approach to animals or plants. The first calculation of the rate of increase for a nonhuman species did not appear until

1940 (Leslie and Ranson, 1940). Birch's (1953) paper was the first to use the methodology to compare the effects of experimental treatments. There have since been many applications of LTREs in animal and plant ecology. Some examples, noteworthy because of their experimental designs or because of the subtlety of the effects they investigate, include Frank et al. (1957; effects of density on *Daphnia pulex*), Watson (1964; effects of host plant nutrient status on the mite *Tetranychus telarius*), King (1967; a three-way factorial study of the effects of food, temperature, and clonal age on the rotifer *Euchlanis dilatata*), Birch et al. (1963; comparing the temperature response of several genotypes of *Drosophila serrata*), Krainacker et al. (1987; the effects of host plant on the Mediterranean fruit fly *Ceratitis capitata*), Reissen and Sprules (1990; demographic costs of antipredator defenses in *Daphnia pulex*), and Lynch (1992; a study comparing the effects of resource depletion on *Daphnia ambigua* and *Ceriodaphnia quadrangula*).

These examples are all laboratory studies of easily cultured organisms with lifetimes short enough to permit cohort life table studies. However, LTREs are not restricted to the laboratory or to short-lived organisms. Particularly with the development of size- and stage-specific demographic methods (Caswell, 1989a), field LTREs have become common, particularly in plants. Examples include Werner and Caswell (1977; comparison of populations of *Dipsacus sylvestris* in eight different old fields), Menges (1987; comparison of the populations of the endangered Furbish's lousewort *Pedicularis furbishiae* along a river system), Kawano et al. (1987; a study of four species of woodland herbs along an environmental gradient), and Horvitz and Schemske (1995; a study of spatial and temporal variation in the demography of the understory herb *Calathea ovandensis*). Animal studies are more difficult, but for examples see Horst (1976; a field study of density effects in the mayfly *Hexagenia limbata*), Levin and Huggett (1990; a field study of two genotypes of the polychaete *Streblospio benedicti*), or Brault and Caswell (1993; a study of the effects of social group structure in the killer whale *Orcinus orca*).

There is a growing literature of toxicological studies using LTREs, e.g., Marshall, 1962 (gamma radiation); Hummon and Hummon, 1975 (DDT); Winner and Farrell, 1976 (copper); Daniels and Allan, 1981 (Dieldrin); Allan and Daniels, 1982 (Kepone); Fitzmayer et al., 1982 (Simazine); Gentile et al., 1982 (heavy metals); Walton et al., 1982 (pH), Rao and Sarma, 1986 (DDT); Levin et al., 1994 (sewage and organic enrichment); see Sibly (Chapter 7 in this volume) for additional references. The full power of the LTRE approach, as seen in the examples cited above, has not yet been applied to ecotoxicology, but there is no reason it cannot be.

It is important to recognize what an LTRE provides, and what it does not. An LTRE measures the population-level effect of the treatment under the conditions of the experiment. Those conditions will probably differ from the conditions in any specific field situation. Thus, the LTRE results are a kind of *population-level bioassay* of the effects of the treatment. Like ordinary, individual-level bioassays, they are projections that tell something about the present, not predictions that tell about the future. They do not permit direct prediction of the growth rate of the population in a field situation unless the rest of the vital rate effects can be extrapolated. Such predictions could be useful, even critical, in some circumstances, but they are extremely difficult. After all, predicting population dynamics in unperturbed environments has frustrated ecologists for as long as there have been ecologists. Predicting the result of perturbations is just that much more difficult.

IV. DECOMPOSING TREATMENT EFFECTS: WHY AND HOW?

Using λ, or other indices, to integrate the diverse and stage-specific effects of environmental factors into their population-level consequences raises another problem. The relation between λ and treatment (Figure 1) shows how the treatments affect population growth, but it obscures the source of those effects. Suppose a treatment increases mortality, reduces growth, and limits reproduction, and suppose that these effects lead to a certain reduction in λ. Which of the vital rate changes is most important? In this section, I will show how to decompose treatment effects on λ into contributions from each of the vital rates, to answer this question.

Decomposition analysis pinpoints the vital rates responsible for the population-level effect of the treatment. It has been applied to a variety of organisms, treatments, and experimental designs:

- Effects of genetic differences in larval development mode in the polychaete *Streblospio benedicti* (Levin et al., 1987; Caswell, 1989c; Levin and Huggett, 1990)
- Interacting effects of food level and DDT exposure in the rotifer *Brachionus patulus* (Caswell, 1989a, using data of Rao and Sarma, 1986)
- Interacting effects of food level and induced predator defenses in the cladoceran *Daphnia pulex* (Walls et al., 1991)
- Effects of fire and fire exclusion on the tropical savanna grasses *Andropogon semiberbis* and *A. brevifolius* (Silva et al., 1991; Canales et al., 1994)
- Effects of pod social structure and of geographical location in killer whales *Orcinus orca* (Brault and Caswell, 1993)
- Effects of pollutants (oil, sewage, and algae) on the polychaetes *Streblospio benedicti* and *Capitella* sp. 1, and differences between those species in their response to pollutants (Levin et al., 1994)
- Contrasts of rare and common species of the plant genus *Eupatorium* (Byers and Meagher, 1995)
- Effects of spatial and temporal environmental variance on plant populations (Caswell and Dixon, unpublished)
- Effects of dieldrin exposure on the cladoceran *Daphnia pulex* (Caswell and Martin, unpublished)

The take-home message of every example that has been published so far is that **it is not safe to assume that the most obvious effect of a toxicant on the vital rates is the source of that toxicant's effect on** λ. In each case, there are large vital rate effects that make only trivial contributions to effects on λ and other, much smaller vital rate effects whose contributions are much larger.

The details of the decomposition analysis depend on the design of the LTRE: on whether treatments are fixed, random, or quantitative, on how many treatment factors there are and how they are cross-classified, and so on. The general approach is described first, followed by specific designs. Some simple examples, not all ecotoxicological, are described to illustrate the concepts.

A. THE GENERAL DECOMPOSITION PRINCIPLE

This decomposition method was introduced by Caswell (1989b). The basic idea is to write a linear model for the treatment effects on λ, similar in spirit to that behind analysis of variance. The terms in this linear model depend on how λ varies in response to changes in the treatments, and this variation is described by the sensitivity analysis of λ (Equation 10).

Consider a simple one-way design with treatments T_I to T_N producing population growth rates $\lambda^{(1)}$ to $\lambda^{(N)}$, as shown in Figure 1. (Note: I have adopted the convention of using superscripts on matrices to denote treatments, leaving subscripts free for later use to denote entries of matrices.) Begin by choosing a *reference matrix* $\mathbf{A}^{(r)}$. The vital rates in this matrix serve as a baseline against which the treatment effects will be measured. Two common choices for a reference matrix are the mean matrix $\mathbf{A}^{(\cdot)} = \Sigma_i \mathbf{A}^{(i)}/N$ and a matrix representing a special "control" condition.

Expanding λ as a function of the a_{ij} around the reference matrix $\mathbf{A}^{(r)}$ gives the basic decomposition result: the population growth rate in any treatment m can be written

$$\lambda^{(m)} \approx \lambda^{(r)} + \sum_{ij} (a_{ij}^{(m)} - a_{ij}^{(r)}) \left. \frac{\partial \lambda}{\partial a_{ij}} \right|_{\mathbf{A}*} \tag{11}$$

where

$$\mathbf{A}* = (\mathbf{A}^{(m)} + \mathbf{A}^{(r)})/2 \tag{12}$$

Each term in the summation in Equation 11 is the contribution of one of the vital rates (given by a_{ij}) to population growth in the mth treatment, measured relative to the reference condition.

The sensitivities in Equation 11 must be evaluated somewhere; Equation 12 indicates that they are evaluated at a matrix $\mathbf{A}*$ which is "midway" between the two matrices, $\mathbf{A}^{(m)}$ and $\mathbf{A}^{(r)}$, being compared. This is not absolutely essential, and other matrices, such as $\mathbf{A}^{(m)}$ or $\mathbf{A}^{(r)}$, could be used instead. Sibly (Chapter 7 in this volume), for example, always evaluates the sensitivities at the reference matrix (a control treatment). However, this choice throws away some information about the response of λ to the changes in the a_{ij}. It uses a single value for the slope of this response, ignoring the curvature of the function.

Information on this curvature is contained in the differences in the sensitivities from point to point, and it seems ill-advised to waste this information. Moreover, the Mean Value Theorem of calculus guarantees that, for each treatment m, there is a matrix somewhere between $\mathbf{A}^{(m)}$ and $\mathbf{A}^{(r)}$ that will make the approximation Equation 11 exact, and using the mean matrix, as in Equation 12, gives good results.

Equation 11 makes it clear that treatment effects on λ depend on both the effect on each of the vital rates and the sensitivity of λ to changes in each of those vital rates. Since the sensitivities usually vary over many orders of magnitude, examining only the effects on the a_{ij} can be very misleading.

B. FIXED EFFECTS DESIGNS

As in analysis of variance, a treatment is said to be fixed if the specific levels included in the study are of interest, rather than being considered a sample drawn from some universe of possible treatments. The basic decomposition formula Equation 11 applies to a one-way fixed effect design. It can easily be extended to factorial, cross-classified fixed effects designs. Consider a two-way design (Figure 2), with $\lambda^{(ij)}$ denoting the population growth rate at level i of the first factor and level j of the second. Let $\lambda^{(\cdot j)}$, $\lambda^{(i\cdot)}$, and $\lambda^{(\cdot\cdot)}$ denote the population growth rates from the matrices averaged over all levels of the first treatment ($\mathbf{A}^{(\cdot j)}$), the second treatment ($\mathbf{A}^{(i\cdot)}$), and both treatments ($\mathbf{A}^{(\cdot\cdot)}$), respectively.

Using the grand mean $\mathbf{A}^{(\cdot\cdot)}$ as the reference matrix, we can write

$$\lambda^{(ij)} = \lambda^{(\cdot\cdot)} + \alpha^{(i)} + \beta^{(j)} + (\alpha\beta)^{(ij)} \tag{13}$$

where α and β are the main effects of the two treatments, and $(\alpha\beta)$ is the interaction effect. Each of these effects can be decomposed

$$\alpha^{(i)} \approx \sum_{k,l} (a_{kl}^{(i\cdot)} - a_{kl}^{(\cdot\cdot)}) \frac{\partial \lambda}{\partial a_{kl}} \tag{14}$$

$$\beta^{(j)} \approx \sum_{k,l} (a_{kl}^{(\cdot j)} - a_{kl}^{(\cdot\cdot)}) \frac{\partial \lambda}{\partial a_{kl}} \tag{15}$$

$$\alpha\beta^{(ij)} \approx \sum_{k,l} (a_{kl}^{(ij)} - a_{kl}^{(\cdot\cdot)}) \frac{\partial \lambda}{\partial a_{kl}} - \alpha^{(i)} - \beta^{(j)} \tag{16}$$

As in the one-way design, the partial derivatives are all calculated midway between the two matrices being compared for the treatment effect being

Trt. A	Treatment B 1	2	3	A Mean
1	$T^{(11)}$	$T^{(12)}$	$T^{(13)}$	$T^{(1.)}$
2	$T^{(21)}$	$T^{(22)}$	$T^{(23)}$	$T^{(2.)}$
3	$T^{(31)}$	$T^{(32)}$	$T^{(33)}$	$T^{(3.)}$
B Mean	$T^{(.1)}$	$T^{(.2)}$	$T^{(.3)}$	$T^{(..)}$

Trt. A	Treatment B 1	2	3	A Mean
1	$A^{(11)}$	$A^{(12)}$	$A^{(13)}$	$A^{(1.)}$
2	$A^{(21)}$	$A^{(22)}$	$A^{(23)}$	$A^{(2.)}$
3	$A^{(31)}$	$A^{(32)}$	$A^{(33)}$	$A^{(3.)}$
B Mean	$A^{(.1)}$	$A^{(.2)}$	$A^{(.3)}$	$A^{(..)}$

Trt. A	Treatment B 1	2	3	A Mean
1	$\lambda^{(11)}$	$\lambda^{(12)}$	$\lambda^{(13)}$	$\lambda^{(1.)}$
2	$\lambda^{(21)}$	$\lambda^{(22)}$	$\lambda^{(23)}$	$\lambda^{(2.)}$
3	$\lambda^{(31)}$	$\lambda^{(32)}$	$\lambda^{(33)}$	$\lambda^{(3.)}$
B Mean	$\lambda^{(.1)}$	$\lambda^{(.2)}$	$\lambda^{(.3)}$	$\lambda^{(..)}$

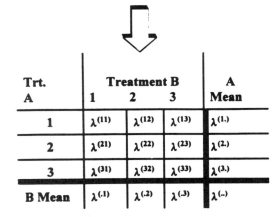

Figure 2 The structure of a factorial LTRE with two treatments (A and B), each with three levels. Treatment $T^{(ij)}$ produces a population projection matrix $A^{(ij)}$, which in turn produces a population growth rate given by the eigenvalue $\lambda^{(ij)}$.

evaluated. Extensions to three-way or higher designs requires only patience in writing down each of the main effects and interactions.

Caswell (1989b) applied this factorial decomposition to a toxicological study by Rao and Sarma (1986) on effects of DDT exposure and food level on a rotifer. Not surprisingly, λ was reduced by DDT exposure, and increased by higher food levels. Of more interest is the existence of an interaction between the two treatments: population growth rate was more sensitive to DDT exposure at low food levels than at high food levels. Examination of the contributions (see Caswell, 1989a for the details) shows that most of the interaction effect is due to effects on survival probability during the first half of the life cycle. Survival in the high food treatment is poorer than expected from the additive model at low DDT levels, but higher than expected at high DDT levels. The contributions of fertility effects to the interaction term are smaller, but they reveal a potentially interesting pattern, with early and late fertility exhibiting different patterns. Fertility contributions late in the life cycle behave much like the survival contributions: high food levels counteract the negative effect of high DDT concentrations. But fertility contributions in the first 5 d of life have the opposite behavior. High food levels make the negative effect of high DDT levels worse, not better. The mechanism behind this effect is unknown, but the ability to pinpoint such subtle age-specific contributions is the advantage of this decomposition analysis.

Another example of a factorial decomposition, using a stage-classified model, can be found in Walls et al. (1991). They studied the population-level effects of the costs of induced morphological defenses against predators. The embryos and first instars of the cladoceran *Daphnia pulex* respond to the presence of the predator *Chaoborus* by growing a neck spine. These spines defend against predators, but at some demographic cost (Riessen and Sprules, 1990). It has been hypothesized that this cost explains why the spines are not produced in the absence of the predator. This effect is conceptually similar to the cost of detoxifying a toxicant (Sibly, Chapter 7 in this volume).

Walls et al. (1991) analyzed a 2×2 cross-classified design with high and low food levels, and with and without exposure to water in which *Chaoborus* had been swimming (the defensive spines are induced by chemical exudates of the predator, so there was no actual predation going on in the experiments). The low food treatment increased survival probability for the late instars, but this increase made only minute contributions to the effect of low food on λ. Most of the food effect was due to reductions in fertility in instars 5–11. The main effect of *Chaoborus* exposure was a complicated mixture of growth effects: reduction in growth in instar 1 (when the spines are formed), an increase in growth in instar 2, a reduction in instar 3, and a slight increase in instar 4. The interaction effect, which was statistically significant, was due to a greater reduction at low food in growth and fertility in the first instars than was predicted by the additive model.

Without the decomposition analysis, these experiments would reveal effects of food, toxicant (DDT or *Chaoborus* extract) and their interaction on population growth. But the sources of those effects would be hidden; there would be no way of knowing, for example, that whereas the effect of food on *Daphnia* population growth rate is due to fertility reductions in instars 5–11, the effect of antipredator defenses comes through a complicated, oscillating effect on individual growth rate.

i. *Alternative Parameterizations of Stage-Classified Models*

The basic decomposition Equation 11 and its extension to factorial designs were written in terms of the treatment effects on the matrix entries a_{ij}. Sometimes, however, it is useful to do the decomposition in terms of alternative parameterizations of the vital rates. Two commonly occurring alternatives are stage-specific growth and survival probabilities, and parameterizations that include age at maturity as well as survival and reproduction.

a. *Stage-Specific Vital Rates*

In size- or stage-classified models in which individuals can grow or develop at most one stage per projection interval, the population projection matrix will have positive elements only in the first row (reproduction), the diagonal (P_i = the probability of surviving and remaining in stage i), and the subdiagonal (G_i = the probability of surviving and growing from stage i to stage $i + 1$):

$$\mathbf{A} = \begin{bmatrix} P_1 & F_2 & F_3 & F_4 & \cdots \\ G_1 & P_2 & 0 & 0 & \cdots \\ 0 & G_2 & P_3 & 0 & \cdots \\ 0 & 0 & G_3 & P_4 & \cdots \\ \vdots & \vdots & \vdots & \ddots & \ddots \end{bmatrix} \tag{17}$$

Because the diagonal and subdiagonal terms include both survival and growth, the interpretation of their contributions to effects on λ can be ambiguous.

One way to parameterize the P_i and G_i in terms of the underlying survival probabilities and growth rates is

$$P_i = \sigma_i (1 - \gamma_i) \tag{18}$$

$$G_i = \sigma_i \gamma_i \tag{19}$$

where σ_i and γ_i are the survival and growth probabilities for stage i. (Note that in some cases, such as the example given below, the fertilities F_i may also contain survival and/or growth terms.) Analyzing the LTRE in terms of contributions of σ_i and γ_i eliminates the confusion of growth and survival that arises when the analysis is done in terms of P_i and G_i. This is accomplished by recognizing that the σ_i, γ_i, and F_i parameterize the matrix just as well as do the matrix entries themselves. Thus, the basic decomposition of a particular treatment m can be written

$$\lambda^{(m)} \approx \lambda^{(r)} + \sum_i (\sigma_i^{(m)} - \sigma_i^{(r)}) \frac{\partial \lambda}{\partial \sigma_i} + \sum_i (\gamma_i^{(m)} - \gamma_i^{(r)}) \frac{\partial \lambda}{\partial \gamma_i} + \sum_i (F_i^{(m)} - F_i^{(r)}) \frac{\partial \lambda}{\partial F_i} \quad (20)$$

$$m = 1, \ldots, N$$

As before, each term in the summation is the contribution of the treatment effect to the effect on λ, but now the effects are described in terms of the underlying vital rates rather than the matrix entries themselves.

Equation 20 requires the sensitivities of λ to changes in the lower-level vital rates; these are obtained by differentiating Equation 18 and Equation 19

$$\frac{\partial \lambda}{\partial \gamma_i} = \frac{\partial \lambda}{\partial P_i} \frac{\partial P_i}{\partial \gamma_i} + \frac{\partial \lambda}{\partial G_i} \frac{\partial G_i}{\partial \gamma_i}$$

$$= \frac{\partial \lambda}{\partial P_i}(-\sigma_i) + \frac{\partial \lambda}{\partial G_i}\sigma_i \quad (21)$$

$$\frac{\partial \lambda}{\partial \sigma_i} = \frac{\partial \lambda}{\partial P_i} \frac{\partial P_i}{\partial \sigma_i} + \frac{\partial \lambda}{\partial G_i} \frac{\partial G_i}{\partial \sigma_i}$$

$$= \frac{\partial \lambda}{\partial P_i}(1 - \gamma_i) + \frac{\partial \lambda}{\partial G_i}\gamma_i \quad (22)$$

In general (Caswell, 1989a), the sensitivity of λ to changes in any lower-order vital rate x is given simply by

$$\frac{\partial \lambda}{\partial x} = \sum_{i,j} \frac{\partial \lambda}{\partial a_{ij}} \frac{\partial a_{ij}}{\partial x} \quad (23)$$

The analysis of effects of food and induced predator defenses in *Daphnia pulex* (Walls et al., 1991) discussed above is actually an example of this type of analysis, but a simpler example is presented here, which makes the approach clear.

An example — Brault and Caswell (1993) developed a stage-classified model for killer whales (*Orcinus orca*) in the coastal waters of Washington State and British Columbia, based on the extensive demographic data of Olesiuk et al. (1990). Four stages (yearlings, immatures, matures, and postreproductives) were identified, leading to a matrix of the form

$$A = \begin{pmatrix} 0 & F_2 & F_3 & 0 \\ G_1 & P_2 & 0 & 0 \\ 0 & G_2 & P_3 & 0 \\ 0 & 0 & G_3 & P_4 \end{pmatrix} \tag{24}$$

In terms of the underlying vital rates σ_i, γ_i, and m (the mean calf production per female), the parameters are given by

$$G_1 = \sigma_1^{1/2}$$

$$P_1 = 0$$

$$G_2 = \gamma_2 \sigma_2$$

$$P_2 = (1 - \gamma_2)\sigma_2$$

$$G_3 = \gamma_3 \sigma_3$$

$$P_3 = (1 - \gamma_3)\sigma_3$$

$$P_4 = \sigma_4$$

$$F_2 = \sigma_1^{1/2} G_2 m / 2$$

$$F_3 = \sigma_1^{1/2}(1 + P_3)m / 2$$

The nonzero value of F_2 represents the reproduction of juveniles that mature and reproduce for the first time during the year. The probability of remaining in the first class, P_1, is zero because this stage lasts for one year by definition.

The sensitivities of λ to the σ_i, γ_i, and m are given by

$$\frac{\partial \lambda}{\partial \sigma_1} = \frac{\partial \lambda}{\partial G_1}\left(\frac{1}{2\sigma_1^{1/2}}\right) + \frac{\partial \lambda}{\partial F_2}\left(\frac{F_2}{2\sigma_1}\right) + \frac{\partial \lambda}{\partial F_3}\left(\frac{F_3}{2\sigma_1}\right) \tag{25}$$

$$\frac{\partial \lambda}{\partial \sigma_2} = \frac{\partial \lambda}{\partial G_2}\gamma_2 + \frac{\partial \lambda}{\partial P_2}(1 - \gamma_2) + \frac{\partial \lambda}{\partial F_2}\left(\frac{F_2}{\sigma_2}\right) \tag{26}$$

$$\frac{\partial \lambda}{\partial \sigma_3} = \frac{\partial \lambda}{\partial G_3}\gamma_3 + \frac{\partial \lambda}{\partial P_3}(1 - \gamma_3) + \frac{\partial \lambda}{\partial F_3}\left(\frac{\sigma_1^{1/2}(1 - \gamma_3)m}{2}\right) \tag{27}$$

$$\frac{\partial \lambda}{\partial \sigma_4} = \frac{\partial \lambda}{\partial P_4} \tag{28}$$

$$\frac{\partial \lambda}{\partial \gamma_2} = \frac{\partial \lambda}{\partial G_2}\sigma_2 - \frac{\partial \lambda}{\partial P_2}\sigma_2 + \frac{\partial \lambda}{\partial F_2}\left(\frac{F_2}{\gamma_2}\right) \tag{29}$$

$$\frac{\partial \lambda}{\partial \gamma_3} = \frac{\partial \lambda}{\partial G_3}\sigma_3 + \frac{\partial \lambda}{\partial P_3}\sigma_3 - \frac{\partial \lambda}{\partial F_3}\left(\frac{\sigma_1^{1/2}\sigma_3 m}{2}\right) \tag{30}$$

$$\frac{\partial \lambda}{\partial m} = \frac{\partial \lambda}{\partial F_2}\left(\frac{F_2}{m}\right) + \frac{\partial \lambda}{\partial F_3}\left(\frac{F_3}{m}\right) \tag{31}$$

Notice that changes in the σ_i and γ_i affect not only P_j and G_j, but also the fertility terms F_j.

This population of orcas is divided into Northern and Southern subpopulations (Bigg et al., 1990). The environments of these populations differ in a variety of unknown factors. Those differences will affect the σ_i, γ_i, and m, and hence λ. In this case, the population growth rates ($\lambda^{(N)} = 1.0248$ and $\lambda^{(S)} = 1.0249$) are almost identical. Thus, the net effect of all the environmental differences affecting all the vital rates is very close to zero. Not only is the difference not significant, but the two values of λ would be this similar less than 1% of the time under the null hypothesis of no location effect (Brault and Caswell, 1993). However, that the net effect is zero does not mean that there may not be important contributions from the various vital rates; it only means that positive and negative contributions must cancel out almost exactly. Thus, it is still possible — and appropriate — to decompose the effect on λ into contributions.

Using the southern population as the reference population, the effect of location on λ is $\lambda^{(N)} - \lambda^{(S)} = -8.007 \times 10^{-5}$. The contributions of the differences in the lower-level vital rates are shown in Figure 3. The linear approximation fits well; the sum of the contributions predicts a difference in λ of -8.032×10^{-5}.

There are positive contributions from an advantage in survival, especially juvenile survival, in the north. There is a negative contribution from an advantage in adult reproductive output in the south. There is also a negative contribution from the difference in adult growth (γ_3). Remember that γ_3 is the rate of growth out of the adult stage into the postreproductive stage; it measures the rate of senescence. Thus, the negative contribution of γ_3 reflects a slower rate of senescence in the south.

This analysis makes two points. First, the effect of location on λ is small, not because geography has no effects on the vital rates, but because there are positive and negative contributions that cancel out nearly exactly. Each individual contribution is still small, but the largest of them are two orders of

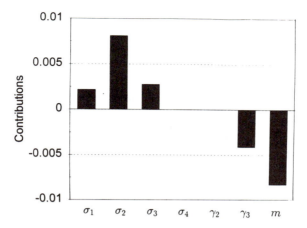

Figure 3   The contributions of geographic differences in survival (σ_i), growth (γ_i), and reproductive output (m) to the difference in λ between northern and southern subpopulations of orcas (*Orcinus orca*) in the coastal waters of Washington and British Columbia.

magnitude larger than the resulting difference in λ. Statistically insignificant differences in λ do not imply that nothing interesting is happening at the level of the vital rates. Second, expressing the contributions in terms of the lower-level vital rates instead of the matrix entries themselves removes a potential source of confusion. Brault and Caswell (1993), analyzing the same contrast directly in terms of the matrix entries, missed the effect of adult growth rate, attributing it instead to an adult survival advantage in the south.

b. Analyzing Effects on Age at Maturity

Age at maturity (α) is known to be demographically important, especially in increasing populations (Lewontin, 1965; Caswell and Hastings, 1980; Caswell, 1982b; Ebert, 1985). However, it does not appear explicitly as an entry in a population matrix, so the sensitivity of λ to changes in α cannot be calculated from the basic Equation 10. It is, however, possible to parameterize a stage-classified model so that age at maturity does appear as an explicit parameter, using the *life cycle graph*.

The life cycle graph (Caswell, 1982a; see Caswell, 1989a for details) contains nodes for each stage in the life cycle, and arrows between nodes denoting the transitions open to individuals in each stage. In the simple life cycle graph of Figure 4, there are two stages (juveniles and adults). The arrows indicate that juveniles can develop into adults, that adults may survive as adults, and that adults contribute new juveniles by reproduction. The coefficients on the arrows indicate the transition probabilities or reproductive outputs, each multiplied by λ raised to a power indicating the time required for the transition. Thus, in Figure 4, juveniles survive to maturity with probability P_1, and require α time units to do so. Adults survive with a probability P_2 and produce F surviving offspring each projection interval. This is the simplest life

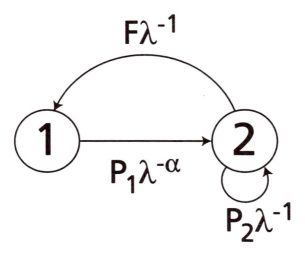

Figure 4   A life cycle graph for the simple two-stage (juvenile and adult) model used to analyze effects on age at maturity in *Capitella* sp.1. The age at maturity is α.

cycle in which effects of juvenile survival, adult survival, fertility, and development time can be studied.

There is a population projection matrix corresponding to any life cycle graph, and it is possible to write down the characteristic equation for that matrix directly from the life cycle graph (Caswell, 1982a). For the graph in Figure 4, the characteristic equation is

$$\lambda^{\alpha+1} - P_2\lambda^{\alpha} - P_I F = 0 \tag{32}$$

The largest solution to this equation is the population growth rate.

Assume that the survival to maturity, P_I, is the result of surviving for the necessary α time steps, with a per-time-step survival probability of σ_1: hence

$$P_I = \sigma_1 \tag{33}$$

For notational consistency, we will set $P_2 = \sigma_2$. This model reduces the complete demographic information to the four parameters σ_1, σ_2, F, and α. The advantage of using the life cycle graph is that the sensitivity of λ to changes in each of these four parameters can be calculated by implicit differentiation of Equation 32, keeping in mind that P_I is also a function of α. The resulting sensitivities are

$$\frac{\partial \lambda}{\partial \sigma_1} = \frac{\alpha \sigma_1^{\alpha-1} F}{(\alpha+1)\lambda^{\alpha} - \alpha\sigma_2\lambda^{\alpha-1}} \tag{34}$$

$$\frac{\partial \lambda}{\partial \sigma_2} = \frac{\lambda}{(\alpha + 1)\lambda - \alpha \sigma_2} \tag{35}$$

$$\frac{\partial \lambda}{\partial F} = \frac{\sigma_1^{\alpha}}{(\alpha + 1)\lambda^{\alpha} - \alpha \sigma_2 \lambda^{\alpha - 1}} \tag{36}$$

$$\frac{\partial \lambda}{\partial \alpha} = \frac{\lambda^{\alpha + 1} \log \lambda - \sigma_2 \lambda^{\alpha} \log \lambda - F\sigma_1^{\alpha} \log \sigma_1}{(\alpha + 1)\lambda^{\alpha} - \sigma_2 \alpha \lambda^{\alpha - 1}} \tag{37}$$

In an LTRE in which $\sigma_1, \sigma_2, \alpha$, and F have been measured, treatment effects on λ can be decomposed into contributions from each parameter. The basic decomposition result becomes

$$\lambda^{(m)} \approx \lambda^{(r)} + (\sigma_1^{(m)} - \sigma_1^{(r)})\frac{\partial \lambda}{\partial \sigma_1} + (\sigma_2^{(m)} - \sigma_2^{(r)})\frac{\partial \lambda}{\partial \sigma_2} + (F^{(m)} - F^{(r)})\frac{\partial \lambda}{\partial F} + (\alpha^{(m)} - \alpha^{(r)})\frac{\partial \lambda}{\partial \alpha}$$

$$m = 1, \ldots, N \tag{38}$$

where the derivatives are evaluated at the mean of the two matrices being compared. The decomposition analysis of factorial experiments and their interactions proceeds as described above, but expressed in terms of the parameters $\sigma_1, \sigma_2, \alpha$, and F.

An example — The polychaete *Capitella* sp.1 responds with rapid population growth to many kinds of pollution and other forms of disturbance. It dominates the benthic biomass at sewage outfalls and in oil-contaminated sediments. Its predilection for such areas has led to its use as an indicator of marine pollution. As part of a study of the responses of estuarine polychaetes to pollutants (Bridges et al., 1994; Levin et al., 1994), *Capitella* sp.1 was exposed in the laboratory to four treatments: control sediments, sediment with sewage added, sediment with hydrocarbons (Number 2 fuel oil) added, and sediment with blue-green algae added. The three treatments are intended to mimic the effects of pollution by sewage, oil, and eutrophic enrichment.

In each treatment, the parameters for the simple two-stage life cycle graph (Figure 4) were estimated (see Levin et al., 1994 for details) for each treatment. The results (Table 1) show that *Capitella* sp.1 clearly thrives in sewage: its fertility is increased more than fourfold and its age at maturity reduced by a factor of three, and these effects are offset by only a small reduction in survival probability. Algal enrichment more than doubles fertility, and halves age at maturity, but it also reduces juvenile and adult survival. Oil pollution has little effect on survival, but slightly reduces fertility and slightly increases age at maturity.

Table 1   Parameter Values (σ_1, σ_2, F, α) for a Simple Two-Stage Model
for the Polychaete *Capitella* sp. 1 in Different Treatments

Treatment	σ_1	σ_2	F	α	λ	P
Control	0.9958	0.9998	48.08	6	1.7914	
Sewage	0.9747	1.0000	215.01	2	4.0576	<0.0005
Oil	0.9964	0.9999	31.24	7	1.6253	0.6637
Algae	0.8755	0.9039	104.32	3	2.5526	0.0360

Note: λ is the population growth rate determined from the two-stage model, and P is the
significance level of a two-tailed randomization test comparing λ for each treatment
with λ for the control.

Data from Levin et al. (1994).

These effects translate into greatly increased values of λ in the sewage and algae treatments, and a slight reduction of λ in the oil treatment. A two-tailed randomization test shows that the first two of these effects are significant, whereas the last one is not.

In this study, the control serves as a natural reference treatment. The contributions of each of the four parameters to the overall effect on λ were calculated relative to this reference, using Equation 38. The contributions are shown in Figure 5. The largest contributions in the sewage and algae treatments come from accelerated maturation. In the sewage treatment, the enormous fertility increase contributes only about a third as much as the accelerated maturity to the effect on λ. In the algae treatment, the positive effects of increased fertility and accelerated maturation are partly offset by the effect of the small reduction in juvenile survival. In the oil treatment, the negative contributions from reduced fertility and delayed maturity are about equal. Levin et al. (1994) discuss these results in more detail, and compare them to another opportunistic polychaete (*Streblospio benedicti*), which responds quite differently to the same treatments.

Without an LTRE approach, the effects of sewage, oil, and algae on *Capitella* could be measured only at the level of the vital rates. Knowing that oil reduces fertility and that algae reduces survival does not determine their effects at the population level. Those can be deduced only from a demographic model that includes the entire life cycle, and from which λ can be calculated. Knowing the effects on λ, however, still leaves the mechanisms unclear. The decomposition analysis provides this information; based on this analysis we know that the increases in λ in the sewage and algae treatments are largely due to effects on age at maturity, and that the small (12%) reduction in juvenile survival has a negative effect comparable in magnitude to the positive effect of the 215% increase in fertility. Such insights greatly increase the power of LTREs.

C. REGRESSION DESIGNS

Of particular interest for ecotoxicology are designs with quantitative treatments, especially designs in which the treatments are levels of exposure to a toxicant. LTREs with quantitative treatments can be analyzed by generalizing

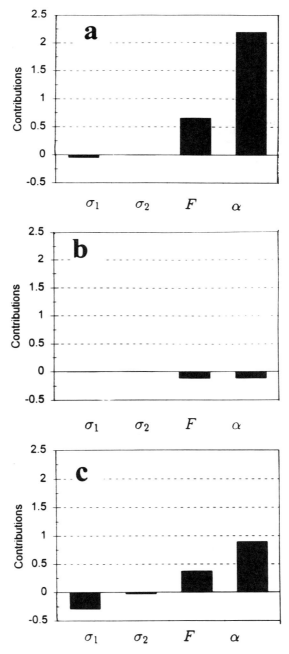

Figure 5 Contributions of juvenile survival (σ_1), adult survival (σ_2), fertility (F) and age at maturity (α) to treatment effects on λ for *Capitella* sp. 1. Contributions are measured relative to the control treatment: (a) sewage treatment, (b) hydrocarbon treatment, (c) blue-green algae treatment.

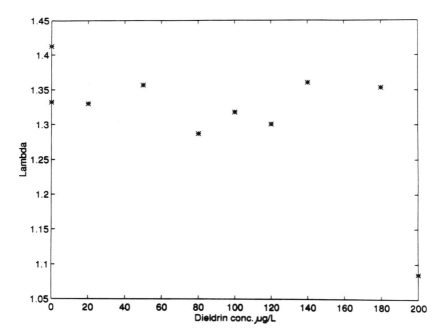

Figure 6   Population growth rate (λ) as a function of dieldrin concentration for *Daphnia pulex* (Data from Daniels, R.E. and J.D. Allan, 1981.)

the approach familiar from regression analysis. A complete discussion of this approach is given by Caswell and Martin (1994); the method is outlined briefly here, with an application to an experiment on the response of *Daphnia pulex* to dieldrin (Daniels and Allan, 1981), see Figure 6.

i. Decomposing Regression Effects

We begin by considering the definition of "effect" in regression designs. It is, of course, possible to treat an experiment with quantitative treatments as a one-way fixed design, comparing each treatment to a selected reference (e.g., the control). Sibly (Chapter 7 in this volume) does so, for example. In a regression design, however, we want to describe the response to a continuously varying exposure level; there are in principle a continuum of treatments. In an ordinary linear regression analysis, this problem is solved by measuring the "effect" of the independent variable x on the dependent variable y by the *slope of the regression*. Whether the effect is significant or not is tested by comparing the slope to zero. Whether the effect is large or small is evaluated by the magnitude of the slope. Whether the effect is linear or not is determined by seeing if the slope is independent of x.

Accordingly, the effect of the treatment variable x on λ is defined as the slope, $\partial\lambda/\partial x$. Unless λ is a linear function of x (and there is no reason to expect that it will be) the slope will vary with x. Figure 7 shows an example. The

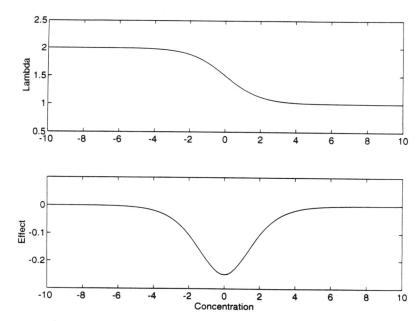

Figure 7 The definition of "treatment effect" for LTRE studies with quantitative treatments. Population growth rate is shown in the upper figure as a function of concentration, for a hypothetical situation. The lower figure shows the effect of concentration, measured by the slope of the upper figure.

intuitive interpretation of this figure would be that changes in x have little effect at low concentrations, large negative effects at intermediate concentrations, and small effects at high concentrations. This intuitive interpretation is captured by using the slope of the function as the measure of effect.

The basic data for a regression LTRE consist of a set of vital rates $a_{ij}(x)$ that are functions of the treatment variable x. These vital rates generate a set of population projection matrices $\mathbf{A}^{(x)}$, from which population growth rates $\lambda(x)$ or $r(x)$ can be derived as a function of x. The effect of x on λ can be decomposed into contributions from each of the vital rates by

$$\frac{d\lambda}{dx} = \sum_{i,j} \frac{\partial \lambda}{\partial a_{ij}(x)} \frac{\partial a_{ij}(x)}{\partial x} \tag{39}$$

The sensitivity $\partial\lambda/\partial a_{ij}(x)$ comes from the matrix $\mathbf{A}^{(x)}$. The vital rate sensitivity $\partial a_{ij}(x)/\partial x$ comes from the functional relationship between the vital rates and x.

The terms in Equation 39 are the contributions of the treatment effect (at level x) on one of the vital rates to the effect (again, at level x) on λ. This contribution can be positive or negative, depending on the signs of $\partial a_{ij}(x)/\partial x$. The contribution of $a_{ij}(x)$ will be small if $a_{ij}(x)$ is not very sensitive to x, or if λ is not very sensitive to $a_{ij}(x)$, or both.

These contributions can be combined in various ways, to describe the effects of different groups of vital rates. Summing across the first row, for example, gives the integrated contribution from reproduction:

$$\sum_{j} \frac{\partial \lambda}{\partial a_{1j}(x)} \frac{\partial a_{1j}(x)}{\partial x} \qquad (40)$$

Similar sums over the subdiagonal in an age-classified matrix would give an integrated contribution from survival. Adding together the survival and fecundity contributions gives an approximation to the derivative of λ with respect to x (this is what Equation 39 says). Integrating this function from the value of λ at $x = 0$ gives λ as a function of x. This function can be compared with the observed response of λ to x; how well the two agree will depend on how well the functions $a_{ij}(x)$ capture the response of the vital rates to x. Fitting a linear function to a nonlinear response, for example, will naturally give a poor agreement. It is important to remember, however, that the $\lambda(x)$ curve obtained from integrating the decomposition analysis is not the same as fitting a curve to the observed data on λ as a function of x. It incorporates information on the mechanisms causing the response at the level of the vital rates.

Note that it is also possible to extend Equation 39 to express the contributions in terms of lower-level parameters (e.g., survival probabilities or development rates) that determine the a_{ij}. Suppose that there are a set of such parameters $z_k(x)$ that are functions of x. The decomposition formula becomes

$$\frac{d\lambda}{dx} = \sum_{k} \frac{\partial z_k(x)}{\partial x} \sum_{i,j} \frac{\partial \lambda}{\partial a_{ij}} \frac{\partial a_{ij}}{\partial z_k} \qquad (41)$$

ii. An Example

As an example, we consider data from Daniels and Allan (1981), who measured age-specific life tables for *Daphnia pulex* exposed to the pesticide dieldrin. Figure 6 shows the response of λ to dieldrin. There is relatively little effect until concentrations reach 180 µg/l, above which λ falls off rapidly. Between 0 and 180 µg/l, there are small variations in λ. At this point, it is not clear whether they represent noise or some aspect of the response of *D. pulex* to this toxicant. We will see later that the decomposition analysis suggests that they are a real effect.

a. Expressing the Vital Rates as a Function of Treatment

A crucial step in Equation 39 is expressing the vital rates $a_{ij}(x)$ as a function of x. This can be done in a variety of ways: simple interpolation between the data points, linear or nonlinear regression, survival analysis, or nonparametric regression. Caswell and Martin (1994) compare these methods for the data from Daniels and Allan (1981); here, I report on the results using a nonparametric regression model which seems to give the best picture of the treatment response in this experiment.

Nonparametric regression, or a generalized additive model, fits a smooth function to a set of data points, without assuming a particular functional form, linear or nonlinear (Hastie and Tibshirani, 1990). Examples include the familiar running mean, kernel smoothers, smoothing splines, and running line smoothers. Each of these methods includes a parameter that adjusts how much smoothing is done. This smoothing parameter must be specified to fit the nonparametric model; one way to choose it is to find the value which, in a particular sense (called cross-validation), makes the fit of the curve close to the data points.

We used a locally weighted running line smoother, known as "loess" (Hastie and Tibshirani, 1990). Suppose that the data is a set of pairs (x_i, y_i). We want to calculate a set of smoothed values of y for a set of x values. For each such point x_0, we take the k nearest neighbors among the data points and fit a weighted linear regression, where the weights depend on how far from x_0 each data point lies (see Hastie and Tibshirani, 1990 for details). The smoothed, or predicted value of y_0 at the point x_0 is simply the value estimated from this local linear regression.

The smoothing parameter in this model is k. If $k = 2$, the procedure simply interpolates a straight line between adjacent data points, without smoothing at all. If k is large enough that each neighborhood includes all the data points, the procedure returns a single weighted linear regression line through the points. Somewhere between these extremes of no smoothing and complete smoothing, there is a "best" value of k. We used cross-validation (Hastie and Tibshirani, 1990, pp. 43–45) to select a value of k that minimizes the average prediction error.

b. Effects on the Vital Rates

Figure 8 shows the effects of dieldrin on survival, measured by the slope of survival probability vs. concentration, as a function of age and concentration. The largest negative effects occur at concentrations from 50 to 150 µg/l and from ages 20 to 40 d. There are some small negative effects at early ages (0 to 15 d) at the highest concentrations, and some positive effects (survival increasing with dieldrin concentration) at ages 40 to 50 d and concentrations of 0 to 50 µg/l. Figure 9 is a similar plot, showing a mixture of positive and negative fertility effects. There are large negative effects at ages 20 to 40 d and concentrations from 100 to 150 µg/l. At earlier ages (7 to 15 d), there are negative effects at low concentrations (0 to 50 µg/l), positive effects at intermediate concentrations (50 to 150 µg/l), and negative effects at the highest concentrations. There are also some positive effects at ages 20 to 40 d and concentrations of 0 to 50 µg/l.

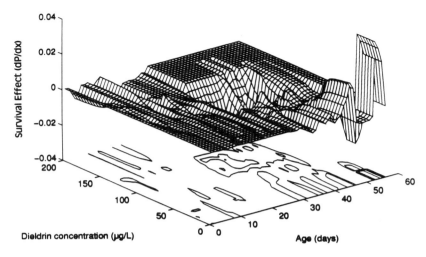

Figure 8   Effects of dieldrin on age-specific survival for *Daphnia pulex*, measured by the slope of a nonparametric regression of survival probability on concentration.

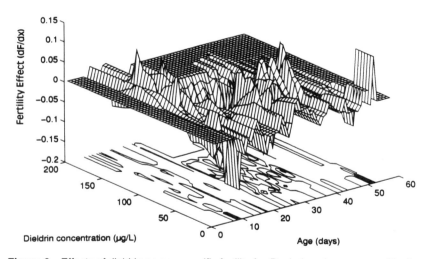

Figure 9   Effects of dieldrin on age-specific fertility for *Daphnia pulex*, measured by the slope of a nonparametric regression of fertility on concentration.

c. Contributions to Effects on λ

These effects are clearly diverse and age specific; what do they imply about λ? Figures 10 and 11 show the contributions of these effects to the effect on population growth, calculated according to Equation 39. Comparing these figures with the corresponding survival and fertility effect figures shows immediately that the contributions to population growth are limited to the first 15 or so days of life (note the difference in the age axes between Figures 8 and 9 and Figures 10 and 11).

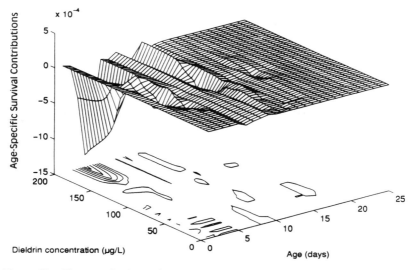

Figure 10   The contributions of age-specific survival effects to the effect of dieldrin concentration on λ for *Daphnia pulex*.

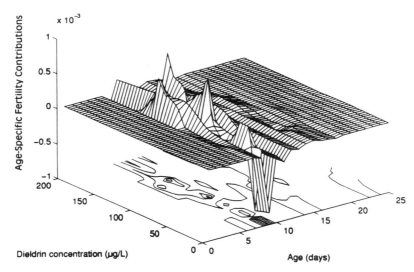

Figure 11   The contributions of age-specific fertility effects to the effect of dieldrin concentration on λ for *Daphnia pulex*.

The most important contributions of the survival effects are negative, and occur at early ages (0 to 5 d) and at the highest concentrations (150 to 200 µg/l). In contrast, the fertility contributions are a mixture of positive and negative terms. In particular, there are large positive contributions from ages 5 to 10 d at concentrations of 50 to 150 µg/l. The improvement of fertility with increasing dieldrin exposure, evident in Figure 9 at these ages, is actually important for population growth.

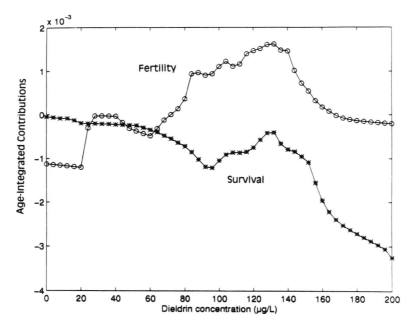

Figure 12   The survival and fertility contributions, integrated over age, to the overall effect of dieldrin concentration on λ in *Daphnia pulex*.

Figure 12 shows the results of integrating these contributions over age, separately for survival and fertility. There are two points to note about this figure. First, on a purely procedural level, remember that survival and fertility effects (Figures 8 and 9) are measured in totally different units. It is impossible to judge their relative importance by comparing their magnitude. Here, the effects have been translated into their contributions to λ, and the values are directly comparable. More interesting is the fact that the survival and fertility contributions, integrated over all ages, are of opposite signs. As far as survival is concerned, λ is a *decreasing* function of dieldrin concentration, becoming more so at the highest concentrations. But, as far as fertility is concerned, λ is an *increasing* function of dieldrin exposure over the range of concentrations from 80 to 160 µg/l. The positive effects on fertility are comparable in magnitude to the negative effects on survival at all but the highest concentrations, when the fertility contributions go to zero and the survival concentrations become most negative.

Figure 13 shows the results of adding the survival and fertility contributions in Figure 12. This is the overall contribution of both survival and fertility, integrated over all ages, to the slope of λ. Since this is the slope of λ, by integrating this curve we can reconstruct λ itself as a function of concentration. The result is shown in Figure 14, with the observed values for comparison.

It appears from this figure that some of the variation in λ from 0 to 160 µg/l is a real response, not merely noise. In particular, there is a slight increase in λ from 100 to 150 µg/l, which we have explained as a result of the increase in

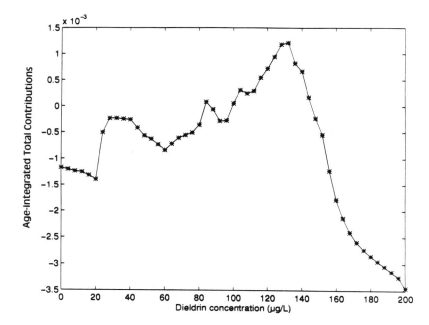

Figure 13   The total contributions, integrated over age and summed over survival and fertility, to the overall effect of dieldrin concentration on λ in *Daphnia pulex*. This is an estimate of the slope of the function relating λ and concentration.

fertility over that range of concentrations. We have also explained the abrupt drop-off in λ at the highest concentrations in the experiment: it is the result of a balance between positive fertility effects and negative survival effects that breaks down when the fertility contributions go to zero. The explanation of the positive fertility effects is an open problem. Daniels and Allan (1981) suggest that it may result from unintentional temperature variation during the experiment, leading to earlier age at maturity in the higher treatments. I have heard several conjectures of possible toxicant responses that could also produce such an effect.

This example shows clearly the richness of the additional insight provided by the decomposition analysis of a LTRE. Daniels and Allan's study is a well-designed and well-executed LTRE. Without the decomposition analysis, however, there is no way to detect the different contributions of survival and fertility effects, and no way to know if the variation in λ between 0 and 160 µg/l is simply noise or if it appears as a result of effects of dieldrin on the vital rates.

V. DISCUSSION

A. DEMOGRAPHIC BIOASSAY

In this chapter, I have tried to present a general approach to demographic bioassay. The first step is the recognition that translating individual effects into

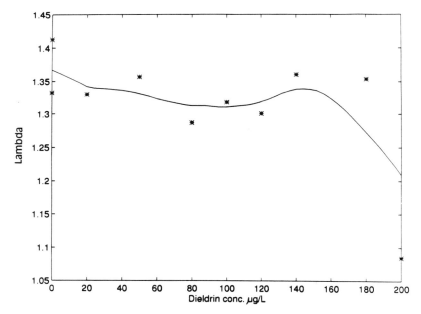

Figure 14   The predicted value of population growth rate λ, calculated from the integral
of the total contribution curve, compared with the observed values of λ
(asterisks), for data from Daniels and Allan (1981) on *Daphnia pulex* exposed
to dieldrin.

population consequences is at the heart of population-level ecotoxicology. The
physiology and development of living organisms are complex. This guarantees
that individual-level toxicant effects will be diverse and stage specific. Thus,
demographic models are an essential tool for translating from the individual to
the population level.

A demographic model describes population dynamics. Depending on the
model, a variety of indices can be produced: population growth rate, stable
stage distributions, reproductive value, sensitivities and elasticities, equilibria,
stability, oscillations, extinction probabilities, and so on. In a bioassay, these
indices are interpreted as projections of current conditions, and hence measures
of treatment effects, not as predictions of actual future dynamics.

Except for the generalization, largely unexplored, to indices other than
population growth rate, this much of the approach is familiar from the work of
Birch and many others. Its emphasis is synthetic, integrating the diverse and
stage-specific individual effects into population-level consequences. The second step, however, is new; it complements the synthetic approach of modeling
with an analytical approach breaking the population effect down into contributions from each of the vital rates (Figure 15).

Decomposition analysis is similar in principle to the methods of multifac-
torial experimental design that have proven so powerful since their introduction by Fisher early in this century. Those methods calculate main effects and

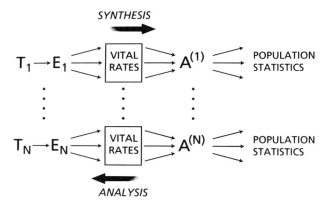

Figure 15   The basic LTRE structure revisited, showing the synthetic (combining diverse effects into single statistics) and analytic (decomposing those single statistics into contributions from the various rates) directions.

interactions that measure the contribution of each factor or combination of factors to an overall response. In the same way, the decomposition analyses I have presented here pinpoint the vital rates that are most responsible for the population-level effects of the toxicant.

The experience with these methods so far suggests a few generalizations. First, the largest contributions to effects on population growth rate do not come from the vital rates with the largest responses to the toxicant. Second, the overall effect on population growth is often a balance of positive and negative contributions from the same toxicant. Third, even closely related species can respond differently — because of different vital rates — to the same toxicant. The generality of these conclusions, and other comparative generalizations, remains a problem for future research.

B. UNRESOLVED ISSUES

This scheme for demographic bioassay leaves unresolved a number of interesting issues, all of which deserve further research.

- **More mechanism**: the decomposition analysis of λ reveals which vital rates contribute the most to population responses. However, the analysis is based on direct measurements of the vital rates under different treatment conditions, and thus does not include the mechanisms determining those rates. A more sophisticated analysis would model the response of the vital rates themselves, as determined by the nature of the toxicant and the physiology of the organism.
- **Additional demographic indices**: although λ is appealing because it integrates all the vital rates into a single statistic, it is not the only possible demographic endpoint. Even within the context of linear matrix population models, other indices can be calculated from the vital rates, including the

stable stage distribution, the damping ratio (which determines the speed with which the stable stage distribution is approached), and the period of transient oscillations. The net reproductive rate (the expected lifetime offspring production) can be calculated for stage-classified as well as age-classified models, and recent theoretical work by Cushing (1994, 1995) has related this statistic to various aspects of the bifurcation behavior of both linear and nonlinear models. The utility of such indices in ecotoxicology remains unexplored.

- **More dynamics**: measuring population response in terms of λ, treated as a projection of conditional future behavior, ignores other possible determinants of dynamics, especially nonlinear ones such as density dependence. It would be valuable to extend demographic bioassay to include more realistic dynamics, and find similar ways to measure the response of those dynamics to changes in the vital rates. Takada and Nakajima (1992), for example, have explored the sensitivity analysis of the equilibrium density in certain kinds of density-dependent matrix models, and further work in these directions would be valuable. However, these refinements will come at a cost, because analyzing toxicant effects in terms of more sophisticated dynamic models will require more sophisticated data to estimate the parameters in those models.

- **Spatial structure**: the spatial pattern of the population and of the toxicological stress may be important in determining population response. More research is needed on how to incorporate metapopulation structure and the effects of dispersal into ecotoxicological analysis.

- **Acute, as opposed to chronic, impacts**: the analyses discussed in this chapter all assume that the effect of the toxicant is a chronic, long-term change in the vital rates. The demographic models and their sensitivity analysis predict the effects of those vital rate changes. An alternative scenario involves acute mortality from a single toxicant impact, without any chronic change in the vital rates. It is important to be able to estimate the effects of such acute stresses on population dynamics and persistence, and to quantify the process of recovery from the acute effect. Little research seems to have been done in this area, although some leads are emerging (Caswell, unpublished).

ACKNOWLEDGMENTS

My research on life table response experiments has benefited from discussions and collaborations with many people, including Solange Brault, Phil Dixon, Lisa Levin, Linda Martin, Juan Silva, and Mari Walls. I am also grateful for the comments and discussion with participants in the workshop. This research was supported by grants from the National Science Foundation (DEB-9211945 and OCE-9115562), the U.S. Environmental Protection Agency (R818404-01-0), and the Office of Naval Research (URIP N00014-92-J-1527). Woods Hole Oceanographic Institution Contribution 8876.

REFERENCES

Allan, J. D. and R. E. Daniels, 1982. Life table evaluation of chronic exposure of *Eurytemora affinis* (Copepoda) to kepone. Mar. Biol. 66, 179–184.

Bigg, M.A., P.F. Olesiuk, G. M. Ellis, J. K. B. Ford and K. C. Balcomb, 1990. Social organization and genealogy of resident killer whales (*Orcinus orca*) in the coastal waters of British Columbia and Washington State. Rep Int. Whaling Comm. Spec. Issue 12, 383–405.

Birch, L. C., 1953. Experimental background to the study of the distribution and abundance of insects. Ecology 34, 698–711.

Birch, L.C., T. Dobzhansky, P.O. Elliot, and R.C. Lewontin, 1963. Relative fitness of geographic races of *Drosophila serrata*. Evolution 17, 72–83.

Brault, S. and H. Caswell, 1993. Pod-specific demography of killer whales (*Orcinus orca*). Ecology 74, 1444–1454.

Bridges, T.S., L.A. Levin, D. Cabrera and G. Plaia, 1994. Effects of sediment amended with sewage, algae, or hydrocarbons on growth and reproduction in two opportunistic polychaetes. J. Exp. Mar. Biol. Ecol. 117, 99–119.

Byers, D.L. and T.R. Meagher, 1995. Demographic characteristics of rarity in plants, as determined through a comparative study in the genus *Eupatorium*. (submitted).

Canales, J., M.C. Trevisan, J.F. Silva and H. Caswell, 1994. A demographic study of an annual grass (*Andropogon brevifolius* Schwarz) in burnt and unburnt savanna. Acta Oecol. 15, 261–273.

Caswell, H., 1978. A general formula for the sensitivity of population growth rate to changes in life history parameters. Theor. Popul. Biol. 14, 215–230.

Caswell, H., 1982a. Stable population structure and reproductive value for populations with complex life cycles. Ecology 63, 1223–1231.

Caswell, H., 1982b. Life history theory and the equilibrium status of populations. Am. Nat. 120, 317–339.

Caswell, H., 1989a. Matrix Population Models: Construction, Analysis, and Interpretation. Sinauer Associates, Sunderland, MA.

Caswell, H., 1989b. The analysis of life table response experiments. I. Decomposition of treatment effects on population growth rate. Ecol. Model. 46, 221–237.

Caswell, H., 1989c. Life history strategies. In: Ecological Concepts, edited by J. M. Cherrett, Blackwell, Oxford, pp. 285–308.

Caswell, H. and A. Hastings, 1980. Fecundity, developmental time, and population growth rate: an analytical solution. Theor. Popul. Biol. 17, 71–79.

Caswell, H. and L.V. Martin, (in prep). Analysis of life table response experiments. IV. Regression designs.

Collyard, S.A., G.T. Ankley, R.A. Hoke and T. Goldstein, 1994. Influence of age on the relative sensitivity of *Hyalella azteca* to diazinon, alkylphenol ethoxylates, copper, cadmium, and zinc. Arch. Environ. Contam. Toxicol. 26, 110–113.

Crowley, P. H., R. M. Nisbet, W. S. C. Gurney and J. H. Lawton, 1987. Population regulation in animals with complex life-histories: formulation and analysis of a damselfly model. Adv. Ecol. Res. 17, 1-59.

Cushing, J.M., 1995. Nonlinear matrix models for structured populations. In: Mathematical Population Dynamics. III. Mathematical Methods and Modelling of Data, edited by A. Arino, S. Axelrod, M. Kimmel and M. Langlais. Wuerz Publishing.

Cushing, J.M., 1994. Nonlinear matrix equations and population dynamics. In: Structured Population Models, edited by H. Caswell and S.D. Tuljapurkar (in preparation).

Daniels, R. E. and J. D. Allan, 1981. Life table evaluation of chronic exposure to a pesticide. Can. J. Fish. Aquat. Sci. 38, 485–494.

DeAngelis, D.L. and L.J. Gross, 1992. Individual-Based Models and Approaches in Ecology. Chapman and Hall, New York.

Ebert, T. A., 1985. Sensitivity of fitness to macroparameter changes: an analysis of survivorship and individual growth in sea urchin life histories. Oecologia 65, 461–467.

Fitzmayer, K. M., J. G. Geiger and M. J. Van Den Avyle, 1982. Effects of chronic exposure to simazine on the cladoceran, *Daphnia pulex*. Arch. Environ. Contam. Toxicol. 11, 603–609

Frank, P. W., C. D. Boll and R. W. Kelly., 1957. Vital statistics of laboratory cultures of *Daphnia pulex* DeGeer as related to density. Physiol. Zool. 30, 287–305.

Gentile, J. H., S. M. Gentile, N. G. Hairston, Jr. and B. K. Sullivan, 1982. The use of life-tables for evaluating the chronic toxicity of pollutants to *Mysidopsis bahia*. Hydrobiologia 93, 179–187.

Hallam, T.G., C.E. Clark and G.S. Jordan, 1983. Effects of toxicants on populations: a qualitative approach. J. Math. Biol. 18, 25–37.

Hallam, T.G., R.R. Lassiter, J. Li and L.A. Suarez, 1990a. Modelling individuals employing an integrated energy response: application to *Daphnia*. Ecology 71, 938–954.

Hallam, T.G., R.R. Lassiter, J. Li and W. McKinney, 1990b. Toxicant-induced mortality in models of *Daphnia* populations. Environ. Toxicol. Chem. 9, 597–621.

Hastie, T.J. and R.J. Tibshirani, 1990. Generalized Additive Models. Chapman and Hall, New York.

Horst, T.J., 1976. Population dynamics of the burrowing mayfly, *Hexagenia limbata*. Ecology 57, 199–204.

Horvitz, C.C. and D.W. Schemske, 1995. Spatiotemporal variation in demographic transitions of a tropical understory herb: projection matrix analysis. Ecol. Monogr. 65, 155–192.

Hummon, W. D. and M. R. Hummon, 1975. Use of life table data in tolerance experiments. Cah. Biol. Mar. 16, 743–749

Kawano, S., T. Takada, S. Nakayama and A. Hiratsuka, 1987. Demographic differentiation and life-history evolution in temperate woodland plants. In: Differentiation Patterns in Higher Plants, edited by K.M. Urbanska, Academic Press, New York, pp. 153–181.

Keyfitz, N., 1968. Introduction to the Mathematics of Population. Addison-Wesley, Reading, MA.

King, C.E., 1967. Food, age, and the dynamics of a laboratory population of rotifers. Ecology 48, 111–128.

Kooijman, S.A.L.M., 1993. Dynamic Energy Budgets in Biological Systems: Theory and Applications in Ecotoxicology. Cambridge University Press, Cambridge.

Kooijman, S.A.L.M. and J.A.J. Metz, 1984. On the dynamics of chemically stressed populations: the deduction of population consequences from effects on individuals. Ecotoxicol. Environ. Saf. 8, 254–274.

Krainacker, D.A., J.R. Carey and R.I. Vargas, 1987. Effect of larval host on life history traits of the mediterranean fruit fly, *Ceratitis capitata*. Oecologia 73, 583–590.

Leslie, P.H. and R.M. Ranson, 1940. The mortality, fertility and rate of natural increase of the vole (*Microtus agrestis*) as observed in the laboratory. J. Anim. Ecol. 9, 27–57.

Levin, L. A. and D. V. Huggett, 1990. Implications of alternative reproductive modes for seasonality and demography in an estuarine polychaete. Ecology 71, 2191–2208.

Levin, L. A., H. Caswell, K. D. DePatra and E. L. Creed, 1987. Demographic consequences of larval development mode: planktotrophy vs. lecithotrophy in *Streblospio benedicti*. Ecology 68, 1877–1886.

Levin, L.A., T. Bridges, H. Caswell, D. Cabrera, G. Plaia and C. DiBacco, 1994. Demographic responses of estuarine polychaetes to sewage, algal, and hydrocarbon contamination: life table response experiments. Ecol. Appl. (in press).

Lewontin, R. C., 1965. Selection for colonizing ability. In: The Genetics of Colonizing Species, edited by H. G. Baker and G. L. Stebbins, Academic Press, New York, pp. 77–94.

Lotka, A.J., 1936. The geographic distribution of intrinsic rate of natural increase in the United States, and an examination of the relation between several measures of net reproductivity. J. Am. Stat. Assoc. 31, 273–294.

Lynch, M., 1992. The life history consequences of resource depression in *Ceriodaphnia quadrangula* and *Daphnia ambigua*. Ecology 73, 1620–1629.

Marshall, J. S., 1962. The effects of continuous gamma radiation on the intrinsic rate of natural increase on *Daphnia pulex*. Ecology 43(4), 598–607.

McCauley, E., W.W. Murdoch, R.M. Nisbet and W.S.C. Gurney, 1990. The physiological ecology of *Daphnia*: development of a model of growth and reproduction. Ecology 71, 703–715.

McKendrick, A.G., 1926. Applications of mathematics to medical problems. Proc. Edinburgh Math. Soc. 40, 98–130.

Menges, E.S., 1987. Population viability analysis for an endangered plant. Conserv. Biol. 4, 52–62.

Metz, J.A.J. and O. Diekmann, 1986. The Dynamics of Physiologically Structured Populations. Springer–Verlag, New York.

Metz, J.A.J., A.M. DeRoos and F. van den Bosch, 1988. Population models incorporating physiological structure: a quick survey of the basic concepts and an application to size-structured population dynamics in waterfleas. In: Size-Structured Populations, edited by B. Ebenman and L. Persson, Springer-Verlag, New York, pp. 102–126 .

Nisbet, R.M. and W.S.C. Gurney, 1986. The formulation of age-structure models. In: Mathematical Ecology, edited by T.G. Hallam and S.A. Levin, Springer-Verlag, New York, pp. 95–115.

Nisbet, R. M., W. S. C. Gurney, W. W. Murdoch and E. McCauley, 1989. Structured population models: a tool for linking effects at the individual and population level. Biol. J. Linn. Soc. 37, 79–99.

O'Brien, P., H. Feldman, E. V. Grill and A. G. Lewis, 1988. Copper tolerance of the life history stages of the splashpool copepod *Tigriopus californicus* (Copepoda, Harpacticoida). Mar. Ecol. 44: 59–64.

Olesiuk, P. F., M. A. Bigg and G. M. Ellis, 1990. Life history and population dynamics of resident killer whales (*Orcinus orca*) in the coastal waters of British Columbia and Washington State. Rep. Int. Whaling Comm. Spec. Issue 12, 209–243.

Pearl, R. and S.L. Parker, 1922. Experimental studies on the duration of life. III. The effect of successive etherizations on the duration of life of *Drosophila*. Am. Nat. 56, 275–280.

Pearl, R. and S.L. Parker, 1924. Experimental studies on the duration of life. X. The duration of life of *Drosophila melanogaster* in the complete absence of food. Am. Nat. 56, 275–280.

Pearl, R., J.R. Miner and S.L. Parker, 1927. Experimental studies on the duration of life. XI. Density of population and life duration in *Drosophila*. Am. Nat. 61, 289–318.

Pearl, R., 1928. The Rate of Living. Alfred A. Knopf, New York.

Rao, T. R. and S. S. S. Sarma, 1986. Demographic parameters of *Brachionus patulus* Muller (Rotifera) exposed to sublethal DDT concentrations at low and high food levels. Hydrobiologia 139, 193–200.

Riessen, H.P. and W.G. Sprules, 1990. Demographic costs of antipredator defenses in *Daphnia pulex*. Ecology 71, 1536–1546.

Silva, J.G., J. Raventos, H. Caswell and M.C. Trevisan, 1991. Population responses to fire in a tropical savanna grass *Andropogon semiberbis*: a matrix model approach. J. Ecol. 79, 345–356.

Takada, T. and H. Nakajima, 1992. An analysis of life history evolution in terms of the density-dependent Lefkovitch matrix model. Math. Biosci. 112, 155–176.

Walls, M., H. Caswell and M. Ketola, 1991. Demographic costs of *Chaoborus*-induced defenses in *Daphnia pulex*. Oecologia 87, 43–50.

Walton, W. E., S. M. Compton, J. D. Allan and R. E. Daniels, 1982. The effect of acid stress on survivorship and reproduction of *Daphnia pulex* (Crustacea: Cladocera). Can. J. Zool. 60, 573–579.

Watson, T.F., 1964. Influence of host plant condition on population increase of *Tetranychus telarius* (Linnaeus) (Acarina: Tetranychidae). Hilgardia 35, 273–322.

Werner, P. A. and H. Caswell, 1977. Population growth rates and age versus stage-distribution models for teasel (*Dipsacus sylvestris* Huds.). Ecology 58, 1103–1111.

Williams, K. A., D. W. J. Green, D. Pascoe and D. E. Gower, 1986. The acute toxicity of cadmium to different larval stages of *Chironomus riparius* (Diptera: Chironomidae) and its ecological significance for pollution regulation. Oecologia 70, 362–366

Winner, R. W. and M. P. Farrell, 1976. Acute and chronic toxicity of copper to four species of *Daphnia*. J. Fish. Res. Board Can. 33, 1685–1691.

CHAPTER 10

Toxicants as Selective Agents in Population and Community Dynamics

Ronald K. Chesser and Derrick W. Sugg

I. INTRODUCTION

Although all sciences depend upon descriptive data and the technologies that aid their collection, synthesis of these data is equally important. Theory plays an important role in the synthetic process; however, its value is not limited to prediction. Modeling nature, at all levels of organization, provides much insight into the workings of our world. As with theory, models developed to explore theoretical issues are important not only for predictive purposes, but because they offer understanding. In this manuscript, we explore some of the theoretical issues of ecotoxicology by modeling some of the processes in ecosystems.

One may ask why ecotoxicology is so descriptive. The answer probably lies partly in our recent interest in some of the issues dealt with by ecotoxicologists. After all, the theory of natural selection was based on keen observations and descriptions made by Charles Darwin (1845), and only later synthesized into a coherent theory (Darwin, 1859, 1871). Another consideration is that ecotoxicology is the melding of two disciplines (ecology and toxicology), crossing the boundaries of many sciences (i.e., chemistry, geology, and biology). Ecotoxicologists often fail to recognize that these sciences and disciplines already have well-developed theories that can aid in our new pursuits. Novel questions associated with ecotoxicology have been the impetus for the new description and techniques that have proliferated in the literature. Description is a natural result of curiosity and, without that curiosity, it is unlikely that we would have even recognized the problems that ecotoxicologists are so concerned with today. As has been so eloquently and succinctly stated by Levin (1989), "theory in the absence of data is sterile, data without theory is uninterpretable." It is important that ecotoxicologists draw on all of the strengths,

1-56670-1127-9/95/$0.00+$.50

whether they be synthetic or descriptive, of other fields if it is to become a viable discipline, much as conservation biology has done. This approach is particularly important because much of the very existence of ecotoxicology is owed to the needs of regulatory bodies dealing with crises.

What role does theory have to play in ecotoxicology? One obvious answer to this question is that it will allow scientists, and regulators, to make predictions about the impacts of toxicants. However, too much emphasis may be placed upon the predictive powers of our theories (Ehrlich, 1986, 1989). It is through theories that we may observe, and ultimately understand, issues that are too complex to glean from descriptive data. Theory can indicate new types of data that must be collected, examine processes for their tractability and usefulness in explaining observed phenomena, and define the bounds under which these phenomena are operating. Ecotoxicologists should welcome theories that address any of these problems, regardless of their predictive abilities.

Obviously, ecosystems are complex entities that will require detailed information to be understood. Although there is considerable data on the composition and functions of ecosystems, a detailed understanding of these systems is lacking. Ironically, we may have too much data to develop a tractable model of ecosystem processes. It is imperative that we simplify the systems to avoid clouding the real issues with unimportant details (Levin, 1989). One method of attaining this goal is to devise standards that are useful for comparing the functions of different ecosystems (Kimball and Levin, 1985) while still maintaining the crucial functions shared by all. Evans (1956) suggested that one obvious simplification is to understand the flow of energy in ecosystems. The utility of this theme has been shown in some classical studies by Lindeman (1942) and Odum (1957). Indeed, many agencies view the flow of materials and energy to be central to understanding ecosystem processes (Schlesinger, 1989). Such a simplification will lead to a loss of predictability, but it does not preclude future improvements in the theories. The laws of thermodynamics have played, and will continue to play, an important role in our understanding of these complex systems. We are likely to benefit from even the most simplistic models as long as we recognize their strengths and weaknesses (Pimm and Gilpin, 1989).

Obvious areas for improvement of previous models of ecosystem functions are the relaxation of assumptions of equilibrium, constancy, and stability; these improvements may be more important than incorporating more complexity and increasing predictive power (Levin, 1989). Some attention must be paid to complexity; after all, simple Lotka-Volterra models poorly describe the massive amounts of energy actually affecting organisms in a complex community (Pimm and Gilpin, 1989). Herein, we develop a model for ecosystems that retains the importance of energy while relaxing the requirements for equilibrium conditions. The ecosystem is viewed as having very complex, nested interactions among the species, but we have limited the number of species to maintain tractability. We have also avoided the trap of closed communities by allowing immigration and recolonization; however, to simplify the system we only allow inclusion of the original member species. Finally, we introduce toxicants as a general selective agent. Our goals are not to provide a predictive model concerning the fate of ecosystems that have been impacted by a general toxicant. Instead, we are interested in understanding the roles that species redundancy, adaptation, and immigration play in the ability of an ecosystem to cope with such an insult. Our model is not perfect, it is even over parameterized for complete understanding, but it may offer some insight that could not be obtained by more classical approaches. This model is intended only to provide plausible functions for ecosystem dynamics. Other functions are possible, and they may alter the response of the ecosystem. More specific models can be developed to explore these details.

II. METHODS

The model described herein simulates the dynamics of a simple community. Each ecosystem starts as an empty patch within a larger system containing the propagules for colonizing the patch. The propagule sources are viewed as essentially infinite in size. Each simulation is repeated ten times with the population sizes, energetics, diversity (Shannon-Wiener Index; Krebs, 1985), and gene frequencies calculated each iteration. The arithmetic averages of these parameters for all ten repetitions are used to summarize the results. Separate simulations were used to establish the baseline dynamics of the ecosystem (no selection), and to determine the changes resulting from a toxicant insult (selection).

A. CONSTANTS

Several constants describing the nature of the ecosystem were used to implement the model. Each ecosystem consisted of 16 species broadly characterized to represent phototrophs $(n = 8)$, herbivores $(n = 4)$, omnivores $(n = 2)$, carnivores $(n = 1)$, and detritivores $(n = 1)$. Additionally, each ecosystem had a variable quantity of external energy at a given time $(\tilde{E}_t)$ that could be acquired and utilized by phototrophs. The external energy at a given time was determined from:

$$\tilde{E}_t = E\left(0.8 + \frac{0.1\sin(t - \pi)}{12}\right) + 0.5 \tag{1}$$

which resulted in a sin-wave with a wavelength of six time intervals. For all simulations, the maximum external energy (E) was set at 25,100 units. A vector, with elements for each species in the ecosystem, was used to describe the amount of external energy needed to support one individual of a given species ($\vec{\varepsilon}$; Table 1). Another vector, of the same size, was used to describe the efficiency with which a species was able to convert external energy to reproductive effort ($\vec{\lambda}$; Table 1). Because only phototrophs could use external energy, the values in these vectors that corresponded to the remaining eight species were zero.

Table 1   Constant Values for 16 Species in
a Model of Ecosystem Dynamics

Species	ε	λ	r
Phototroph	6.2750	0.1594	0.0833
Phototroph	6.2750	0.1594	0.0833
Phototroph	1.5688	0.6375	0.3333
Phototroph	1.5688	0.6375	0.3333
Photoroph	0.3138	3.1873	0.6667
Photoroph	0.3138	3.1873	0.6667
Photoroph	62.7500	0.0159	0.0333
Photoroph	62.7500	0.0159	0.0333
Herbivore 1	0.0000	0.0000	0.3333
Herbivore 2	0.0000	0.0000	0.3333
Herbivore 3	0.0000	0.0000	0.1667
Herbivore 4	0.0000	0.0000	0.1667
Omnivore 1	0.0000	0.0000	0.1111
Omnivore 2	0.0000	0.0000	0.1667
Carnivore	0.0000	0.0000	0.0833
Detritivore	0.0000	0.0000	0.9900

Note: Columns contain the species designation,
amount of external energy (ε) necessary to
support one individual of that species, the
relative factor for conversion (λ) of energy
into population growth, and the intrinsic
growth rates (r) of the species.

Central to the determination of ecosystem dynamics are the interactions among species. A matrix of values was used to describe the potential flow of energy from one species to another, termed the Z-matrix hereafter (Table 2). Z-values represent the potential flow of energy from individuals of the species i (represented by the row) to individuals of the species j (represented by the column). Positive values indicate a potential gain for species j, while negative values indicate a potential loss of energy in the presence of species i. A schematic representation of interactions among trophic levels, as it pertains to energy flow, is given in Figure 1. Another matrix was used to describe the flow of materials from one species to another, termed the C-matrix (Table 2). The Z-matrix was used for the calculation of carrying capacities, while the C-matrix was used for control of toxicant flow. Finally, a vector of population growth rates was used with carrying capacities to calculate the change in population sizes (Table 1).

B. INITIAL VALUES

Each simulation was carried out for a total of 1220 time intervals, with each interval representing the generation time of the shortest lived species (in this case, the detritivore). To initiate the simulation, several values were chosen to seed the first interval. Population sizes ($\vec{N}$) for all species except the detritivore were set as random, uniform numbers in the range from zero to ten. The initial population size of the detritivore was set at 10,000. The genotype frequencies of all species ($\vec{\rho}$) were assigned random, uniform numbers in the range of zero to one. Finally, the toxicant levels ($\vec{\tau}$) and the deviation from

Table 2   Example Matrices for Interactions Among Species (Z-matrix) and Flow of Materials (C-matrix) in a Model of Ecosystem Dynamics

Z-matrix

Species	Phototroph 1	Phototroph 2	Phototroph 3	Phototroph 4	Phototroph 5	Phototroph 6	Phototroph 7	Phototroph 8	Herbivore 1
Phototroph 1	-0.1000	-0.1000	-0.4000	-0.4000	-2.0000	-2.0000	-0.0100	-0.0100	0.0200
Phototroph 2	-0.1000	-0.1000	-0.4000	-0.4000	-2.0000	-2.0000	-0.0100	-0.0100	0.0400
Phototroph 3	-0.0250	-0.0250	-0.1000	-0.1000	-0.5000	-0.5000	-0.0025	-0.0025	0.0100
Phototroph 4	-0.0250	-0.0050	-0.1000	-0.1000	-0.5000	-0.5000	-0.0025	-0.0025	0.0050
Phototroph 5	-0.0050	-0.0050	-0.0200	-0.0200	-0.1000	-0.1000	-0.0005	-0.0005	0.0100
Phototroph 6	-0.0050	-0.0050	-0.0200	-0.0200	-0.1000	-0.1000	-0.0005	-0.0005	0.0194
Phototroph 7	-1.0000	-1.0000	-4.0000	-4.0000	-20.0000	-20.0000	-0.1000	-0.1000	0.0000
Phototroph 8	-1.0000	-1.0000	-4.0000	-4.0000	-20.0000	-20.0000	-0.1000	-0.1000	0.0000
Herbivore 1	-0.1000	-0.1000	-0.8000	-0.5000	-2.0000	-4.0000	-0.1000	-0.0100	-0.2060
Herbivore 2	-0.1000	-0.1000	-0.5000	-0.8000	-4.0000	-2.0000	-0.1000	-0.1000	-0.2000
Herbivore 3	-0.0500	-0.0500	-0.8000	-0.5000	-10.0000	-10.0000	-0.0100	-0.0050	-0.5000
Herbivore 4	-1.0000	-1.0000	-1.8000	-2.0000	-8.0000	-5.0000	-0.5000	-0.5000	-1.2760
Omnivore 1	0.0000	0.0000	-0.1000	-0.1000	-5.0000	-8.0000	-0.0100	-0.0100	-0.5000
Omnivore 2	-0.3000	-0.3000	-0.5000	-0.5000	-5.0000	-5.0000	-0.2000	-0.2000	-1.0000
Carnivore	0.0000	0.0000	0.0000	0.0000	0.0000	0.0000	0.0000	0.0000	-10.0000
Detritivore	0.0005	0.0005	0.0019	0.0018	0.0098	0.0098	0.0001	0.0001	0.0000

Z-matrix

Species	Herbivore 2	Herbivore 3	Herbivore 4	Omnivore 1	Omnivore 2	Carnivore	Detritivore
Phototroph 1	0.0400	0.0400	0.0100	0.0100	0.0020	0.0000	336.4172
Phototroph 2	0.0200	0.0100	0.0100	0.0100	0.0020	0.0000	336.4172
Phototroph 3	0.0050	0.0010	0.0010	0.0020	0.0050	0.0000	63.9193
Phototroph 4	0.0100	0.0200	0.0020	0.0010	0.0050	0.0000	63.9193
Phototroph 5	0.0194	0.0025	0.0050	0.0050	0.0050	0.0000	6.0555
Phototroph 6	0.0100	0.0200	0.0050	0.0050	0.0050	0.0000	6.0555
Phototroph 7	0.0000	0.0000	0.1200	0.1000	0.2500	0.0150	3263.2460
Phototroph 8	0.0000	0.0000	0.1000	0.1000	0.2100	0.0100	3263.2460
Herbivore 1	-0.2000	-0.2000	-0.0100	0.1000	0.0500	0.0010	168.2086
Herbivore 2	-0.2060	-0.1000	-0.0100	0.1000	0.0500	0.0010	168.2086
Herbivore 3	-0.5000	-0.5129	-0.5000	0.0010	-0.5000	0.1000	336.4172
Herbivore 4	-1.2760	-0.3000	-0.3195	-0.1798	-0.2000	0.1224	336.4172
Omnivore 1	-0.5000	-0.8000	-0.5000	-0.3790	-0.2500	0.0000	33.6417
Omnivore 2	-1.0000	-0.3000	-0.1740	-0.1500	-0.2160	-0.0100	42.3886
Carnivore	-10.0000	-13.6400	-3.8597	-10.2000	-6.7406	-0.1170	1345.6690
Detritivore	0.0000	0.0000	0.0000	0.0000	0.0000	0.0000	-0.1209

Table 2 (continued)

C-matrix

Species	Phototroph 1	Phototroph 2	Phototroph 3	Phototroph 4	Phototroph 5	Phototroph 6	Phototroph 7	Phototroph 8	Herbivore 1
Phototroph 1	0.0779	0.0000	0.0000	0.0000	0.0000	0.0000	0.0000	0.0000	0.0779
Phototroph 2	0.0000	0.0935	0.0000	0.0000	0.0000	0.0000	0.0000	0.0000	0.1871
Phototroph 3	0.0000	0.0000	0.0742	0.0000	0.0000	0.0000	0.0000	0.0000	0.1484
Phototroph 4	0.0000	0.0000	0.0000	0.0612	0.0000	0.0000	0.0000	0.0000	0.0612
Phototroph 5	0.0000	0.0000	0.0000	0.0000	0.0112	0.0000	0.0000	0.0000	0.1120
Phototroph 6	0.0000	0.0000	0.0000	0.0000	0.0000	0.0088	0.0000	0.0000	0.1698
Phototroph 7	0.0000	0.0000	0.0000	0.0000	0.0000	0.0000	0.0934	0.0000	0.0000
Phototroph 8	0.0000	0.0000	0.0000	0.0000	0.0000	0.0000	0.0000	0.1029	0.2641
Herbivore 1	0.0000	0.0000	0.0000	0.0000	0.0000	0.0000	0.0000	0.0000	0.0000
Herbivore 2	0.0000	0.0000	0.0000	0.0000	0.0000	0.0000	0.0000	0.0000	0.0000
Herbivore 3	0.0000	0.0000	0.0000	0.0000	0.0000	0.0000	0.0000	0.0000	0.0000
Herbivore 4	0.0000	0.0000	0.0000	0.0000	0.0000	0.0000	0.0000	0.0000	0.0000
Omnivore 1	0.0000	0.0000	0.0000	0.0000	0.0000	0.0000	0.0000	0.0000	0.0000
Omnivore 2	0.0000	0.0000	0.0000	0.0000	0.0000	0.0000	0.0000	0.0000	0.0000
Carnivore	0.0000	0.0000	0.0000	0.0000	0.0000	0.0000	0.0000	0.0000	0.0000
Detritivore	0.1068	0.1068	0.1077	0.1067	0.1140	0.1140	0.1650	0.1650	0.0000

C-matrix

Species	Herbivore 2	Herbivore 3	Herbivore 4	Omnivore 1	Omnivore 2	Carnivore	Detritivore
Phototroph 1	0.1558	0.2226	0.1558	0.1558	0.0229	0.0000	0.1311
Phototroph 2	0.0935	0.0668	0.1871	0.1871	0.0275	0.0000	0.1573
Phototroph 3	0.0742	0.2120	0.0594	0.1187	0.2182	0.0000	0.0949
Phototroph 4	0.1224	0.3498	0.0980	0.0490	0.1801	0.0000	0.0783
Phototroph 5	0.2173	0.0400	0.2240	0.2240	0.1647	0.0000	0.0068
Phototroph 6	0.0875	0.2500	0.1750	0.1750	0.1287	0.0000	0.0053
Phototroph 7	0.0000	0.0000	0.2241	0.1868	0.3433	0.0000	0.1524
Phototroph 8	0.0000	0.0000	0.2058	0.2058	0.3177	0.0000	0.1679
Herbivore 1	0.0000	0.0000	0.0000	0.5129	0.1886	0.0128	0.0216
Herbivore 2	0.2641	0.0000	0.0000	0.5129	0.1886	0.0128	0.0216
Herbivore 3	0.0000	0.4139	0.0000	0.0023	0.0000	0.5649	0.0190
Herbivore 4	0.0000	0.0000	0.5040	0.0000	0.0000	0.4827	0.0133
Omnivore 1	0.0000	0.0000	0.0000	0.9978	0.0000	0.0000	0.0022
Omnivore 2	0.0000	0.0000	0.0000	0.0000	0.9934	0.0000	0.0066
Carnivore	0.0000	0.0000	0.0000	0.0000	0.0000	0.8968	0.1032
Detritivore	0.0000	0.0000	0.0000	0.0000	0.0000	0.0000	0.0141

Note: The **Z**-matrix represents the pathways and relative magnitudes for the flow of energy among the species. The **C**-matrix represents the pathways and relative magnitudes for the flow of toxicants, and other materials, through the ecosystem.

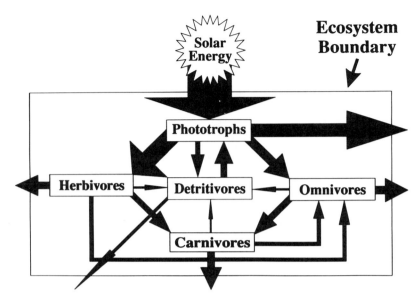

Figure 1   Schematic representation of energy flow in a hypothetical ecosystem. Energy enters the system from an external source and some proportion is captured by phototrophs and some is lost. Energy can flow from phototrophs to other trophic levels with the potential for loss at each level. The width of the arrows designates the magnitude of energy involved and will vary depending upon the composition of the ecosystem (i.e., number of species, numbers of individuals) and external factors (i.e., time of year, weather).

Hardy-Weinberg expectations $(\vec{F})$ were set to zero, and the mean population fitnesses $(\vec{\omega})$ were set to 1.0 for each species. The above parameters were recalculated each time interval.

A number of other parameters were determined randomly at the beginning of each simulation and held constant. The intrinsic dispersal rate of each species $(\vec{D})$ was a random, uniform number between zero and 0.5. The dominance factors for genotypes $(\vec{h})$ and the selection coefficients $(\vec{s})$ were random, uniform numbers between zero and one.

C. CARRYING CAPACITIES

To determine the vector of carrying capacities $(\vec{K})$, it was first necessary to determine the external energy necessary to support the extant populations in the ecosystem. The external energy used at any time interval (E_t) was determined from:

$$E_t = \sum_{i=1}^{\eta} N_{i,t-1}\varepsilon_i \qquad (2)$$

where η is the number of species in the ecosystem and t is the time interval. Subtracting this quantity from the total external energy at a given time interval yields the external energy available to phototrophs for growth and reproduction. This resulting quantity was then divided evenly among the phototrophs to yield the net external energy $(\tilde{E}_t)$ at a given time interval:

$$\tilde{E}_t = \frac{E_t - E_t}{\eta_{phototrophs}} = \frac{E\left(0.8\,\dfrac{0.1\sin(t-\pi)}{12}\right) + 0.5 - \displaystyle\sum_{i=1}^{\eta} N_{i,t-1}\varepsilon_i}{\eta_{phototrophs}} \tag{3}$$

Finally, the carrying capacity of each species was determined by the net gain or loss in energy from species interactions, as mediated through the Z-matrix, and the gain in energy from the external source:

$$K_{i,t} = \lambda_i(\tilde{E}_t + N_{i,t-1}\varepsilon_i) + \sum_{j=1}^{\eta} Z_{j,i}N_{j,t-1} \tag{4}$$

Equations 1 to 4 describe the processes that occur at the transition between time intervals; however, there are four states within a time interval that must be accounted for. These stages result from the interdependencies of some of the variables. Although the orders of these stages will depend on specific assumptions (*i.e.*, juvenile or adult dispersal), herein the stages are distribution of toxicant, selection, population growth, and immigration. These states, and the variables changed during these states, are described below (Sections II.D to II.G).

D. TOXICANT FLOW

As with energy, toxicants must flow through the ecosystem. While the Z-matrix determines the effect of one species on the ability of another species to acquire and utilize energy, a new matrix is needed to determine actual flow patterns. The C-matrix serves this purpose by having values only for possible flow of materials from one species to another (i.e., energy and toxicants can flow directly from a phototroph to a herbivore, but not the other way around). Additionally, these materials may be retained by a species such that $C_{i,i} > 0$. If there is no loss of material from the system, then the sum of the rows in the C-matrix will all be 1. Any row sum less than one indicates a loss of material from the system. With this in mind, the toxicant concentration in a species during one iteration can be determined by the toxicant concentrations in other species during the previous iteration. Because the ecosystem composition is dynamic, the actual flow of toxicants can change with each time interval. For that reason, the actual flow must be calculated using a new matrix, v. Because flow of materials can occur only from extant species, each population must be checked for size before its effect on overall flow can be determined. If a population is extinct, then all of its toxicant is diverted to the detritivore, such that $v_{i,\eta} = 1$. The actual flow of toxicant from species i to species j becomes:

$$v_{i,j,t} = \frac{C_{i,j}}{\sum_{j=1;N_i \neq 0}^{\eta} C_{i,j}} \tag{5}$$

Next, the toxicant level in species i can be determined by summing the product of the toxicant level and the flow variable from all species in the ecosystem.

$$\tau_{i,t}^{(d)} = \sum_{j=1}^{\eta} \tau_{j,t-1} V_{j,i,t} \tag{6}$$

and the total toxicant in the system can be obtained by summing the toxicant concentrations across all species.

E. SELECTION

Selection in this model was viewed to act on three possible genotypes, each having a separate fitness (ω_1, ω_2, and ω_3, respectively). These genotype fitnesses for each individual at a given time interval were determined as follows:

$$\omega_{1,i,t} = 1 \qquad \omega_{2,i,t} = 1 - h_i s_i \tau_{i,t}^{(d)} \qquad \omega_{3,i,t} = 1 - s_i \tau_{i,t}^{(d)} \tag{7}$$

The superscript (d) indicates that the toxicant level is at an intermediate state after distribution of toxicants (before selection) during a given time interval. The mean population fitness at a given time interval was determined from:

$$\overline{\omega}_{i,t} \frac{P_{i,t-1}[(\omega_{1,i,t}^2 - 2\omega_{2,i,t}^2 + \omega_{3,i,t}^2)(1 + F_{i,t-1} - P_{i,t-1}F_{i,t-1}) + 2(\omega_{2,i,t}^2 + \omega_{3,i,t}^2)]}{\overline{\omega}_{i,t-1}} \tag{8}$$

and the mortality of individuals changed the population size as follows:

$$N_{i,t}^{(s)} = N_{i,t-1} \frac{(p_{i,t-1} - p_{i,t-1}^2)[F_{i,t-1}(\omega_{1,i,t} - 2\omega_{2,i,t} + \omega_{3,i,t}) + 2(\omega_{2,i,t} - \omega_{3,i,t})] + p_{i,t-1}^2 + (1 - p_{i,t-1})\omega_{3,i,t}}{\overline{\omega}_{i,t-1}} \tag{9}$$

where the superscript (s) indicates an intermediate stage (after selection) in the transition of a variable. Because selection also changed genotype frequencies and deviations from Hardy-Weinberg expectations, these parameters were also recalculated as follows:

$$p_{i,t}^{(s)} = p_{i,t-1} \frac{(p_{i,t-1} + F_{i,t-1} - p_{i,t-1}F_{i,t-1})(\omega_{1,i,t} - \omega_{2,i,t}) + \omega_{2,i,t}}{\overline{\omega}_{i,t-1}}$$

$$F_{i,t} = 1 - \frac{\omega_{2,i,t}(1 - F_{i,t-1})}{\overline{\omega}_{i,t-1}} \tag{10}$$

F. POPULATION GROWTH

Because toxicants can affect fecundity as well as mortality, the growth rate of each population must be determined for each time interval:

$$r_{i,t} = \overline{\omega}_{i,t} r_{i,0} \tag{11}$$

where $r_{i,0}$ is the initial intrinsic growth rate for species i determined randomly as described above. The population size for species i after reproduction was thus represented as:

$$N_{i,t}^{(r)} = N_{i,t}^{(s)} \left[1 + r_{i,t} \left[1 - \frac{N_{i,t}^{(s)}}{K_{i,t}} \right] \right] \tag{12}$$

where the superscript (r) indicates that populations are at a new intermediate stage (after reproduction) during the transition between time intervals.

G. IMMIGRATION

During each time interval, there was the potential for individuals to move into the ecosystem from some other propagule pool. Because it is unlikely that immigrants of a species will successfully invade an ecosystem that does not have the resources to support additional members of that species, immigration rates must account for the proximity of population sizes to carrying capacities as well as intrinsic dispersal rates. Additionally, some randomness is expected. Therefore, the immigration rate (I) of species i at a given time interval was determined from

$$I_{i,t} = XD_i \rho \left[1 - \frac{N_{i,t}^{(r)}}{K_{i,t}} \right] \tag{13}$$

where D_i is the intrinsic dispersal rate explained above and ρ is a random, uniform number between zero and one. Low, medium, and high immigration rates are obtained by assigning the values of 1, 10, and 20, respectively, to X. With this expression, immigration had a maximum value when population size was zero and carrying capacity was greater than zero. Population size after immigration was

$$N_{i,t} = N_{i,t}^{(r)} + I_{i,t} \qquad (14)$$

The introduction of new individuals into the population resulted in changes in the toxicant levels and genotype frequencies of those species. The expressions for these parameters after immigration were as follows:

$$p_{i,t} = p_{i,t}^{(s)} - \frac{I_{i,t}(p_{i,t}^{(s)} - \rho)}{N_{i,t}}$$

$$\tau_{i,t} = \tau_{i,t}^{(d)} - \left(1 - \frac{I_{i,t}}{N_{i,t}}\right) \qquad (15)$$

where $N_{i,t}$ was the population size after immigration.

III. ECOSYSTEM STABILITY

Issues brought to light by the large-scale destruction of our environments and the potential for climate change on a global scale have punctuated our lack of knowledge concerning ecosystems. Whether the destructive factors are the result of land clearing or the introduction of toxicants, it is clear that a better understanding of ecosystem-level phenomena is necessary to address these problems. Central to our understanding of ecosystems is the issue of ecosystem stability (e.g., Elton, 1933). Ecosystem stability can be viewed in terms of either the structure or function of the system. The stability can be assessed in terms of the system's susceptibility to perturbations or its resiliency after change (see Schlesinger, 1989). These measures of stability apply to equilibrium as well as dynamic systems. Given the multitude of possible measures of ecosystem stability, it is not surprising that scientists have not come to a consensus about the appropriate measures. We do not propose to answer the question of which measure is the best; we simply choose a scenario that we feel will lead to some understanding of how ecosystems may respond to perturbations. We feel that the availability and flow of energy in the ecosystem play important roles in determining its stability, either for resistance to change or resiliency. This view comes from the idea that all systems tend toward entropy without the input of energy. This does not rule out the importance of other resources, such as essential nutrients, but simply assumes that these resources impact the ability of organisms to acquire and utilize energy. Another view we feel is important for furthering our understanding of ecosystem stability is that static equilibria are difficult to maintain in complex systems. A dynamic ecosystem is dependent upon extrinsic and intrinsic forces (Levin, 1989) that act on the availability of energy. Finally, we view the problems of assessing structure and function as being intimately related, bound in such a way that general approaches to ecosystem stability are unduly complicated by separate treatment of these components. This is not to say that the differences in these two are not important; indeed, we present results that are intended to contrast the two. We feel that changes in the functional properties of an ecosystem are in some measure proportional to changes in diversity.

Results for the simulations that do not incorporate selection illustrate the dynamic nature of the model, regardless of whether one is concerned with population sizes or available energy and diversity (Figure 2). Although no steady-state equilibrium is attained in the time frame used here, one can see that the major oscillations are diminishing over time. However, the minor oscillations persist. The reasons for these oscillations are the extrinsic and intrinsic factors that impinge upon the ecosystem. Solar energy, and other external sources of energy, maintain the ecosystem in its fight against entropy. Individual species provide the machinery with which the ecosystem converts energy to a quasi-stable state. Other resources, such as nutrients, play a mediating role through their impact on the acquisition and utilization of energy.

The minor oscillations in the component parts of the ecosystem (e.g., population sizes, energy, and diversity) result from extrinsic factors. The magnitude of these oscillations depends on the nature of the energy source. It is obvious that temperate zones experience a regular cycle of available solar energy; however, tropical zones may also show regular cycles in the availability of water. Because water influences the ability of organisms to acquire and utilize energy, the cyclic nature of this resource will also lead to fluctuations in the amount of available energy. It is the extrinsic forces that prevent a true stable equilibrium for ecosystems. We feel that such dynamic properties are representative of most, if not all, ecosystems (see also Levin, 1989; Schlesinger, 1989).

The availability of external energy has its greatest impact on the primary producers. External energy directly affects their carrying capacities, which in turn impacts their population sizes. However, the population sizes of one species will affect the carrying capacities of many others. Thus, the cyclic nature of external sources of energy (and nutrients) cascades through the ecosystem (Figure 2). In general, rapidly reproducing species are more susceptible to outside forces, and producers are more susceptible than consumers (Figure 3); however, population sizes are obviously correlated with position in the trophic structure. Such results are to be expected given the important role population growth rates play in extinction dynamics (e.g., Leigh, 1981; Goodman, 1987).

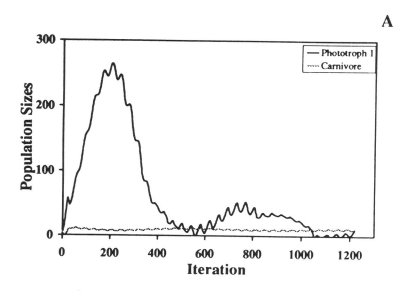

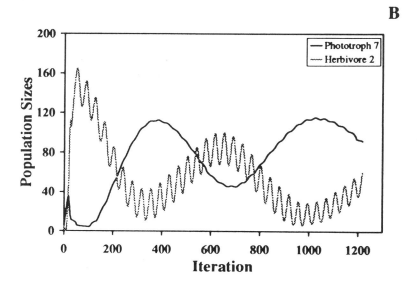

Figure 2   Dynamics of a model ecosystem without selection. Plot A depicts the population sizes of a phototroph ($r = 0.0833$, $D = 0.1070$) and a carnivore ($r =$
0.0833, $D = 0.2236$). Plot B depicts the population sizes of a phototroph ($r =$
0.0333, $D = 0.2347$) and a herbivore ($r = 0.3333$, $D = 0.2126$). Plot C depicts
the available external energy ($\bar{E}$), which may be negative when the number of
phototrophs in the ecosystem requires more energy than is supplied by the
external source. Plot D depicts the Shannon–Wiener index of diversity.

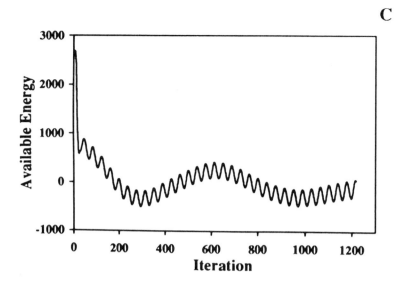

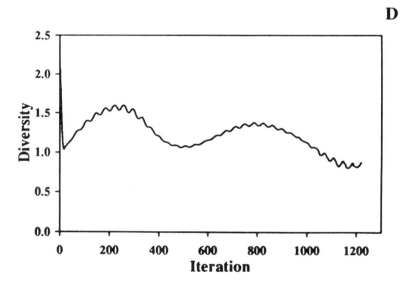

Figure 2 (continued).

Changes in population sizes of phototrophs, resulting from external sources, influence the carrying capacities of all the species in the ecosystem. These changes alter the flow of energy in the system, representing the intrinsic cycles. Although population sizes increase with greater availability of energy, the availability of energy decreases with increasing population sizes. One species may increase because there is a large amount of energy flowing into that

A

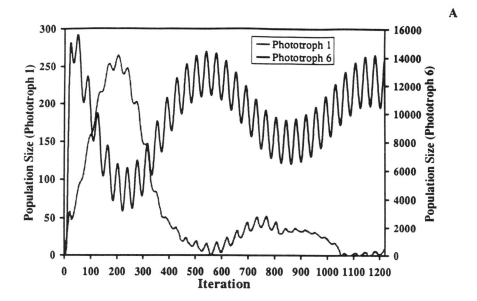

B

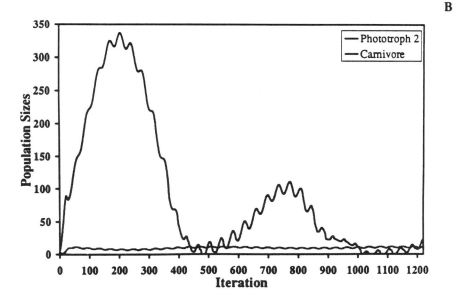

Figure 3   The effect of population growth rates and trophic level on population dynamics in a model ecosystem without selection. Plot A contrasts phototroph 1 ($r = 0.0833$, $D = 0.1070$) with phototroph 6 ($r = 0.6667$, $D = 0.2812$) to show the rapid responses of highly fecund species. Plot B contrasts a phototroph ($r = 0.0833$, $D = 0.2319$) and a carnivore ($r = 0.0833$, $D = 0.2236$) to show producers are more responsive to changes in the availability of external energy than consumers.

portion of the ecosystem from internal and/or external sources, but an increase in their numbers may provide energy for a number of species that consume them (see also Pimm and Gilpin, 1989). These consumers increase in numbers while decreasing the numbers of the species they consume. These relationships take the general form of the traditional competition equation (Lotka, 1925; Volterra, 1926), but they are further complicated by the variable nature of the carrying capacities.

IV. PERTURBATIONS

We view the introduction of toxicants into an ecosystem as simply a selective agent. Of course, a toxicant can indiscriminately kill individuals, regardless of their genotype, but such effects are beyond the scope of this paper which deals more with adaptive responses. We have simplified the mode of selection to represent a single-locus, two-allele system. We have also assumed that selection is directional, as opposed to stabilizing or disruptive, for all species. Finally, we assume that selection is acting on all species, not just the members of a single group (i.e., insecticides aimed at pest species). Other scenarios are possible, but the general intent of this chapter is to elucidate the perturbations that can occur in an ecosystem with the introduction of a selective agent, and not to dwell on specific modes of selection or toxicant action.

Perturbations of the ecosystem change both the structure (number of species and population sizes) and the function (energy flow) of an ecosystem. As the population sizes of a species increase or decrease, the amount of energy flowing through that node in the ecosystem also changes. The relative importance of the various pathways for the flow of energy changes dynamically with changing diversity and relative numbers of each species. Extinctions can lead to the loss of some pathways while increasing the flow of energy through others when some species are released from competition. Perturbations result in changes in the intrinsic cycles by first impacting the carrying capacities (either a direct decrease or indirect increase/decrease). These changes in carrying capacities result in concomitant changes in population size. As with the normal dynamics of the system, changes in population sizes will have an effect on the carrying capacities of other species. Even with the resulting large changes in structure, the basic functioning of the ecosystem (the processing of energy to maintain some order) may return to normal cycles rapidly (Figure 4). This may explain why there is little change in major ecosystem function with declining diversity (Rapport et al., 1985). By comparing unperturbed to perturbed ecosystems (Figures 2C, 2D, and 4), one can also see that, after perturbation, diversity drops more sharply when energy decreases during each extrinsic cycle. Such a result indicates that the simplified system is more sensitive to change. Although the number of species may decrease, the relative abundances of particular species are also changing. Simple measures of diversity, such as the Shannon-Wiener Index (Krebs, 1985), are sensitive to abundance

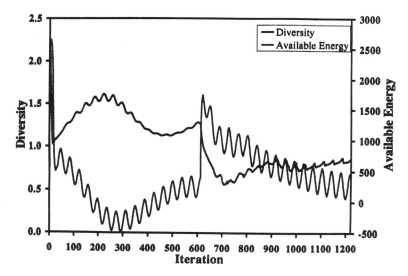

Figure 4   The effect of selection, resulting from a general toxicant, on the diversity and availability of energy in a model ecosystem. The toxicant was introduced at iteration 610. Both indices responded quickly; however, available energy resumed a fairly regular cycle after a few iterations while diversity remained affected for the remainder of the simulations. Of particular interest are the sharp drops in diversity with decreasing energy during the minor cycles, a phenomenon not observed prior to the perturbation.

and evenness (Peet, 1974; Magurran, 1988), but may say nothing of functional roles and relative importance.

Major perturbations in an ecosystem can lead to extinctions, both short-termed and long-termed. These extinctions result in the collapse of some pathways for the flow of energy. However, the relative abundances of some species will increase and much of the available energy will flow through previously existing pathways. These effects impact both the structure (diversity) and function (energetics) of the ecosystem. The result is a system that is much simplified in terms of both structure and function. Additionally, the system is relatively more variable than unperturbed systems (Figure 5). While the energetics of the system may undergo major fluctuations, these changes are of short duration. Conversely, the diversity of perturbed ecosystems remain substantially less than that of "normal" ecosystems for a considerable time period. This increase in variation is not simply a function of simplifying the system, but also results from the "load" experienced by extinct species continually trying to recolonize the ecosystem. These species are artificially maintained in the system by immigration alone, and they have an affect on the richness of the environment. Additionally, they utilize energy that would be available to other species that were relatively unaffected by the perturbation. Thus, structure and function become less predictable after perturbations because of the secondary impacts of recolonization.

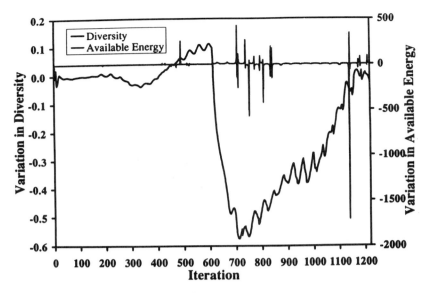

Figure 5   Coefficients of variation for diversity and available energy in a model ecosystem with the introduction of a general toxicant. Both indices show considerable variation when compared with simulations not involving selection. Deviations in available energy are short–lived, whereas those for diversity continue for the remainder of the simulations.

V. SPECIES REDUNDANCIES

As with ecosystem stability, the breadth or complexity of an ecosystem can be viewed in the context of either structure or function. All species utilize energy; however, they differ in how much is necessary for growth, reproduction, and maintenance. They also differ in the form in which energy must be acquired. Very similar uses of energy, particularly in how it is acquired from and disseminated to other components of the ecosystem, lead to some functional overlap in the ecosystem (a similar case can be made for nutrients). In the broadest sense, all phototrophs are redundant species because they acquire energy from an external source and pass some of that energy on to higher trophic levels. Narrowing the scope of redundancy leads to a plethora of examples of bacteria that play a critical role in the cycling of nutrients (i.e., nitrogen). It is these overlaps in function that provide the redundancies in an ecosystem.

With the view of the Hutchinsonian niche (Hutchinson, 1959), it is unlikely that two species will perform exactly the same function in an ecosystem. However, here again the intimate relationships between structure and function are obvious. The greater the number of species in an ecosystem, the greater likelihood that particular functions will be performed by different species. The exact pathways of energy from two phototrophs to a carnivore may be quite different in magnitude and complexity, but energy still gets to the carnivore when either is present (Pimm and Gilpin, 1989). This phenomenon is quite

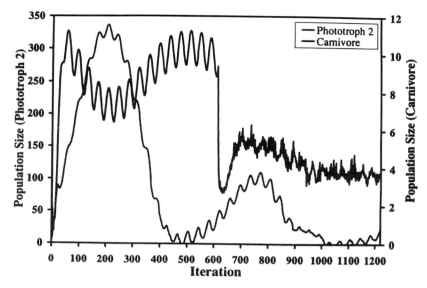

Figure 6   Population dynamics in a model ecosystem with the introduction of a toxicant. The phototroph ($r = 0.0833$, $D = 0.2689$, $h = 0.5905$, $s = 0.5404$) responds dramatically and shows considerable disruption of the normal dynamics. The carnivore ($r = 0.0833$, $D = 0.2384$, $h = 0.5874$, $s = 0.5273$) increases considerably after the introduction of the toxicant and maintains nearly normal dynamics. The different responses of these two species is due largely to the number of pathways for energy flow that lead to each, with consumers having many more pathways than producers.

obvious from the results that low-r phototrophs are more sensitive to ecosystem perturbations than low-r carnivores (Figure 6). Although any given phototroph may be susceptible to extinction, energy can still flow through other phototrophs to herbivores and then to the carnivore. Herein, we have forced considerable redundancy by manipulation of the Z-matrix because incorporating the thousands of species necessary to obtain an accurate representation is impractical. Additionally, models of ecosystem dynamics tend to oversimplify these patterns because no single patch in an ecosystem is likely to contain all the potential redundancies, but immigration may allow these species to invade when a competitor or predator is impacted by a perturbation.

The obvious question from these results concerns how much of the change in an ecosystem after perturbation is "compensation" to maintain some stability and how much is simply a consequence. Obviously, the decrease in population sizes, or extinctions, of some species will relax others from competition or predation. Alternatively, the species that rely almost exclusively on impacted species will be similarly impacted. The relaxation of competition is simply a consequence of the perturbation that frees energy for some species. The result is a simplified system with fewer pathways for energy flow, even if the total amount is relatively unchanged. The key to compensation is the issue of stability; is the resulting system less sensitive to further perturbation? In the presence of high levels of immigration, and therefore recolonization, new perturbations are likely to free energy and allow sink populations to rebound. Under this scenario, the new ecosystem would probably be quite able to cope with new perturbations (see also Long, 1981). However, in the absence of these potential colonizers further perturbations are likely to lead to further simplification of the system, indicating a more susceptible state. Stability must be viewed in terms of proportional changes. A simple system may be stable simply because less energy is necessary to fight entropy (i.e., fewer interactions; May, 1973); however, the loss of one species can lead to a proportionately greater change in structure and function than in a complex system. This is especially likely when complex food webs collapse down to simple food chains (see Pimm, 1982, 1987). A simple ecosystem may be more resistant to change, but it is also likely to be less resilient (i.e., change back to the original state).

In general, there are advantages and disadvantages associated with redundancy in an ecosystem. Having redundant species in an ecosystem may increase the rate at which the impacts on the flow of energy can be alleviated. However, this is simply a response to the perturbation and leads to a more simplified system. At the same time, the release from competition with redundant species will lead to an increase in the population sizes of the remaining species. They may prevent the recolonization and establishment of their competing, redundant species at these higher population sizes (see also Levin, 1989). Such changes in the ecosystem may limit its ability to "compensate" when faced with new insults.

VI. IMMIGRATION

In the model presented here, immigration plays an important role in the ecosystem in both the ecological and genetic contexts. Immigration is continual, but it is limited by available resources. In the purely ecological context, a constant influx of individuals into an ecosystem may artificially maintain populations that would go extinct under other circumstances. Such scenarios are believed to play a major role in the dynamics of many natural populations (see Brown and Kodric-Brown, 1977; Pulliam, 1988; Levin, 1989; Pulliam and Danielson, 1991). At the ecosystem level, these artificially maintained populations have an effect on other species by changing their carrying capacities. For example, a population consisting of a single carnivore is doomed to extinction because it cannot propagate, but during its lifetime it will consume a measurable quantity of energy that other species of carnivores and omnivores could use. Because these types of scenarios are possible in natural, undisturbed ecosystems, one may not be disturbed by the detrimental roles redundant species may play in perturbed systems. However, in a system that has been dramatically simplified by a perturbation, the effects of recolonizing species can cascade throughout the entire system (Levin, 1989). These reintroductions will continually change the pathways of energy flow as a few, but proportionally large number of, species go through cycles of extinction and recolonization. Alternatively, these recolonizations may also act as a means by which individual species can resample the environment. When conditions are favorable, they may be able to re-establish themselves, increase the complexity of the ecosystem, and increase the ability of the ecosystem to compensate for new perturbations.

In the context of genetics, immigration rates also play an important role in the structure and function of an ecosystem. Individuals that move into the environment are likely to have different gene frequencies, by chance, than those within the ecosystem. After a perturbation, this influx of genes is likely to slow the adaptive process by introducing unfavorable genes (Figure 7). However, if a species goes extinct, individuals recolonizing the ecosystem may, by chance, have gene frequencies that are favorable under the current selection pressures. Thus, as with the ecological consequences, the genetic consequences of immigration are double-edged.

The constant drifting of gene frequencies resulting from immigration leads to considerable variation in gene frequencies. Rapidly reproducing species adapt the most rapidly (Figure 7A), whereas less fecund species are more susceptible to stochastic changes in gene frequencies (Figure 7B). There is a secondary trade-off associated with the influx of new genes. Although increasing the frequencies of favorable genes maximizes the fitness of populations in the face of the environmental insult, it may do so at the cost of limiting adaptation to future selective pressures. Conversely, large variations in gene frequencies resulting from drift maintain the potential for future adaptation, but at the cost of suboptimal survivorship and reproduction (e.g., shifting balance; Wright, 1970, 1977).

VII. GENERALIZATIONS

The approach we have taken treats the impact of toxicant agents as simply a type of natural selection. There are undoubtedly situations in which toxicants impact growth and/or reproduction regardless of genotypes. Such scenarios are of obvious importance from an ecological standpoint, but the focus of this chapter is to concentrate on adaptive processes in the context of ecosystem stability. Treating toxicants as selective agents does not lead to any major shifts in our ideas concerning population dynamics (Fisher, 1958), but the fact that they are becoming more prevalent in our environment, and that they affect many species, does suggest that higher levels of organization may respond in unpredictable manners. Gilpin (personal communication) discussed Volterra's work concerning the application of pesticides to a simple predator-prey model, the results of which were counter-intuitive. As systems become more complex, and the number of interdependencies of species increases, the ability to make good prediction based on traditional models is limited, and perhaps impossible for more complex models. It is in this regard that ecotoxicology has the potential to make a great impact on ecology; not by changing our ideas of selection and adaptation, but by forcing us to deal with ecosystems in their full complexity.

A

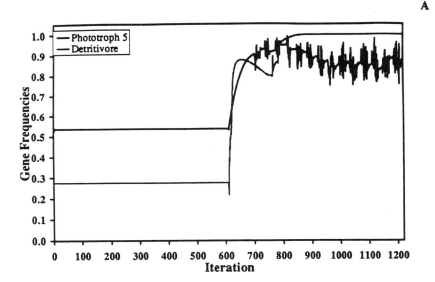

B

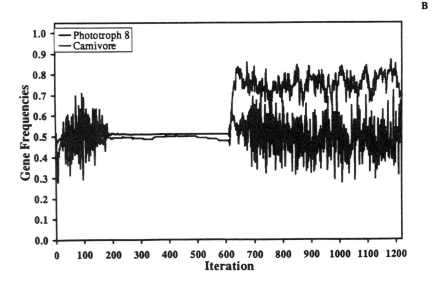

Figure 7   The dynamics of gene frequencies in a model ecosystem with the introduction of a toxicant. Plot A shows that rapidly reproducing species, such as phototroph 5 ($r = 0.6667$, $D = 0.2545$, $h = 0.4038$, $s = 0.5483$) and the detritivore ($r = 0.9900$, $D = 0.2625$, $h = 0.5086$, $s = 0.4764$) respond quickly and are relatively unaffected by drift from immigration. Plot B depicts two species, phototroph 8 ($r = 0.0333$, $D = 0.2573$, $h = 0.5451$, $s = 0.5860$) and the carnivore ($r = 0.0833$, $D = 0.2625$, $h = 0.5086$, $s = 0.4764$) with low population growth rates. These species respond slowly to the selection pressure and are very susceptible to the drifting effect of immigration.

One area that ecotoxicology must address is the problem of remediation in contaminated environments. Obviously cleanup of affected sites will be a major component of many remediation efforts, but other possibilities exist. One may feel that removal of toxicants is too expensive or otherwise impractical. Under such situations, how should one manage an **ecosystem** to preserve its structure and function? Should one isolate the impacted area so as to prevent the transport of the toxicant by either biotic or abiotic components? As we have discussed, immigration has both beneficial and detrimental effects on the ecosystem. Limiting the efflux, and consequently the influx, of individuals may have short-term rewards (more rapid adaptation). However, the complexity of the ecosystem is likely to diminish, making it more susceptible to future perturbations and bringing into question its long-term survival.

Given the goal of long-term persistence of an ecosystem, is reintroduction of extinct species a viable alternative to letting the ecosystem respond naturally? Again, there are benefits and costs to this strategy. These introduced species will utilize resources that other species, quite capable of adapting to the insult, could use. Such a scenario may place further limits on the ecosystem to stabilize. The alternative is an ecosystem that will reach a quasi-equilibrium more rapidly, but at the cost of lost complexity. To regain the complexity that will preserve the original ecosystem and ensure resiliency to future perturbations, it is necessary that some species fill the functional voids left by extinctions. Continual attempts at reintroduction are likely to result in some propagules that are capable of adapting to the present situation in the ecosystem, and their presence may benefit the long-term survival of the ecosystem.

The literature is replete with models of adaptive responses of organisms (see Crow and Kimura, 1970; Falconer, 1989) and gene dynamics (e.g., Chesser, 1991a, 1991b; Caballero and Hill, 1992a, 1992b; Chesser et al., 1993; Sugg and Chesser, 1994). The majority of such models assume that such perturbations have negligible effects on other aspects of the ecosystem that may influence an organism's ability to function. Herein, we have shown that adaptive outcomes may be quite complex and be functions of interacting components of the ecosystem. These complexities may indeed be relevant to ecotoxicology because of the general effects of many toxicants, the persistent nature of many toxicants, and the transfer and biomagnification of some toxicants.

Selection is of obvious importance to individual species; however, other pressures acting on a species having to function within an ecosystem may limit adaptation. Alternatively, the response of energy flow and utilization to environmental changes may be rapid, but in this context selection is not playing an important role (although more complex models may incorporate genes for adapting to altered species interaction). It is the combination of these phenomena that may leave ecosystems less able to cope with future challenges.

ACKNOWLEDGMENTS

This work was supported by funds from Contract DE-AC09-765R00-819 between the United States Department of Energy and the University of Georgia's Savannah River Ecology Laboratory.

REFERENCES

Brown, J.H. and A. Kodric-Brown, 1977. Turnover rates in insular biogeography: effect of immigration on extinction. Ecology 58, 445–449.

Caballero, A. and W.G. Hill, 1992a. A note on the inbreeding effective population size. Evolution 46, 1969–1972.

Caballero, A. and W.G. Hill, 1992b. Effective size in nonrandom mating populations. Genetics 130, 909–916.

Chesser, R.K., 1991a. Gene diversity and female philopatry. Genetics 127, 437–447.

Chesser, R.K., 1991b. Influence of gene flow and breeding tactics on gene diversity within populations. Genetics 129, 573–583.

Chesser, R.K., O.E. Rhodes, Jr., D.W. Sugg and A.F. Schnabel, 1993. Effective sizes for subdivided populations. Genetics 135, 1221–1232.

Crow, J.F. and M. Kimura, 1970. An Introduction to Population Genetics Theory. Harper & Row, New York.

Darwin, C., 1845. The Voyage of the Beagle. John Murray, London.

Darwin, C., 1859. The Origin of Species. John Murray, London.

Darwin, C., 1871. The Descent of Man, and Selection in Relation to Sex. John Murray, London.

Ehrlich, P.R., 1986. Which animal will invade? In: Ecology of Biological Invasions of North America and Hawaii, edited by H.A. Mooney and J.A. Drake, Springer-Verlag, New York, pp. 79–95.

Ehrlich, P.R., 1989. Discussion: management — is ecological theory any good in practice? In: Perspectives in Ecological Theory, edited by J. Roughgarden, R.M. May and S.A. Levin, Princeton University Press, Princeton, NJ, pp. 306–318.

Elton, C., 1933. The Ecology of Animals. Methuen, London.

Evans, F.C., 1956. Ecosystem as the basic unit in ecology. Science 123, 1127–1128.

Falconer, D.A., 1989. Introduction to Quantitative Genetics, 3rd edition. Longman, London.

Fisher, R.A., 1958. The Genetical Theory of Natural Selection. Dover Press, New York.

Goodman, D., 1987. The demography of chance extinction. In: Viable Populations for Conservation, edited by M.E. Soulé, Cambridge University Press, Cambridge, England, pp. 11–34.

Hutchinson, G.E., 1959. Homage to Santa Rosalia or why are there so many kinds of animals? Am. Nat. 95, 137–145

Kimball, K.D. and S.A. Levin, 1985. Limitations of laboratory bioassays and the need for ecosystem level testing. BioScience 35, 165–171.

Krebs, C.J., 1985. Ecology: The Experimental Analysis of Distribution and Abundance. Harper & Row, New York.

Leigh, E.G., 1981. The average lifetime of a population in a varying environment. J. Theor. Biol. 90, 213–239.

Levin, S.A., 1989. Challenges in the development of a theory of community and ecosystem structure and function. In: Perspectives in Ecological Theory, edited by J. Roughgarden, R.M. May and S.A. Levin, Princeton University Press, Princeton, NJ, pp. 242–255.

Lindeman, R.L., 1942. The trophic-dynamic aspect of ecology. Ecology 23, 399–418.

Long, J., 1981. Introduced Birds of the World. David and Charles, London.

Lotka, A.J., 1925. Elements of Physical Biology. Williams & Wilkins, Baltimore, MD.

Magurran, A.E., 1988. Ecological Diversity and its Measurement. Princeton University Press, Princeton, NJ, USA.

May, R., 1973. Stability and Complexity in Model Ecosystems. Princeton University Press, Princeton, NJ.

Odum, H.T., 1957. Trophic structure and productivity of Silver Springs, Florida. Ecol. Monogr. 27, 55–112.

Peet, R.K., 1974. The measurement of species diversity. Annu. Rev. Ecol. Syst. 5, 285–307.

Pimm, S.L., 1982. Food Webs. Chapman and Hall, New York, NY.

Pimm, S.L., 1987. Determining the effects of introduced species. Trends Ecol. Evol. 2, 106–108.

Pimm, S.L. and M.E. Gilpin, 1989. Theoretical issues in conservation biology. In: Perspectives in Ecological Theory, edited by J. Roughgarden, R.M. May and S.A. Levin, Princeton University Press, Princeton, NJ, pp. 287–305.

Pulliam, H.R., 1988. Sources, sinks, and population regulation. Am. Nat. 132, 652–661.

Pulliam, H.R. and B.J. Danielson, 1991. Sources, sinks, and habitat selection: a landscape perspective on population dynamics. Am. Nat. 137 (Suppl.), S50–S66.

Rapport, D.J., H.A. Regier and T.C. Hutchinson, 1985. Ecosystem behavior under stress. Am. Nat. 125, 617–640.

Schlesinger, W.H., 1989. Discussion: ecosystem structure and function. In: Perspectives in Ecological Theory, edited by J. Roughgarden, R.M. May and S.A. Levin, Princeton University Press, Princeton, NJ, pp. 268-274.

Sugg, D.W. and R.K. Chesser. 1994. Effective population sizes with multiple paternity. Genetics 137,1147-1155.

Volterra, V., 1926. Fluctuations in the abundance of a species considered mathematically. Nature 118, 558–560.

Wright, S., 1970. Random drift and the shifting balance theory of evolution. In: Biomathematics: Mathematical Topics in Population Genetics, edited by K. Kojima, Springer-Verlag, New York, pp. 1-31.

Wright, S., 1977. Evolution and the Genetics of Populations, Vol. 3. Experimental Results and Evolutionary Deductions. University of Chicago Press, Chicago.

Effects of Environmental Stressors on Interspecific Interactions of Aquatic Animals

Gary J. Atchison, Mark B. Sandheinrich, and Michael D. Bryan

I. OVERVIEW

Interspecific interactions, such as predation and competition, form linkages among individuals, populations, and communities. In order to survive, grow, and reproduce, organisms must balance many demands in diverse and changing environments. Animals must forage effectively, avoid predation, find appropriate habitat, attract mates, minimize competition for scarce resources, and interact in many ways with the biotic and abiotic components of their environment. How stressors affect the outcomes of these interactions and thus affect population, community, and ecosystem dynamics should be a key question in ecotoxicology, but little effort has been expended on this issue. Ecotoxicologists primarily focus on individuals and populations with laboratory studies that identify contaminant concentrations that decrease rates of survival, growth, and reproduction. However, do standard chronic laboratory tests provide an accurate measure of these effects?

Laboratory fish production index (LFPI) tests were the first widely used chronic tests (Mount and Stephan, 1967). Current tests still use the concept of production as the empirical endpoint but are limited in their assessment of toxic effects on ecological interactions. Fish are maintained in clean tanks, with little or no cover, no predators and an abundance of high quality, nonevasive, and easily handled food. In the real world, life is more challenging. Most toxicant effects on growth observed in these laboratory tests are due to the stress-altered physiology of the organism—reduced motivation to feed, reduced assimilation, or increased maintenance rates. But growth is dependent, in part, on the ability of a fish to search for and capture evasive prey. Survival not only includes physiological tolerance to stressors, but also requires resource partitioning to minimize competition and avoidance of predation. Stressors can affect foraging ability, availability of prey, and the ability to avoid predators. Consider how we would design experiments if we asked questions about the effects of stressors on an organism's fitness (lifetime contribution of offspring) instead of production.

The limited number of toxicological studies of interspecific interactions generally use empirical endpoints, such as toxicant effects on the number of prey consumed or the capture efficiency of predators (Sandheinrich and Atchison, 1990). The intent of these tests is to establish water quality criteria, to contribute to risk assessment through the identification of a no-observed-effect concentration (NOEC) or lowest-observed-effect concentration (LOEC), or to define the dose-response pattern. Results are rarely placed within an ecological context; the focus is generally on the prey *or* the predator, but not on predation ecology and the effects of altered predation on higher and lower trophic levels. Additionally, these studies often test very simple predator-prey systems with few species (commonly fish are the predator) which do not represent all of the diverse foraging modes found in aquatic systems.

Behavioral ecologists have approached the study of predator-prey interactions in a very different way. Instead of using a strictly empirical methodology, they have frequently taken a mechanistic approach and developed models (e.g., optimal foraging, bioenergetics, and prey-switching behavior) to better understand the underlying mechanisms that determine the outcome of predation. A mechanistic approach has been defined as the use of autecology (behavioral ecology, physiological ecology, etc.) as the basis for constructing a theoretical framework to explain community structure (Schoener, 1986). Rate variables used in these models (e.g., handling time, capture rate, caloric intake, respiration rate, search volume) are determined through laboratory and field experiments, and may be altered by toxicants. Although the use of a mechanistic approach in ecotoxicology has been previously advocated (Sandheinrich and Atchison, 1990), there has been little effort to apply these models to understanding toxicant effects on interspecific interactions and subsequent effects, both direct and indirect, on community structure and function.

Our objectives are to: (1) demonstrate how studies in behavioral ecology can contribute to ecotoxicology; we will focus on interspecific interactions, primarily predation and competition, in aquatic ecosystems, and (2) briefly review current knowledge of the theoretical behavioral ecology of these specific interactions, review aquatic ecotoxicological research that is relevant to these interactions, and provide suggestions for further research.

II. GENERAL BACKGROUND

The study of predation and competition is an active area of basic ecological research. Aspects of predation and competition that are of specific interest to this chapter include predator search behavior (Bell, 1991), general

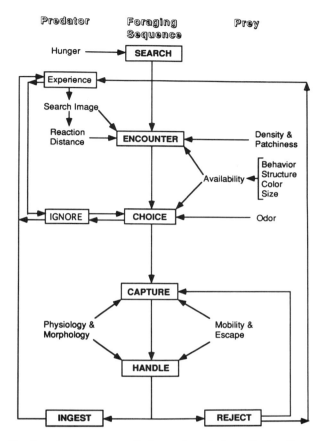

Figure 1   The foraging sequence with the various factors affecting the predator and the prey.

decision-making (Krebs and Kacelnik, 1991), foraging theory (Stephens and Krebs, 1986; Kamil et al., 1987; Hart, 1993), predation strategies (Taylor, 1984; Kerfoot and Sih, 1987; Endler, 1991), competition for resources (Milinski and Parker, 1991), resource partitioning (Werner et al., 1977; Mittelbach, 1988), predation risk, and avoidance of predation (Sih, 1992; Huntingford and Wright, 1993; Lima and Dill, 1990; Milinski, 1993; Werner and Anholt, 1993). We will focus on several of these topics.

III. PREDATION

The foraging sequence is complex (Figure 1), with many factors potentially affecting the central sequence. Many of these factors are physiological and biochemical in nature and will not be discussed in detail here, although they may well be affected by stressors. Our main focus will be those factors that affect the outcome of the foraging sequence and that are population or community oriented.

In order for a predator to acquire food in the wild, it must successfully execute the following sequence of behaviors: prey search, identification, choice, pursuit and capture, handling, and ingestion. Characteristics of both predator and prey influence the outcome of each stage of this predation sequence. Chemical stressors can affect any one or all of these behaviors, thereby reducing food intake per unit time for the predator, and altering predation pressures on the prey. We will first focus on the characteristics of the predator that influence foraging success.

A. HUNGER AND MOTIVATION TO FEED

The control of appetite by the brain is not well understood but probably involves several areas (Heath, 1987). Little has been done to determine whether stressors affect appetite, although a common consequence of stress is the cessation of feeding (Sandheinrich and Atchison, 1990).

B. SEARCHING BEHAVIOR

Predators use refined search strategies to locate specific prey (Bell, 1991). A predator's success in finding prey is dependent upon the match of the appropriate strategy to a particular type of prey. This strategy is based upon past foraging experience, environmental conditions and complexity, characteristics of the prey (i.e., apparent size, color, evasiveness), and various senses associated with finding prey (e.g., vision, olfaction, gustation, lateral line system, etc.). Thus, prey encounter rates are primarily determined by: (1) the search strategy of the predator; (2) acuity of the predator's visual and other sensory systems employed to acquire prey; and (3) prey behavior, density, and distribution in the environment. Environmental stressors may affect search strategy, learning, and sensory systems in predatory fishes, as well as the appearance and abundance of their prey. An effect on one or more of these factors will generally reduce prey encounter rates and may ultimately reduce growth of predators through reductions in rates of caloric intake. There are many search strategies used by aquatic predators, but none as well studied or applied to toxicology as often as planktivory. Planktivorous foraging by fishes has been extensively studied by ethologists and ecologists (Werner, 1974; Werner and Hall, 1974; O'Brien et al., 1976; Vinyard, 1980; Gardner, 1981; Luecke and O'Brien, 1981; Mittelbach, 1981, 1983; Bartell, 1982; Eggers, 1982; Janssen, 1982; O'Brien et al., 1985, 1986, 1989, 1990; Wetterer and Bishop, 1985; Evans and O'Brien, 1988; Wanzenbock and Schiemer, 1989; Walton et al., 1992) and is well understood. Thus, planktivores provide an excellent model for further discussion of contaminant effects on predation in aquatic systems.

White crappie (*Pomoxis annularis*), bluegill (*Lepomis macrochirus*), and other planktivores move in a saltatory or stop-and-go manner while searching for prey (Janssen, 1982; O'Brien et al., 1986; Evans and O'Brien, 1988; Bryan, 1993). Fish using this strategy search for prey only while stopped and, when no prey is found, move approximately half their maximum reaction distance (defined as the maximum distance at which a fish can locate a specific prey) to a new position, where they stop to search again for prey (O'Brien et al., 1986, 1990; Evans and O'Brien, 1988). From their new position, these fish can search a new volume of water (unsearched area) that partially overlaps with the previous search volume. Reaction distance is included in most models of fish predation on zooplankton (Werner and Hall, 1974; Eggers, 1977; Confer et al., 1978; Wright and O'Brien, 1984). Independent measurement of reaction distance in centrarchid planktivorous fish indicated a linear relationship between prey length and reaction distance (Werner and Hall, 1974; Confer and Blades, 1975). However, in their work with bluegill, Sandheinrich and Atchison (1990) and Bryan (1993) found a great deal of variability in this measure, presumably due to the saltatory search strategy used by bluegill. Measurement of reaction distance must be tailored to the individual test species and the search strategy employed in prey capture.

Work by O'Brien and co-workers indicates that planktivorous fish maximize their search efficiency by managing their search time. Assessing contaminant effects on search strategy is not, however, as simple as measuring effects of contaminants on reaction distance to prey (e.g., Nyman, 1981; Sandheinrich and Atchison, 1989), one of few foraging variables previously investigated by ecotoxicologists. Reaction distance is only a small part of the complex strategies that zooplanktivorous fishes employ while foraging (see Janssen, 1982; Evans and O'Brien, 1988; O'Brien et al., 1976, 1985, 1989, 1990; Ehlinger, 1989). For example, bluegill alter the length of their search pauses and the distance of their repositioning movements to maximize the efficiency of their saltatory search for specific prey types and sizes (Janssen, 1982), and for environments of variable complexity (Ehlinger, 1989). Regardless of effects on maximum reaction distance, stressors may decrease search efficiency, and hence prey encounter rates, by altering the amount of time allocated to actual searching (i.e., pauses) and that devoted to moving to new unsearched areas (i.e., runs). Maximum reaction distance is less important than these other components that define optimal search strategy in bluegill and other planktivores. We are assessing the effects of cadmium on the allocation of search time by juvenile bluegill (Bryan, unpublished data). Preliminary analyses indicate that exposure to cadmium significantly alters time allocation associated with searching for daphnid prey, and reduces prey attack and capture rates.

C. PREY CHOICE

In addition to affecting prey search, environmental stressors may also change prey choice. In such cases, predation pressures will be intensified for some prey populations, while being lessened for others. Differential risk of predation for different species may ultimately contribute to changes in prey community structure.

Few studies have investigated the effects of contaminant exposure on prey choice or patch switching behavior. Farr (1978) showed that exposure of gulf killifish (*Fundulus grandis*) to methyl parathion significantly altered their prey choice. Bryan (1993) found cadmium reduced the percentage of bluegills' diet composed of large *Daphnia*, but only when foraging in an arena where large *Daphnia* were relatively scarce. He concluded that cadmium did not affect prey selection directly, but rather indirectly through effects on prey search strategy. In this same study, patch switching efficiency of bluegill, as measured by caloric gain per unit time, was not affected by cadmium exposure. No other studies of this type have been done and, therefore, we cannot compare results among chemical groups or predators. Additional work is clearly needed in this area to determine how contaminants affect prey choice by key aquatic predators.

D. CAPTURE

Perhaps no other components of the predation sequence are more important than successful prey capture, handling, and ingestion. Regardless of prey encounter rates, opportunities for a meal must be capitalized upon as they arise if maximum growth rates are to be achieved by predators. Effects of environmental stressors on prey capture have been primarily quantified by ecotoxicologists through two measures: (1) capture efficiency (i.e., number of prey captured/number of prey attacked), and (2) prey handling time.

Bryan (1993) observed no effect of cadmium on capture efficiency in juvenile bluegill foraging on *Daphnia*, presumably due to the ease with which bluegill capture and ingest these nonevasive, easily handled prey (Eggers, 1977; Drenner et al., 1978; Vinyard, 1980; Ehlinger, 1989, 1990). Studies reporting toxicant-altered capture efficiency (Finger et al., 1985; Mathers et al., 1985; Sandheinrich and Atchison, 1989) involved larger or more evasive prey. Sandheinrich and Atchison (1989) suggested that the sensitivity of capture efficiency to toxicants may be related to the ratio of predator size to prey size. We further suggest that prey evasiveness is important. The harder a particular prey is to capture and consume, the more likely toxicants may affect foraging success.

Sandheinrich and Atchison (1989) also reported that copper increased the handling times of bluegill feeding on large *Daphnia* and larger invertebrate prey (e.g., *Enallagma* sp. and *Hyalella azteca*). It should be noted, however, that increased handling time was, in part, the result of repetitive spitting and recapturing of prey. Zebrafish (*Brachydanio rerio*) exposed to zinc (Cairns and Loos, 1967) and lead (Nyman, 1981), and flagfish (*Jordanella floridae*) exposed to alkyl benzene sulfonate detergent (Foster et al., 1966) repetitively seized and ejected their prey while foraging. Repeated spitting and recapture of prey items was not observed in bluegill exposed to cadmium (Bryan, 1993). Repetitive capture and spitting of prey by fish may be due to a blockage of the gustatory senses by the contaminant (Hidaka, 1970; Sutterlin and Sutterlin, 1970; Zelson and Cagan, 1979; Bryant, 1990). Such predators visually identify their prey, and subsequently pursue, capture, and begin to handle it, but reject the prey because they fail to receive gustatory confirmation that the captured item is edible.

A mechanistic approach to evaluating contaminant effects on predation, through the assessment of stress-altered search strategies, motivation to feed, reaction distance, search volume, etc., can determine the specific component(s) of the predation event affected by contaminants. The ecological relevance of behavioral toxicity data is dependent upon choosing realistic predator-prey combinations and test environments for study. Toxicant-induced changes in behavioral measures can then, and only then, be meaningfully applied to optimal foraging and bioenergetics models (e.g., Werner and Hall, 1974; Mittelbach, 1981, 1983; Rice, 1990; Wildhaber and Crowder, 1990) which may predict toxicant-altered diets and growth of fish in the wild. Additional work in this area is needed. Understanding how xenobiotics affect fish foraging behavior will improve our ability to predict direct effects on fish, and also improve our ability to predict indirect and cascading effects (Kerfoot, 1987; Mills et al., 1987; Carpenter and Kitchell, 1993) of toxicant exposure throughout aquatic communities.

E. OPTIMAL FORAGING THEORY AND ECOTOXICOLOGY

Study of the large and somewhat complicated array of factors influencing predator-prey interactions has been generally restricted to empirical tests of isolated parts of the array. However, the main question for behavioral ecologists and ecotoxicologists is the outcome of the entire process, not just the various components. A mechanistic approach, with research focused on the effects of stressors on the most important components of the predation sequence, may allow the development of testable hypotheses of contaminant-altered feeding and growth through the incorporation of current foraging models.

Werner and Hall (1974) presented evidence that the size-selective nature of bluegill predation on zooplankton was related to the optimal allocation of time spent searching for and handling prey. They developed a model, based upon the relationship of prey search and handling time to energy return, to predict the diet of bluegill foraging in the pelagic zone of a lake. A quantitative test of the model indicated that bluegill chose prey in accordance with optimal foraging theory.

Optimal foraging models assume that individuals most efficient in feeding will be most fit, i.e., optimally foraging fish will acquire more energy, grow faster, and produce more offspring than fish foraging suboptimally. These average-rate-maximizing models construct cost/benefit relationships that incorporate trade-offs between searching for and handling prey. Each prey item provides a "benefit" to the predator in terms of a fixed mean amount of energy, but a "cost" is also associated with each prey in terms of time and energy required to pursue, capture, and consume it. The model predicts whether a forager should attack the prey it encounters or pass it over and continue searching. By using the model, the range of prey sizes eaten and the maximum net energy intake for a given-sized fish and prey distribution can be determined. The optimal foraging model developed by Mittelbach (1981) and used by Turner and Mittelbach (1990) and others predicts the rate of net energy intake (E_n/T) by

$$\frac{E_n}{T} = \frac{(\Sigma(\lambda_i E_i p_i) - C_s)}{1 + \Sigma \lambda_i H_i p_i} \tag{1}$$

where $E_i = Ae_i - C_h H_i$; where A = assimilable fraction of prey energy content, e_i = energy content of prey item of type i (joules), λ_i = encounter rate with prey type and size i, H_i = handling time (seconds) of prey size i, C_h = cost of handling (joules/second), C_s = cost of searching (joules/second), and p_i = probability that prey item i is attacked once encountered. Stephens and Krebs (1986) presented the biological assumptions for the validity of this basic prey model (BPM) and these are further discussed by Hart and Gill (1993). Assumptions are as follows:

• Searching and handling do not occur simultaneously.
• Prey are encountered sequentially and at random.
• The E_is, H_is, and λ_is are constant and independent of p_i, the probability that prey type i will be eaten.
• Encounter without attack does not result in a gain or loss of energy by the forager.
• The forager makes decisions as described by the model and its behavior remains consistent over time.

Mittelbach (1981) expanded upon the earlier work of Werner and Hall (1974) by using the optimal foraging model to predict the diet and net energetic intake of bluegill foraging in the pelagic, littoral, and benthic zones of a lake. Laboratory feeding experiments were performed to determine prey encounter rates and prey handling times for bluegill foraging on different sizes of zooplankton, damselfly naiads, and midge larvae. Predictions of bluegill prey size selection and habitat use, based upon maximization of net energy gain, were then examined for bluegill in a small lake. The ambient prey size-frequency distribution was compared with that of the model and the fishes' stomach contents. The model closely predicted the actual diet and habitat use of bluegill in the lake and confirmed the utility of this mechanistic approach to the study of fish foraging.

These studies suggest that prey choice and energy intake may be impaired by stress effects on the prey encounter rates and handling times. Sandheinrich and Atchison (1990) used the optimal foraging model and the ambient size frequency distribution of invertebrates found in Mittelbach (1981) to predict the diet of bluegill if prey handling times were altered by a toxicant. An equivalent proportional increase in handling time for all prey sizes would change the predicted size-frequency distribution of prey in the bluegill diet and would result in a lower rate of net energy intake. Studies of copper effects on bluegill foraging suggest that handling time increases of 50% to 100% are not unusual and are not equal across all prey size classes (Sandheinrich and Atchison, 1989).

As pointed out, the BPM assumes that the internal state of the organism is constant and does not affect the probability that a particular prey type will be taken. The model seems to work well with planktivorous fishes such as bluegill (Werner, 1984) and threespine stickleback (*Gasterosteus aculeatus*) feeding on various size classes of *Daphnia* (Gibson, 1980). However, Hart and Ison (1991) found that this model did not do a good job of predicting diet of stickleback eating large prey, such as *Ascellus*. They surmised that, when a predator is relatively large compared to the prey, each additional prey consumed would result in a small incremental addition to stomach fullness and thus a relatively small decrease in the hunger of the predator. However, when the size of the prey is relatively large compared to the predator, each prey item consumed may significantly decrease the hunger of the predator; thus, the assumption of internal stasis of the predator is not valid.

Hart and Gill (1993), following the work of Mangel and Clark (1986, 1988) and Mangel (1992), developed a stochastic dynamic programming (SDP) model that is much more complex than the BPM; it takes into account the effect of a predator's changing internal state on decisions made during the foraging bout. Their primary concern was how the physiological state of hunger changed prey selection over time. For foraging decisions, as well as many other decisions, there must be a mechanism whereby the predator's behavior is continually adjusted to new circumstances. In addition to hunger, learning certainly plays a part in this process (Hughes et al., 1992). This too must be taken into account in future optimality models. Work with these models should eventually provide greater insights into the dynamics of foraging and how the foraging process changes from species to species or within a species confronted by different prey arrays over time.

Predation strategies change from fish that eat zooplankton to those that eat macroinvertebrates; they differ even more in piscivorous fish (Breck, 1993). A piscivore's daily ration consists of a few large prey, with handling times that are short relative to digestion times. Capture efficiency (ratio of prey capture to attack) is very great for planktivores but probably much less for piscivores (Crowder et al., 1992). Breck (1993) developed a foraging model for piscivores and discussed its applications to species where capture efficiency is an important variable influencing foraging rates.

Although foraging models may predict the diet of fish, they may not fully predict contaminant effects on growth. Reduction in the rate of net energy intake may be compensated for by an increase in time spent foraging. In addition, growth is affected by more than food intake. Bryan (1993) found that cadmium did reduce food intake but decreased growth was probably also a consequence of increased metabolic rate. Bioenergetics models (see Rice,

1990) complement foraging models by predicting changes in growth as a function of changes in physiological rates and amount of food consumed. Combinations of foraging models and bioenergetics models should be useful in ecotoxicology.

F. RISKS OF PREDATION

Foraging in an optimal manner must be balanced with other demands that animals must meet to maximize fitness—the assumed goal of optimal foraging theory. While animals must eat to survive and grow, they generally must do so within the constraints of predation pressure. As Lima and Dill (1990) pointed out, few failures in the life of an organism are as unforgiving as the failure to avoid a predator. There is an expanding theoretical base for understanding predation risks and how they affect other behavioral decisions that organisms must make. Optimal foraging theory, taken out of context with predation risk, is very incomplete, as are most laboratory feeding studies aimed at understanding how a toxicant affects feeding efficiency.

In addition to understanding how stressors alter the foraging mechanisms, we also must determine if stressed organisms are more vulnerable to predation and how this vulnerability may influence risk-sensitive foraging strategies. Many studies have assessed effects of stressors on vulnerability to predation. Most of these were very simple experiments, with prey first exposed to the stressor and then to a predator (which generally was not exposed to the stressor—an event unlikely to occur in the wild); a refuge for the prey may have been available. One of the first such studies (Goodyear, 1972) showed that exposure of mosquitofish (*Gambusia affinis*) to ionizing radiation increased their risk of predation by largemouth bass (*Micropterus salmoides*). Mosquitofish that strayed from a refuge were vulnerable to the bass. A similar system was used by Kania and O'Hara (1974) to demonstrate that mercury exposure increased predation risk of mosquitofish to largemouth bass. Many species have been used in various test systems; increased predation risk can occur with exposure of prey to thermal stress (Coutant, 1973; Coutant et al., 1974; Yocom and Edsall, 1974), insecticides (Hatfield and Anderson, 1972; Little et al., 1990), pentachlorophenol (Brown et al., 1985), fluorene (Finger et al., 1985), and cadmium (Sullivan et al., 1978).

Several studies exposed both the prey and predator to a stressor. Farr (1978) showed that exposure of two prey species (grass shrimp, *Palaemonetes pugio*, and sheepshead minnows, *Cyprinodon varigatus*) and a predator (gulf killifish, *Fundulus grandis*) to methyl parathion had a significant effect on the predator's choice of prey. Under normal conditions, gulf killifish preferred sheepshead minnow as prey but exposure to the insecticide increased activity of the grass shrimp, thereby making them more visible and available to the predator, which subsequently increased their vulnerability to predation. Tagatz (1976) exposed grass shrimp (*P. vulgaris*) and pinfish (*Lagodon rhomboides*) to the organochlorine insecticide mirex; grass shrimp were significantly more vulnerable to pinfish predation when exposed to the pesticide. Woltering et al. (1978) exposed both largemouth bass and mosquitofish to ammonia and found a decrease in predation risk for the prey; the bass were more sensitive to ammonia than the prey. Hedtke and Norris (1980) found that ammonia increased the vulnerability of juvenile chinook salmon (*Oncorhynchus tshawytscha*) to predation by brook trout (*Salvelinus fontinalis*). Kolar and Rahel (1993) and Rahel and Kolar (1990) showed that hypoxia can cause some benthic animals to become more vulnerable to predation, but has little effect on hypoxia-tolerant prey. Clements et al. (1989) demonstrated that copper increased the vulnerability of two species of caddisflies to a stonefly predator in artificial streams.

Beitinger (1990) found that prey vulnerability to predation was increased in 23 of 29 experiments that he reviewed, and in 16 of 17 experiments in which only the prey were treated. The extent of stressor effects on predator-prey interactions depends on whether the predator *or* prey is more sensitive to the stressor.

Predation is often one of the major processes influencing community structure in aquatic ecosystems. In this brief review, however, we show that relatively few species interactions and stressors have been tested in the laboratory or field. In addition, there are many adaptations that organisms have evolved to reduce risks of predation and very few of these have been investigated for sensitivity to stressors.

Sih (1987) and Endler (1991) described various aspects of the evolution of predator avoidance. Little et al. (1985) emphasized that many of these antipredator adaptations could be affected by stressors. Antipredator behavioral mechanisms common in freshwater animals include: (1) habitat selection, refuging, and migrations to remove prey from areas occupied by predators; (2) schooling or swarming; (3) vigilance and ability to sense the presence of the predator, which includes visual, lateral line or mechanical, and chemosensory mechanisms; (4) ability to avoid detection, including prey color and movement patterns; and (5) prey ability to escape, including evasive swimming and flight to refuge. Foragers must sample their environment to gather information on both the presence of food and predators (Sih, 1992). Learning is involved in predator avoidance (Huntingford and Wright, 1993) and can be affected by the stressor.

Growth can also be an antipredator mechanism. Rapid growth allows an individual to handle a larger size-array of food, thus more effectively using available resources, while simultaneously growing too large for the predator to handle. Slow growth leaves prey vulnerable to predation for longer periods of time. The implications of the presence of predators on ontogenetic habitat and diet shifts of prey fishes have been extensively studied (Werner and Gilliam, 1984; Mittelbach, 1986; Werner and Hall, 1988; Werner and Anholt, 1993).

There are many components to predation risk; these vary greatly among species and size-classes within a species. Lima and Dill (1990) presented a simple model to represent predation risk:

$$P(death) = 1 - e^{-\alpha dT} \tag{2}$$

where α is the rate of encounter between predator and prey, d is the probability of death given an encounter, and T is the time spent vulnerable to encounter. An encounter is defined as "whenever the distance separating prey and predator is less than whichever of their detection radii is greater" (Lima and Dill, 1990). To determine the probability of death (d), one must determine the probabilities of the subcomponents of predation risk that include encounter, attack, capture, and consumption.

Many defense adaptations may be altered by toxicants and make prey more vulnerable to predation. Stressors can cause prey to become hyperactive and consequently increase their visibility to predators, or reduce the effectiveness of schools, the vigilance of prey, or their escape response or speed. Few of these alterations have been investigated and no systematic study has been made to gain a broad understanding of this area relative to ecotoxicology. Current predation risk models could be much more effectively used by ecotoxicologists, both in designing experiments and in predicting toxicant effects on the vulnerability of prey to predators.

G. BALANCING PREDATION RISK AND FORAGING

To maximize its fitness, an organism must do more than just optimize one aspect of its life history, such as foraging or avoiding predation. There must be compromises among a diverse array of conflicting demands. There is a growing body of evidence to support this complex decision-making process (Lima and Dill, 1990; Krebs and Kacelnik, 1991). A great deal of research has recently been devoted to the subject of balancing predation risks with foraging efficiency (Gotceitas, 1990; Gotceitas and Colgan, 1990; Lima and Dill, 1990; Turner and Mittelbach, 1990; Sih, 1992; Milinski, 1993; Pettersson and Bronmark, 1993; Werner and Anholt, 1993). Sih (1992) discussed the following aspects of balancing feeding with predator avoidance and presented a model to test these ideas.

Refuge	Open area
Predator absent	Predators perhaps present
Food scarce	Food abundant

Prey must make choices to (1) stay in the refuge to avoid predation but sacrifice food intake, or (2) venture out into open water to feed and incur higher risks of predation. The uncertainty is mainly involved with the presence of predators in open water. An organism can gain information by sampling open water for food and predators. Milinski (1993) noted several connections between foraging and risk of predation: (1) an animal is more conspicuous by its foraging activities, (2) an animal must concentrate on finding food at the expense of vigilance for predators, and (3) at times food is plentiful where risk of predation is high. Time spent foraging may expose the animal to predators but never foraging is not an option.

A basic rule in this decision process may be to minimize the ratio of mortality rate to foraging rate (Gilliam and Fraser, 1987) or to minimize the ratio of mortality rate to growth rate (Werner and Anholt, 1993). A key to applying this rule, however, is the organism's hunger state (Milinski, 1993). Less hungry organisms should shift the compromise towards predation avoidance, whereas very hungry organisms should chance greater risk of predation. The work of Werner and co-workers (Werner et al., 1983; Werner and Gilliam, 1984; Werner and Hall, 1988; and Werner and Anholt, 1993) has shown that the combination of predation pressure and food availability affects habitat choices of centrarchid fishes. Milinski (1993) reviewed the experimental evidence for such trade-offs. As with the development of optimal foraging theory, the early models were quite simple. They often had an assumption of constant death rates that were habitat specific; the organism's main control over death was habitat selection. However, more recent model development accounts for control of death rate (predation rate) by other means, such as changing activity patterns (Werner and Anholt, 1993).

A toxicant could affect many aspects of these decisions, yet we are not aware of research to explicitly test "trade-off" hypotheses. Toxicants could affect: (1) predator presence or predation effectiveness, (2) prey ability to remain in a refuge (hyperactivity or starvation), (3) prey ability to sense the presence of a predator, or (4) prey ability to escape from predators. The refuge can be structural (habitat selection) or behavioral (schooling). Much theoretical and experimental work has been done to understand the role of both types of refuges on the trade-offs of foraging and predation risk. Study in this area should offer ecotoxicologists many rewarding insights.

IV. COMPETITION FOR RESOURCES

Interspecific competition is another important ecological process influencing the structure of communities (Connell, 1983; Schoener, 1983; Sih et al., 1985). However, in contrast to the numerous studies of toxicant effects on predator-prey interactions (Sandheinrich and Atchison, 1990), little research has focused on the effects of xenobiotics on competition and the concomitant change in aquatic community structure. The few studies in this area may be due, in part, to differences in the nature of the interactions and subsequent development of ecological theory. Predation is a direct interaction between predator and prey, is readily observed in the laboratory and field, and has an extensive foundation of theoretical concepts on the mechanisms governing the process in aquatic systems. Conversely, competition is a more indirect interaction among species via common exploitation of a limited resource and is not easily directly measured or observed. Consequently, theoretical concepts have primarily described competition in a strictly empirical or phenomenological way and have only recently incorporated knowledge of ecophysiology, life history strategies, etc. into cogent explanations of the mechanisms governing resource competition and its influence on aquatic community structure (Rothhaupt, 1990).

Freshwater zooplankton have been the focus of many laboratory and field studies in basic and applied ecology; they are the only group for which there is sufficient evidence to suggest that contaminants affect competition. We briefly examine the patterns of zooplankton community structure observed in uncontaminated and contaminated systems. We then demonstrate how a mechanistic approach, in conjunction with information on contaminant effects on zooplankton filtration rates, can be used to develop testable hypotheses to account for changes in zooplankton communities in stressed systems.

Predictable trends in freshwater zooplankton community structure are apparent. In lakes with abundant fish populations, size-selective predation by planktivores (O'Brien, 1979; Lazarro, 1987) eliminates large zooplankton and results in a zooplankton community dominated by small-bodied species (Brooks and Dodson, 1965; Lynch, 1979; Zaret, 1980). In the absence of intense predation by fish, larger zooplankton dominate and small species become rare (Brooks and Dodson, 1965). Two hypotheses have been proposed to explain the dominance of large zooplankton in the absence of intense planktivory: (1) invertebrate predators prey heavily on small species and reduce their abundance (Dodson, 1974), and (2) small zooplankton are suppressed by competitively superior large zooplankton through one or more processes (e.g., size-efficiency hypothesis, Brooks and Dodson, 1965). For example, large cladocerans typically suppress rotifer populations to low levels by exploitation and interference competition (Gilbert, 1988). Several reviews have summarized information and hypotheses on the relative importance of predation and competition in structuring zooplankton communities (Bengtsson, 1987; Kerfoot and Sih, 1987; DeMott, 1989; Rothhaupt, 1990).

Reviews of the response of freshwater zooplankton communities to acidification or pesticide application indicate that the size distribution of the zooplankton community undergoes a pronounced shift to smaller organisms and large species become rare (Havens and Hanazato, 1993; deNoyelles et al., 1994, Moore and Folt, 1993). Several studies (e.g., Day et al., 1987; Kaushik et al., 1985; Lozano et al., 1992) have attributed this shift to the direct effects of the chemical on large zooplankton mortality with subsequent numerical increases of small zooplankton due to release from invertebrate predation, or exploitation and interference competition. Hence, increases in populations of small zooplankton (especially rotifers) are often the result of indirect chemical effects on the community. Though there are exceptions (e.g., Liber et al., 1994), rotifer populations are generally most tolerant of chemical stress and also most resilient following removal of the stressor (Havens and Hanazato, 1993). Rotifers are followed, in order of tolerance and resilience, by small cladocerans, then copepods, and finally large cladocerans. Havens and Hanazato (1993) suggested there are several reasons why small zooplankton taxa predominate in stressed aquatic systems, including rapid reproductive rates

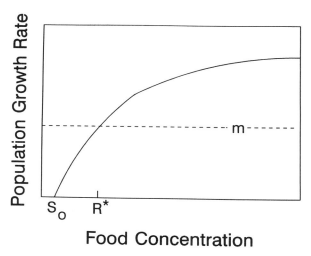

Food Concentration

Figure 2   The Monod growth curve with a minimum food requirement for zero population growth (S_o). With increasing resource density, the population reaches maximal growth rate, r_{max}, asymptotically. R^* is the resource concentration where the growth rate equals the mortality rate (m). (From Rothhaupt, K. O., 1990. Arch. Hydrobiol. 118, 1–29. With permission.)

(r-strategists), development with few transitions through sensitive stages (i.e., fewer molts prior to maturity), or greater number of stress-tolerant species. However, they also concluded that no single explanation accounted for all observed trends in zooplankton responses to anthropogenic stress and that additional research was needed to identify the mechanisms responsible.

Though there is substantial evidence to suggest that toxicants may influence competitive interactions in zooplankton communities, no studies have directly assessed this issue. Rothhaupt (1990) suggested that a mechanistic approach to exploitative competition may be beneficial in understanding the outcome of competitive interactions in zooplankton communities. For zooplankton, if we assume that population growth is dependent upon food resources, then the intrinsic growth rate of the population can be expressed in the form of a modified Monod function (Monod, 1950) by incorporating a minimum food requirement for zero population growth (Rothhaupt, 1990)

$$r = r_{max} \frac{S - S_o}{S - S_O + k_S} \qquad (3)$$

where r = intrinsic growth rate (1/t), r_{max} = maximal growth rate, S = food concentration (e.g., cells/liter), S_o = minimum food requirement for zero population growth, and k_s = half saturation constant.

This equation demonstrates that population growth rate (r) is a function of food concentration (Figure 2). The threshold food level (S_o) is the concentration at which individual body and population growth is zero and energy assimilation is equal to respiration. Threshold food concentrations required for populations to persist (R^*, Tilman, 1982) are greater than S_o if predation, in

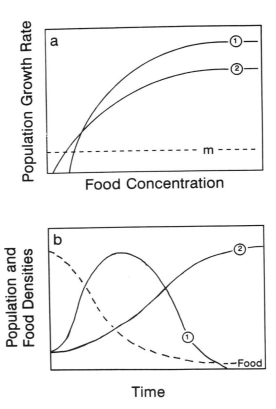

Figure 3   Schematic representation of competition for a common resource between two species with different threshold food concentrations (S_o) and maximum intrinsic growth rates (r_{max}). (A) Numerical response curves for both species: species 1 reaches a higher r_{max} when food is abundant; species 2 has lower threshold requirements; m is the mortality rate. (B) Time course of population densities and resource availability in competition, starting with a high initial resource density. (From Rothhaupt, K.O., 1990. Arch. Hydrobiol. 118, 1–29. With permission.)

addition to senescence or starvation, is a significant source of mortality in the population.

The threshold hypothesis (Lampert, 1977; Lampert and Schober, 1980) suggests that, among species with similar feeding modes (i.e., no resource partitioning) and under steady-state conditions (i.e., nonfluctuating food supply), those species with the lowest food density threshold for reproduction are the best competitors. Under this model, species with greatest r_{max} will dominate when food is abundant. However, when food is limited, the species with the lowest S_o will dominate, regardless of its r_{max} (Figure 3).

Bengtsson (1987) reviewed 20 zooplankton competition studies of 37 species combinations to evaluate the experimental evidence for and against several hypotheses for competitive dominance among zooplankton. The threshold concept was supported in only 36% of the tests, but none of the other three hypotheses assessed (size-efficiency hypothesis, Brooks and Dodson, 1965;

small body size hypothesis, Neill, 1975, Hanski and Ranta, 1983; r_{max} hypothesis, Goulden et al., 1978) had sufficient empirical support to be considered as general explanations of zooplankton competition. However, Rothhaupt (1990) stated that in many of the studies that Bengtsson (1987) found not to agree with the threshold concept, non-steady-state conditions (Goulden et al., 1982; Romanovsky and Feniova, 1985) and resource partitioning (DeMott and Kerfoot, 1982) occurred. Therefore, even though competition in zooplankton communities probably needs to be considered within a multifactor framework (Bengtsson, 1987), there is still sufficient empirical support to initially consider the threshold hypothesis as a mechanism for the decline of large zooplankton and an increase in small zooplankton in contaminant-stressed systems.

For this discussion, the threshold hypothesis should be reconsidered in terms of energy intake rather than strictly in terms of food concentration. If filtration rates for individual species are constant and independent of food concentration (Rigler, 1961), then the net reproductive rate (which is dependent upon net energy intake) will primarily be a function of food concentration. However, when food levels are constant, the energy intake (and hence reproduction rates) will be dependent upon filtration rate. This suggests that toxicant-induced alterations in filtration will reduce the "apparent food concentration" (i.e., energy intake), potentially below levels required for reproduction and maintenance of population size. Differential sensitivity of zooplankton species to toxicant-altered filtration rates would alter competitive relationships under the threshold hypothesis.

Feeding rates of cladocera, copepods, and rotifers are inhibited by a variety of chemicals (Table 1) and are a sensitive indicator of toxicant stress. For example, McNaught (1989) found zooplankton filtration rates more sensitive than net reproductive rates and respiration to a variety of chemicals. Jones et al. (1991) showed that sodium dodecyl sulfate induced changes in filtration at the same concentrations that altered population level responses (reproduction, survival, R_o). If contaminant stress inhibits filtration, the threshold food concentration will shift to a higher value. Kersting (1983) found that pesticide toxicity increased the threshold concentration of *Chlorella* for a population of *Daphnia magna*.

Evidence from studies of the effects of toxic cyanobacteria on zooplankton suggests that: (1) rotifers may be less sensitive than cladocerans to inhibition of feeding by toxicant stress (Fulton and Paerl, 1987; Fulton, 1988); (2) susceptibility of cladocerans to a toxicant is positively related to body size (Lampert, 1981, 1982; Gilbert, 1990); and (3) differential sensitivity to a toxicant can prevent suppression of a rotifer population by a large cladoceran population and result in dominance by the rotifers (Gilbert, 1990). In addition, small cladocerans are better competitors at low food concentrations than large species (Tillman and Lampert, 1984; Romanovsky and Feniova, 1985). Consequently, the shift in dominance from large to small zooplankton in contaminant-stressed systems may be due, in part, to differential species sensitivity of food acquisition rates to chemical toxicity and the ability to reproduce under

Table 1 Chemical Concentrations Inhibiting Filtration Rates of Various Species of Freshwater Zooplankton

Chemical	Species	Maximum relative (% control) inhibition	Range of concentrations (μg/l)	Ref.
Cu	Cyclops bicuspidatus	−29	0.2–21.9	McNaught et al. 1987[b]
	Diaptomus spp.	−90	1.2–21.9	McNaught et al. 1988
	Daphnia magna	−93	10–40	Flickinger et al. 1982
	Brachionus calyciflorus	−50	32[a]	Ferrando et al. 1993
Zn	Diaptomus spp.	−100	0.1–32.0	McNaught et al. 1987[b]
	Daphnia retrocurva	−71	0.1–32.0	McNaught et al. 1988
	Daphnia schodleri	−99	0.1–32.0	McNaught et al. 1987
Mg	Daphnia magna	−87	75,000	Mirza 1968[b]
Nitrite	Daphnia magna	−47	40,000	Mirza 1968[b]
PCB	Cyclops bicuspidatus	−43	25	McNaught et al. 1984[b] McNaught 1982
Pentachloro-phenate	Brachinous calyciflorus	−50	1,850[a]	Ferrando et al. 1993
3,4-Dichloro-aniline	Brachionus calyciflorus	−50	41,200[a]	Ferrando et al. 1993
Lindane	Daphnia pulex	−25	50	Gliwicz and Sieniawska 1986
	Brachionus calyciflorus	−50	8,500[a]	Ferrando et al. 1993
Toxaphene	Diaptomus sicilis	−65	0.2–15	McNaught et al. 1987[b]
	Limnocalanus marcurus	−57	0.2–15	McNaught et al. 1987[b]
Endosulfan	Daphnia magna	−70	150–620	Fernandez-Casalderrery et al. 1994
	Brachionus calyciflorus	−50	2,960[a]	Fernandez-Casalderrery et al. 1992
Diazinon	Daphnia magna	−97	0.45–0.90	Fernandez-Casalderrery et al. 1994
	Brachionus calyciflorus	−50	14,390[a]	Fernandez-Casalderrery et al. 1992
Fenvalerate	Daphnia galeata mendotae	−100	0.05–0.5	Day and Kaushik 1987; Day et al. 1987
	Ceriodaphnia lacustris	−70	0.01–0.5	Day and Kaushik 1987; Day et al. 1987
	Diaptomus orgonensis	−100	0.05–0.5	Day and Kaushik 1987; Day et al. 1987
Pulpmill effluent	Daphnia retrocurva	−87	5% – 10%	Cooley 1977
Dichlobenil	Daphnia magna	−50	1,000	Kersting and Van der Honing 1981
Sodium dodecyl sulfate	Daphnia catawba	−65	5,000–30,000	Jones et al. 1991

[a] EC50

[b] Cited in McNaught (1989).

conditions of reduced food intake. This hypothesis may not adequately account for species shifts in all stressed systems. For example, Sierzen and Frost (1990) were not able to detect a direct effect of acidification (from pH 6.2 to 5.2) on the feeding rates of zooplankton. However, the threshold hypothesis does provide a mechanism for developing predictions of toxicant-altered competition by ranking the relative reduction in the filtering activity of different species due to toxicant exposure. Experiments are needed to directly assess the relationship between toxicant-inhibited feeding and competitive dominance in mixed zooplankton species assemblages.

V. CONCLUSIONS

Modeling interspecific interactions is not greatly different from ecological risk assessment. An animal faces many challenges and decisions about predator avoidance, foraging, and competition. Those that make the best choices presumably are the most fit. Predation efficiency and competitive ability are important elements of fitness, the ability to pass on genetic contributions to future generations. An animal must balance several conflicting risks, most of which are not examined in standard laboratory toxicity tests (e.g., trade-offs between avoiding predation and finding food or mates). Sprague (1981), in a paper entitled "Ethologists are Environmental Drop-outs," called for behavioral ecologists to be more involved in the applied problems of water pollution and contaminants. We now make a similar plea to ecotoxicologists. Incorporate the many concepts and models of behavioral ecology into our field. Traditional, standardized tests can be better designed to provide the data needed to predict the effects of toxicants on interspecific interactions that have important implications for community and ecosystem dynamics.

ACKNOWLEDGMENTS

A portion of this paper was based on work supported by the U.S. Environmental Protection Agency (Grant #R-815921-01-0) and the Iowa State University Agriculture and Home Economics Experiment Station. MBS was supported by a University of Wisconsin-La Crosse Faculty Research Grant. We thank Sandra Brewer and Bill Richardson for review of this manuscript.

REFERENCES

Bartell, S.M., 1982. Influence of prey abundance on size-selective predation by bluegills. Trans. Am. Fish. Soc. 111, 453–461.

Beitinger, T.L., 1990. Behavioral reactions for the assessment of stress in fishes. J. Gt. Lakes Res. 16, 495–528.

Bell, W.J., 1991. Searching Behaviour: The Behavioural Ecology of Finding Resources, Chapman and Hall, New York.

Bengtsson, J., 1987. Competitive dominance among Cladocera: are single-factor explanations enough? Hydrobiologia 145, 245–257.

Breck, J.E., 1993. Foraging theory and piscivorous fish: are forage fish just big zooplankton? Trans. Am. Fish. Soc. 122, 902–911.

Brooks, J.L. and S.I. Dodson, 1965. Predation, body size, and composition of plankton. Science 150, 28–35.

Brown, J.A., P.H. Johansen, P.W. Colgan and R.A. Mathers, 1985. Changes in the predator-avoidance behavior of juvenile guppies (*Poecilia reticulata*) exposed to pentachlorophenol. Can. J. Zool. 63, 2001–2005.

Bryan, M.D., 1993. Effects of cadmium on juvenile bluegill (*Lepomis macrochirus*) foraging behavior and growth. Doctoral Dissertation. Iowa State University, Ames, IA.

Bryant, B.P., 1990. Specific inhibition of the binding of the taste stimulus, L-alanine, by sulfhydryl reagents, in *Ictalurus punctatus*. Comp. Biochem. Physiol. 95A, 533–537.

Cairns, J., Jr. and J.J. Loos, 1967. Changed feeding rate of *Brachydanio rerio* (Hamilton-Buchanan) resulting from exposure to sublethal concentrations of zinc, potassium dichromate, and alkyl benzene sulfonate detergent. Proc. Pa. Acad. Sci. 40, 47–52.

Carpenter, S.R. and J.F. Kitchell (Eds.), 1993. The Trophic Cascade in Lakes, Cambridge University Press, New York.

Clements, W.H., D.S. Cherry and J. Cairns, Jr., 1989. The influence of copper exposure on predator-prey interactions in aquatic insect communities. Freshwater Biol. 21, 483–488.

Confer, J.L. and P.I. Blades, 1975. Reaction distance to zooplankton by *Lepomis gibbosus*. Verh. Int. Verein. Limnol. 19, 2493–2497.

Confer, J.L., G.L. Howick, M.H. Corzette, S.L. Kramer, S. Fitzgibbon and R. Landesberg, 1978. Visual predation by planktivores. Oikos 31, 27–37.

Connell, J.H., 1983. On the prevalence and relative importance of interspecific competition: evidence from field experiments. Am. Nat. 122, 661–696.

Cooley, J.M., 1977. Filtering rate performance of *Daphnia retrocurva* in pulpmill effluent. J. Fish. Res. Board Can. 34, 863–868.

Coutant, C.C., 1973. Effect of thermal shock on vulnerability of juvenile salmonids to predation. J. Fish. Res. Board Can. 30, 965–973.

Coutant, C.C., H.M. Ducharme and J.R. Fisher, 1974. Effects of cold shock on vulnerability of juvenile channel catfish (*Ictalurus punctatus*) and largemouth bass (*Micropterus salmoides*) to predation. J. Fish. Res. Board Can. 31, 351–354.

Crowder, L.B., J.A. Rice, T.J. Miller and E.A. Marschall, 1992. Empirical and theoretical approaches to size-based interactions and recruitment variability in fishes. In: Individual-Based Models and Approaches in Ecology: Populations, Communities and Ecosystems, edited by D.L. DeAngelis and L.J. Gross, Chapman and Hall, New York, pp. 237–255.

Day, K. and N.K. Kaushik, 1987. Short-term exposure of zooplankton to the synthetic pyrethroid, fenvalerate, and its effect on the rates and assimilation of the algae, *Chlamydomonas reinhardii*. Arch. Environ. Contam. Toxicol. 16, 423–432.

Day, K.E., N.K. Kaushik and K.R. Solomon, 1987. Impact of fenvalerate on enclosed freshwater planktonic communities and on in-situ rates of filtration of zooplankton. Can. J. Fish. Aquat. Sci. 44, 1714–1728.

DeMott, W.R., 1989. The role of competition in zooplankton succession. In: Plankton Ecology: Succession in Plankton Communities, edited by U. Sommer, Springer-Verlag, New York, pp. 253–296.

DeMott, W.R. and W.C. Kerfoot, 1982. Competition among cladocerans: nature of the interaction between *Bosmina* and *Daphnia*. Ecology 63, 1949–1966.

DeNoyelles, F., Jr., S.L. Dewey, D.G. Huggins and W.D. Kettle, 1994. Aquatic mesocosms in ecological effects testing: detecting direct and indirect effects of pesticides. In: Aquatic Mesocosm Studies in Ecological Risk Assessment, edited by R.L. Graney, J.H. Kennedy and J.H. Rodgers, Jr., Lewis Publishers, Boca Raton, FL, pp. 577–603.

Dodson, S.I., 1974. Zooplankton competition and predation: an experimental test of the size-efficiency hypothesis. Ecology 55, 605–613.

Drenner, R.W., J.R. Strickler and W.J. O'Brien, 1978. Capture probability: the role of zooplankter escape in the selective feeding of planktivorous fish. J. Fish. Res. Board Can. 35, 1370–1373.

Eggers, D.M., 1977. The nature of prey selection by planktivorous fish. Ecology 58, 46–59.

Eggers, D.M., 1982. Planktivore preference by prey size. Ecology 63, 381–390.

Ehlinger, T.J., 1989. Learning and individual variation in bluegill foraging: habitat-specific techniques. Anim. Behav. 38, 643–658.

Ehlinger, T.J., 1990. Habitat choice and phenotypic-limited feeding efficiency in bluegill: individual differences and trophic polymorphism. Ecology 71, 886–896.

Endler, J.A., 1991. Interactions between predators and prey. In: Behavioural Ecology: An Evolutionary Approach, edited by J.R. Krebs and N.B. Davies, Blackwell Scientific, Boston, MA, pp. 169–196.

Evans, B.I. and W.J. O'Brien, 1988. A reevaluation of the search cycle of planktivorous arctic grayling, *Thymallus arcticus*. Can J. Fish. Aquat. Sci. 45, 187–192.

Farr, J.A., 1978. The effect of methyl parathion on predator choice of two estuarine prey species. Trans. Am. Fish. Soc. 107, 87–91.

Fernandez-Casalderry, A., M. Ferrando and E. Andreau-Moliner, 1992. Filtration and ingestion rates of *Brachionus calyciflorus* after exposure to endosulfan and diazinon. Comp. Biochem. Physiol. 103C, 357–361.

Fernandez-Casalderry, A., M.D. Ferrando and E. Andreau-Moliner, 1994. Effect of sublethal concentrations of pesticides on the feeding behavior of *Daphnia magna*. Ecotoxicol. Environ. Saf. 27, 82–89.

Ferrando, M.D., C.R. Janssen, E. Andreu and G. Persoone, 1993. Ecotoxicological studies with the freshwater rotifer *Brachionus calyciflorus*. III. Short-term exposure to pollutants and their effects on rates of filtration and ingestion. Ecotoxicol. Environ. Saf. 26, 1–9.

Finger, S.E., E.E. Little, M.G. Henry, J.F. Fairchild and T.P. Boyle, 1985. Comparison of laboratory and field assessment of fluorene. I. Effects of fluorene on the survival, growth, reproduction, and behavior of aquatic organisms in laboratory tests. In: Validation and Predictability of Laboratory Methods for Assessing the Fate and Effects of Contaminants in Aquatic Ecosystems. ASTM STP 865, edited by T.P. Boyle, American Society for Testing and Materials, Philadelphia, PA, pp. 120–133.

Flickinger, A.L., R.J.F. Bruins, R.W. Winner and J.H. Skillings, 1982. Filtration and phototactic behavior as indices of chronic copper stress in *Daphnia magna* Straus. Arch. Environ. Contam. Toxicol. 11, 457–463.

Foster, N.R., A. Scheier and J. Cairns, Jr., 1966. Effects of ABS on feeding behavior of flagfish, *Jordanella floridae*. Trans. Am. Fish. Soc. 95, 109–110.

Fulton, R.S., III, 1988. Grazing on filamentous algae by herbivorous zooplankton. Freshwater Biol. 20, 263–271.

Fulton, R.S. III and H.W. Paerl, 1987. Toxic and inhibitory effects of the blue-green alga *Microcystis aeruginosa* on herbivorous zooplankton. J. Plankton Res. 9, 837–855.

Gardner, M.B., 1981. Mechanisms of size selectivity by planktivorous fish: a test of hypotheses. Ecology 62, 571–578.

Gibson, R.M., 1980. Optimal prey-size selection by three-spined sticklebacks (*Gasterosteus aculeatus*): a test of the apparent-size hypothesis. A. Tierpsychol. 52, 291–307.

Gilbert, J.J., 1988. Suppression of rotifer populations by *Daphnia*: a review of the evidence, the mechanisms, and the effects on zooplankton community structure. Limnol. Oceanogr. 36, 1286–1303.

Gilbert, J.J., 1990. Differential effects of *Anabaena affinis* on cladocerans and rotifers: mechanisms and implications. Ecology 71, 1727–1740.

Gilliam, J.F. and D.F. Fraser, 1987. Habitat selection under predation hazard: test of a model with foraging minnows. Ecology 68, 1856–1862.

Gliwicz, M.Z. and A. Sieniawska, 1986. Filtering activity of *Daphnia* in low concentrations of a pesticide. Limnol. Oceanogr. 31, 1132–1138.

Goodyear, C.P., 1972. A simple technique for detecting effects of toxicants or other stresses on a predator-prey interaction. Trans. Am. Fish. Soc. 101, 367–370.

Gotceitas, V., 1990. Foraging and predator avoidance: a test of a patch choice model with juvenile bluegill sunfish. Oecologia 83, 346–351.

Gotceitas, V. and P. Colgan, 1990. The effects of prey availability and predation risk on habitat selection by juvenile bluegill sunfish. Copeia, 409–417.

Goulden, C.E., L.L. Henry and A.J. Tessier, 1982. Body size, energy reserves, and competitive ability in three species of Cladocera. Ecology 63, 1780–1789.

Goulden, C.E., L. Hornig and C. Wilson, 1978. Why do large zooplankton species dominate? Verh. Int. Verein. Limnol. 20, 2457–2460.

Hanski, I. and E. Ranta, 1983. Coexistence in a patchy environment: three species of *Daphnia* in rock pools. J. Anim. Ecol. 52, 263–279.

Hart, P.J.B., 1993. Teleost foraging: facts and theories. In: Behaviour of Teleost Fishes, edited by T.J. Pitcher, Chapman and Hall, New York, pp. 253–284.

Hart, P.J.B. and A.B. Gill, 1993. Choosing prey size: a comparison of static and dynamic foraging models for predicting prey choice by fish. Mar. Behav. Physiol. 23, 91–104.

Hart, P.J.B. and S. Ison, 1991. The influence of prey size and abundance, and individual phenotype on prey choice by the three-spined stickleback, *Gasterosteus aculeatus* L. J. Fish Biol. 38, 359–372.

Hatfield, C.T. and J.M. Anderson, 1972. Effects of two insecticides on the vulnerability of Atlantic salmon (*Salmo salar*) parr to brook trout (*Salvelinus fontinalis*) predation. J. Fish. Res. Board Can. 29, 27–29.

Havens, K.E. and T. Hanazato, 1993. Zooplankton community responses to chemical stressors: a comparison of results from acidification and pesticide contamination research. Environ. Pollut. 82, 277–288.

Heath, A.G., 1987. Water Pollution and Fish Physiology. CRC Press, Boca Raton, FL.

Hedtke, J.L. and L.A. Norris, 1980. Effects of ammonium chloride on predatory consumption rates of brook trout (*Salvelinus fontinalis*) on juvenile chinook salmon (*Oncorhynchus tshawytscha*) in laboratory streams. Bull. Environ. Contam. Toxicol. 24, 81–89.

Hidaka, I., 1970. The effects of transition metals on the palatal chemoreceptors of the carp. Jpn. J. Physiol. 20, 599–609.

Hughes, R.N., M.J. Kaiser, P.A. Mackney and K. Warburton, 1992. Optimizing foraging behaviour through learning. J. Fish Biol. 41(Suppl. B), 77–91.

Huntingford, F.A. and P.H. Wright, 1993. The development of adaptive variation in predator avoidance in freshwater fishes. Mar. Behav. Physiol. 23, 45–61.

Janssen, J., 1982. Comparison of searching behavior for zooplankton in the obligate planktivore, blueback herring (*Alosa aestivalis*) and a facultative planktivore, bluegill (*Lepomis macrochirus*). Can J. Fish. Aquat. Sci. 39, 1649–1654.

Jones, M., C. Folt and S. Guarda, 1991. Characterizing individual, population and community effects of sublethal levels of aquatic toxicants: an experimental case study using *Daphnia*. Freshwater Biol. 26, 35–44.

Kamil, A.C., J.R. Krebs and H.R. Pulliam (Eds.), 1987. Foraging Behavior. Plenum Press, New York.

Kania, H.J. and J. O'Hara, 1974. Behavioral alterations in a simple predator-prey system due to sublethal exposure to mercury. Trans. Am. Fish. Soc. 103, 134–136.

Kaushik, N.K., G.L. Stephenson, K.R. Solomon and K.E. Day, 1985. Impacts of permethrin on zooplankton communities in limno corrals. Can. J. Fish. Aquat. Sci. 42, 77–85.

Kerfoot, W.C., 1987. Cascading effects and indirect pathways. In: Predation: Direct and Indirect Impacts on Aquatic Communities, edited by W.C. Kerfoot and A. Sih, University Press of New England, Hanover, NH, pp. 57–70.

Kerfoot, W.C. and A. Sih (Eds.), 1987. Predation: Direct and Indirect Impacts on Aquatic Communities. University Press of New England, Hanover, NH.

Kersting, K., 1983. Direct determination of the "threshold food concentration" for *Daphnia magna*. Arch. Hydrobiol. 96, 510–514.

Kersting, K. and H. Van Der Honing, 1981. Effect of the herbicide dichlobenil on the feeding and filtering rate of *Daphnia magna*. Verh. Int. Verein. Limnol. 21, 1135–1140.

Kolar, C.S. and F.J. Rahel, 1993. Interaction of a biotic factor (predator presence) and an abiotic factor (low oxygen) as an influence on benthic invertebrate communities. Oecologia 95, 210–219.

Krebs, J.R. and A. Kacelnik, 1991. Decision-making. In: Behavioural Ecology: An Evolutionary Approach, edited by J.R. Krebs and N.B. Davies, Blackwell Scientific, Boston, MA, pp. 105–136.

Lampert, W., 1977. Studies on the carbon balance of *Daphnia pulex* DeGeer as related to environmental conditions. IV. Determination of the "threshold" concentration as factor controlling the abundance of zooplankton species. Arch. Hydrobiol. Suppl. 48, 361–368.

Lampert, W., 1981. Inhibitory and toxic effects of blue-green algae on *Daphnia*. Int. Rev. Hydrobiol. 66, 285–198.

Lampert, W., 1982. Further studies on the inhibitory effect of the toxic blue-green *Microcystis aeruginosa* on the filtering rate of zooplankton. Arch. Hydrobiol. 95, 207–220.

Lampert, W. and U. Schober, 1980. The importance of "threshold" food concentrations. In: Evolution and Ecology of Zooplankton Communities, edited by W.C. Kerfoot, University Press of New England, Hanover, NH, pp. 264–267.

Lazarro, X., 1987. A review of planktivorous fishes: their evolution, feeding behaviors, selectivity and impacts. Hydrobiologia 146, 97–167.

Liber, K., K.R. Solomon, N.K. Kaushik and J.H. Carey, 1994. Impact of 2,3,4,6-tetrachlorophenol (DIATOX®) on plankton communities in limnocorrals. In: Aquatic Mesocosm Studies in Ecological Risk Assessment, edited by R.L. Graney, J.H. Kennedy and J.H. Rodgers, Jr., Lewis Publishers, Boca Raton, FL, pp. 257–294.

Lima, S.L. and L.M. Dill, 1990. Behavioral decisions made under the risk of predation: a review and prospectus. Can. J. Zool. 68, 619–640.

Little, E.E., R.D. Archeski, B.A. Flerov and V.I. Kozlovskaya, 1990. Biological indicators of sublethal toxicity in rainbow trout. Arch. Environ. Contam. Toxicol. 19, 380–385.

Little, E.E., B.A. Flerov and N.N. Ruzhinskaya, 1985. Behavioral approaches in aquatic toxicity investigations: a review. In: Toxic Substances in the Aquatic Environment: An International Aspect, edited by P.M. Mehrle, R.H. Gray and R.L. Kendall, Water Quality Section, American Fisheries Society, Bethesda, MD, pp. 72–98.

Lozano, S.J., S.L. O'Halloran, K.W. Sargent and J.C. Brazner, 1992. Effects of esfenvalerate on aquatic organisms in littoral enclosures. Environ. Toxicol. Chem. 11, 35–47.

Luecke, C. and W.J. O'Brien, 1981. Prey location volume of a planktivorous fish: a new measure of prey vulnerability. Can. J. Fish. Aquat. Sci. 38, 1264–1270.

Lynch, M., 1979. Predation, competition, and zooplankton community structure: an experimental study. Limnol. Oceanogr. 24, 253–272.

Mangel, M., 1992. Rate maximizing and state variable theories of diet selection. Bull. Math. Biol. 54, 413–422.

Mangel, M. and C.W. Clark, 1986. Towards a unified foraging theory. Ecology 67, 1127–1138.

Mangel, M. and C.W. Clark, 1988. Dynamic Modeling in Behavioural Ecology. Princeton University Press, Princeton, NJ.

Mathers, R.A., J.A. Brown and P.H. Johansen, 1985. The growth and feeding behavior responses of largemouth bass (Micropterus salmoides) exposed to PCP. Aquat. Toxicol. 6, 157–164.

McNaught, D.C., 1982. Short cycling of contaminants by zooplankton and their impact on Great Lakes ecosystems. J. Gt. Lakes Res. 8, 360–366.

McNaught, D.C., 1989. Functional bioassays utilizing zooplankton: a comparison. Hydrobiologia 188/189, 117–121.

McNaught, D.C., S.D. Bridgham and C. Meadows, 1987. Effects of contaminants in River Raisin on ecosystem function of the bacteria, phytoplankton and zooplankton. Unpublished data report to U.S. Environmental Protection Agency.

McNaught, D.C., S.D. Bridgham and C. Meadows, 1988. Effects of complex effluents from the River Raisin on zooplankton grazing in Lake Erie. In: Functional Testing of Aquatic Biota for Hazard Estimation. ASTM STP 988, edited by J. Cairns and J.R. Pratt, American Society for Testing and Materials, Philadelphia, PA, pp. 128–137.

McNaught, D.C., D. Griesmer, M. Buzzard and M. Keendy, 1984. PCBs in Saginaw Bay: development of functional indices to estimate inhibition of ecosystem fluxes. EPA-600/S3-84-008, U.S. Environmental Protection Agency, Washington, D.C.

Milinski, M., 1993. Predation risk and feeding behaviour. In: Behaviour of Teleost Fishes, edited by T.J. Pitcher, Chapman and Hall, New York, pp. 285–305.

Milinski, M. and G.A. Parker, 1991. Competition for resources. In: Behavioural Ecology: An Evolutionary Approach, edited by J.R. Krebs and N.B. Davies, Blackwell Scientific, Boston, MD, pp. 137–168.

Mills, E.L., J.L. Forney and K.J. Wagner, 1987. Fish predation and its cascading effects on the Oneida Lake food chain. In: Predation: Direct and Indirect Impacts on Aquatic Communities, edited by W.C. Kerfoot and A. Sih, University Press of New England, Hanover, NH, pp. 118–131.

Mirza, M., 1968. An ecological study on the nature of pollution in Tonawanda and Ellicott Creeks of the Niagara River Basin and the effects of various chemical variables on the feeding and reproductive rates of *Daphnia magna*. Ph.D. thesis, State University of New York, Buffalo, NY.

Mittelbach, G.G., 1981. Foraging efficiency and body size: a study of optimal diet and habitat use by bluegills. Ecology 62, 1370–1386.

Mittelbach, G.G., 1983. Optimal foraging and growth in bluegills. Oecologia 59, 157–162.

Mittelbach, G., 1986. Predator-mediated habitat use: some consequences for species interactions. Environ. Biol. Fishes 16, 159–169.

Mittelbach, G.G., 1988. Competition among refuging sunfishes and effects of fish density on littoral zone invertebrates. Ecology 69, 614–623.

Monod, J., 1950. La technie de culture contiue. Theorie et application. Ann. Inst. Pasteur 79, 390–410.

Moore, M. and C. Folt, 1993. Zooplankton body size and community structure: effects of thermal and toxicant stress. Trends Ecol. Evol. 8, 178–183.

Mount, D.I. and C.E. Stephan, 1967. A method for establishing acceptable toxicant limits for fish — malathion and the butoxyethanol ester of 2,4-D. Trans. Am. Fish. Soc. 96, 185–193.

Neill, W.E., 1975. Experimental studies on microcrustacean competition, community composition and efficiency of resource utilization. Ecology 56, 809–826.

Nyman, H.G., 1981. Sublethal effects of lead (Pb) on size selective predation by fish: applications on the ecosystem level. Verh. Int. Verein. Limnol. 21, 1126–1130.

O'Brien, W.J., 1979. The predator-prey interaction of planktivorous fish and zooplankton. Am. Sci. 67, 572–581.

O'Brien, W.J., B. Evans and C. Luecke, 1985. Apparent size choice of zooplankton by planktivorous sunfish: exceptions to the rule. Environ. Biol. Fishes 13, 225–233.

O'Brien, W.J., B.I. Evans and G.L. Howick, 1986. A new view of the predation cycle of a planktivorous fish, white crappie (*Pomoxis annularis*). Can. J. Fish. Aquat. Sci. 43, 1894–1899.

O'Brien, W.J., B.I. Evans and H.I. Browman, 1989. Flexible search tactics and efficient foraging in saltatory searching animals. Oecologia 80, 100–110.

O'Brien, W.J., H.I. Browman and B.I. Evans, 1990. Search strategies of foraging animals. Am. Sci. 78, 152–160.

O'Brien, W.J., N.A. Slade and G.L. Vinyard, 1976. Apparent size as the determinant of prey selection by bluegill sunfish (*Lepomis macrochirus*). Ecology 57, 1304–1310.

Pettersson, L.B. and C. Bronmark, 1993. Trading off safety against food: state dependent habitat choice and foraging in crucian carp. Oecologia 95, 353–357.

Rahel, F.J. and C.S. Kolar, 1990. Trade-offs in the response of mayflies to low oxygen and fish predation. Oecologia 84, 39–44.

Rice, J.A., 1990. Bioenergetics modeling approaches to evaluation of stress in fishes. Am. Fish. Soc. Symp. 8, 80–92.

Rigler, F.H., 1961. The uptake and release of inorganic phosphorus by *Daphnia magna* Straus. Limnol. Oceanogr. 6, 165–174.

Romanovsky, Y.E. and I.Y. Feniova, 1985. Competition among Cladocera: effect of different levels of food supply. Oikos 44, 243–252.

Rothhaupt, K.O., 1990. Resource competition of herbivorous zooplankton: a review of approaches and perspectives. Arch. Hydrobiol. 118, 1–29.

Sandheinrich, M.B. and G.J. Atchison, 1989. Sublethal copper effects on bluegill, *Lepomis macrochirus*, foraging behavior. Can. J. Fish. Aquat. Sci. 46, 1977–1985.

Sandheinrich, M.B. and G.J. Atchison, 1990. Sublethal toxicant effects on fish foraging behavior: empirical vs. mechanistic approaches. Environ. Toxicol. Chem. 9, 107–119.

Schoener, T.W., 1983. Field experiments on interspecific competition. Am. Nat. 122, 240–285.

Schoener, T.W., 1986. Mechanistic approaches to community ecology: a new reductionism? Am. Zool. 26, 81–106.

Sierszen, M.E. and T.M. Frost, 1990. Effects of an experimental lake acidification on zooplankton feeding rates and selectivity. Can. J. Fish. Aquat. Sci. 47, 772–779.

Sih, A., 1987. Predators and prey lifestyles: an evolutionary and ecological overview. In: Predation: Direct and Indirect Impacts on Aquatic Communities, edited by W.C. Kerfoot and A. Sih, University Press of New England, Hanover, NH, pp. 203–224.

Sih, A., 1992. Prey uncertainty and the balancing of antipredator and feeding needs. Am. Nat. 139, 1052–1069.

Sih, A., P. Crowley, M. McPeek, J. Petranks and K. Strohmeier, 1985. Predation, competition, and prey communities: a review of field experiments. Annu. Rev. Ecol. Syst. 16, 269–311.

Sprague, J., 1981. Ethologists are environmental drop-outs. In: Ecology and Ethology of Fishes, edited by D.L.G. Noakes and J.A. Ward, Dr. W. Junk, Boston, MA, p. 138.

Stephens, D.W. and J.R. Krebs, 1986. Foraging Theory. Princeton University Press, Princeton, NJ.

Sullivan, J.F, G.J. Atchison, D.J. Kolar and A.W. McIntosh, 1978. Changes in the predator-prey behavior of fathead minnows (*Pimephales promelas*) and largemouth bass (*Micropterus salmoides*) caused by cadmium. J. Fish. Res. Board Can. 35, 446–451.

Sutterlin, A.M. and N. Sutterlin, 1970. Taste responses in the Atlantic salmon (*Salmo salar*) parr. J. Fish. Res. Board Can. 27, 1927–1942.

Tagatz, M.E., 1976. Effect of mirex on predator-prey interaction in an experimental estuarine ecosystem. Trans. Am. Fish. Soc. 105, 546–549.

Taylor, R.J. 1984. Predation. Chapman and Hall, New York.

Tillman, U. and W. Lampert, 1984. Competitive ability of differently sized *Daphnia* species: an experimental test. J. Freshwater Ecol. 2, 311–323.

Tilman, D., 1982. Resource Competition and Community Structure. Princeton University Press, Princeton, NJ.

Turner, A.M. and G.G. Mittelbach, 1990. Predator avoidance and community structure: interactions among piscivores, planktivores, and plankton. Ecology 71, 2241–2254.

Vinyard, G.L., 1980. Differential prey vulnerability and predator selectivity: effects of evasive prey on bluegill (*Lepomis macrochirus*) and pumpkinseed (*L. gibbosus*) predation. Can. J. Fish. Aquat. Sci. 37, 2294–2299.

Walton, W.E., N.G. Hairston, Jr. and J.K. Wetterer, 1992. Growth-related constraints on diet selection by sunfish. Ecology 73, 429–437.

Wanzenbock, J. and F. Schiemer, 1989. Prey detection in cyprinids during early development. Can. J. Fish. Aquat. Sci. 46, 995–1001.

Werner, E.E., 1974. The fish size, prey size, handling time relation in several sunfishes and some implications. J. Fish. Res. Board Can. 31, 1531–1536.

Werner, E.E., 1984. The mechanisms of species interactions and community organization in fish. In: Ecological Communities: Conceptual Issues and the Evidence, edited by D.R. Strong, Jr., D. Simberloff, L.G. Abele and A.B. Thistle, Princeton University Press, Princeton, NJ, pp. 360–382.

Werner, E.E. and B.R. Anholt, 1993. Ecological consequences of the trade-off between growth and mortality rates mediated by foraging activity. Am. Nat. 142, 242–272.

Werner, E.E. and J.F. Gilliam, 1984. The ontogenetic niche and species interactions in size-structured populations. Annu. Rev. Ecol. Syst. 15, 393–425.

Werner, E.E., J.F. Gilliam, D.J. Hall and G.G. Mittelbach, 1983. An experimental test of the effects of predation risk on habitat use in fish. Ecology 64, 1540–1548.

Werner, E.E. and D.J. Hall, 1974. Optimal foraging and the size selection of prey by the bluegill sunfish (*Lepomis macrochirus*). Ecology 55, 1042–1052.

Werner, E.E. and D.J. Hall, 1988. Ontogenetic habitat shifts in bluegill: the foraging rate-predation risk trade-off. Ecology 69, 1352–1366.

Werner, E.E., D.J. Hall, D.R. Laughlin, D.J. Wagner, L.A. Wilsmann and F.C. Funk, 1977. Habitat partitioning in a freshwater fish community. J. Fish. Res. Board Can. 36, 360–370.

Wetterer, J.K. and C.J. Bishop, 1985. Planktivore prey selection: the reactive field volume model vs. the apparent size model. Ecology 66, 457–464.

Wildhaber, M.L. and L.B. Crowder, 1990. Testing a bioenergetics-based habitat choice model: bluegill (*Lepomis macrochirus*) responses to food availability and temperature. Can. J. Fish. Aquat. Sci. 47, 1664–1671.

Woltering, D.M., J.L. Hedtke and L.J. Weber, 1978. Predator-prey interactions of fishes under the influence of ammonia. Trans. Am. Fish. Soc. 107, 500–504.

Wright, D.I. and W.J. O'Brien, 1984. The development and field test of a tactical model of the planktivorous feeding of white crappie. Ecol. Monogr. 54, 65–98.

Yocom, T.G. and T.A. Edsall, 1974. Effect of acclimation temperature and heat shock on vulnerability of fry of lake whitefish (*Coregonus clupeaformis*) to predation. J. Fish. Res. Board Can. 31, 1503–1506.

Zaret, T.M., 1980. Predation and Freshwater Communities. Yale University Press, New Haven, CT.

Zelson, P.R. and R.H. Cagan, 1979. Biochemical studies of taste sensation. VIII. Partial characterization of alanine-binding taste receptor sites of catfish *Ictalurus punctatus* using mercurials, sulfhydryl reagents, trypsin and phospholipase C. Comp. Biochem. Physiol. 64B, 141–147.

Ecotoxicology and the Redundancy Problem: Understanding Effects on Community Structure and Function

James R. Pratt and John Cairns, Jr.

I. INTRODUCTION

Ecotoxicologists are concerned about the impact of anthropogenic stressors on ecological systems and so, implicitly, are studying the health of ecosystems (see Chapters 10 and 13). Strategies for ecosystem protection assume that ecosystems can withstand some degree of impact without adversely altering ecosystem function, that is, important ecological processes (Lawton and Brown, 1993). The assumed "assimilative capacity" of ecosystems is thought to occur because systems have redundant parts. Resolving issues of ecological redundancy is precisely the business of ecotoxicology, since human impacts on the environment, especially from toxic chemicals, may remove or debilitate organisms locally, regionally, and globally. Historically, community structure has been monitored to indicate the state of an ecosystem, even though systems ecologists have focused on the processing of matter and energy (i.e., function) for nearly 50 years. Logically, if processes are the fundamental characteristics of ecosystems, then the identities of the participating structures (species) may be unimportant. On the other hand, ecologists have long characterized communities, ecosystems, and landscapes by their dominant species, irrespective of the dominance of particular processes.

There are two contrasting views (hypotheses) of ecological or functional redundancy. (The term *functional redundancy* refers to the principle that function is not altered by changing structure.) The first hypothesis is the so-called *rivet popper hypothesis* (Ehrlich and Ehrlich, 1981) in which the ecosystem is viewed as analogous to an airplane, with individuals and species viewed as equal rivets sustaining the whole. Impacts (toxic chemicals, habitat

1-56670-1127-9/95/$0.00+$.50

destruction) remove individuals and species, leading to impoverishment in terms of rivets. Every rivet that is popped weakens the overall structure, resulting eventually in a catastrophe. In the early stages of rivet popping, little difference can be detected in the integrity of the structure, while continued impact results in the eventual failure of the structure to function.

An alternative view is the *redundant species hypothesis* of Walker (1991) and, subsequently, Lawton and Brown (1993) that suggests that some species are more important than others. In Walker's conception, *drivers* are either dominant or key species, and *passengers* include groups with high redundancy. It is argued that groups with higher redundancy can withstand some simplification without producing functional consequences in the ecosystem. That is, the ecosystem is functionally homeostatic in the face of species loss for some species. A similar concept for testing has arisen from food web theory in which the degree of connectedness of food webs may result in species deletion sensitivity (Orians and Kunin, 1990). A web that is insensitive to deletion of a particular species (a passenger) is presumed to have redundancy.

Testing these hypotheses is important to ecotoxicology, risk assessment, conservation biology, and fundamental ecology. If ecosystems have redundant components, then we should determine precisely how much ecological simplification can occur without damage to our life support system i.e., the free services provided by ecosystems (Pimm, 1993; Cairns and Pratt, 1995). If ecosystems lack redundancy, we should redouble our efforts to assess levels of stressors that produce adverse effects, and we should be more proactive in managing for waste minimization and conservation (perhaps preservation) of biological resources.

In this discussion, we examine the background of the redundancy concept, compare toxicological and ecological approaches to predicting impacts, discuss the interplay between ecology and ecotoxicology, and evaluate the conventional wisdom of ecosystem protection. Examples are drawn largely from our work on aquatic ecosystems. Streams, rivers, and shallow waters are particularly vulnerable since they are a common site of waste disposal, are a small part of the landscape, and provide essential services (drinking water, food, recreation) for human populations.

II. POSSIBLE RELATIONSHIPS BETWEEN STRUCTURE AND FUNCTION

Three very different relationships are possible between ecosystem structure and function (Cairns and Pratt, 1986). First, natural community structure and function may be so intimately related that a change in one inevitably results in a change in the other. If this were the case, only one attribute would need to be measured with the assurance that extrapolation to other attributes was quite reliable. Since persuasive evidence indicates that structural attributes are more easily measured than function, the method of choice would primarily be structural measures. Structural measures are also those with which ecologists

Table 1   Ecosystem Differences Affecting
Expression of Ecotoxicological Impacts

Stream ecosystems	Forested ecosystems
Short-lived species	Long-lived species
Biomass turnover rapid	Biomass turnover slow
Functional plasticity	Functional sensitivity
Pollutant conduits	Pollutant sinks

still have the greatest familiarity and which have the longest history of measurement.

Second, functional measures may be more sensitive to stressors than structural measures. The variability in community structure from place to place is high and possibly problematic in detecting adverse effects at the community and ecosystem levels. Ecological functions integrate the collective activities of many species, and the physiological abilities of individuals can be reduced by stressors without individuals being eliminated from populations. Therefore, measuring collective functions might reveal effects that would be missed by "critter counting."

Third, structural measurements may be more sensitive to both natural and anthropogenic stressors than functional measurements because of the hypothesized redundancy of most natural systems. The redundancy argument is based on the assumption that communities rich in species may have an array of taxa in key functional groups or guilds. Therefore, if a species were eliminated, one or more other species could expand to fulfill the functional role.

III. ECOSYSTEM DIFFERENCES

Most of the species that are targets of concern and conservation are large and terrestrial, even though many extant species are small and aquatic. Large and long-lived species, such as the well-known charismatic megavertebrates and even perennial flowering plants, persist for sufficiently long periods of time to be discovered, even when they are extremely rare, and their habitat requirements can often be deduced even if these are difficult to replace or protect.

At the ecosystem level the responses of terrestrial and aquatic systems may differ considerably (Table 1). Many terrestrial ecosystems are characterized by the complex physical and biological structure of the dominant plant community. Stressors of the terrestrial environment, such as habitat destruction, fragmentation, acidic deposition, and soil pollution, are persistent and clearly affect the functioning of ecosystems. The persistent nature of stressors is exacerbated by the comparatively long life histories of the resident organisms: most species, including even the smallest insects, reproduce during only part of the year in the temperate zone. This makes ecological recovery from stress a long process. The slower responses of species in terrestrial environments have led to predictions about the inability of the terrestrial plant community to adapt quickly to changing climate and suggest that species turnover occurs over comparatively long time scales (Abrams, 1992).

Aquatic ecosystems, especially streams, are often quite different. The predominant species of concern in stream and river systems are short-lived (even sport fishes live only a few years), and flowing waters tend to displace the effects of stressors, such as chemical inputs, and dilute these pollutants downstream. The aquatic environment is characterized by rapid turnover of biomass and species, leading to frequent species replacements even in unstressed systems and assisting the recovery process in damaged systems (Herricks and Cairns, 1976). The dichotomy between aquatic and terrestrial ecosystems is often not clear, since species turnover also may be characteristic of terrestrial ecosystems (Johnson and Mayeaux, 1992), terrestrial ecosystems contain very many short-lived species, lakes can certainly be pollutant sinks, and the oceans are the ultimate sinks for most contaminants. Regardless, neither terrestrial nor aquatic communities are random aggregations of species (Schultze and Mooney, 1993).

The soil is particularly problematic since it includes representative species with both long and short turnover times. The macrobiota of soils exclusive of the terrestrial plant cover performs essential functions in reducing dead organic material directly in the detritus-based food web. While larger organisms such as salamanders and insects may be active for long periods, the soil microbiota is strongly influenced by the availability of water. In this sense, microbial components of the soil actually inhabit a temporally varying freshwater ecosystem. When the soil is dry, these organisms are largely inactive, although the activities of enzyme systems responsible for degrading portions of the detrital milieu may remain active on soil particle surfaces or in the periplasts of Gram-negative bacteria (Overbeck and Chrost, 1990). In this regard, functional activities may be unrelated to the presence of viable organisms, at least temporarily.

IV. PREDICTIVE APPROACHES

A. WHAT HAS ECOTOXICOLOGY LEARNED FROM ECOLOGY?

Environmental toxicology is widely regarded as a child of mammalian toxicology, but over the past decade, increasing attention has been paid to the use of ecological information in decision making. That is, the science of ecotoxicology has matured as increasing attention has been paid to ecologically significant effects of contaminants, as opposed to the attention paid to traditional toxicological measures. On the other hand, ecological measures have not found a welcome audience in the transition from science to application in ecosystem protection (e.g., "ecological" risk assessment). For example, replicated ponds were proposed and used for several years in the United States to assess pesticide risks (Graney et al., 1994), but these systems were quickly abandoned in a policy decision. Protocols for conducting ecotoxicity tests in laboratory microcosms have been developed and used for numerous research studies (see Cairns, 1984a,b; Pratt and Bowers, 1990), but no microcosm study has ever been required to register a new chemical for commercial use in the United States. Ecological studies still are not fully incorporated into risk assessment for many reasons ranging from ecosystem variability to the perceived imprecision of ecological measures (but see Pratt and Bowers, 1992). In summary, ecotoxicology has expanded to include the measurement of ecological effects, but these studies have not been used in the regulation of chemical stressors.

B. TOXICOLOGICAL PARADIGMS

Toxicological studies examine the adverse effects of xenobiotics at the level of the individual. In acute toxicity testing, the endpoint (i.e., measured response) of interest is the level of toxicant producing an effect in 50% of the test organisms (LD50, LC50, EC50). Decision making under chronic exposure is most often based on hypothesis testing using analysis of variance (ANOVA) models which are followed by pairwise tests to define the no-observed-effects concentration or level (NOEC or NOEL) and the lowest-observed-effects concentration or level (LOEC or LOEL). Generally, both acute and chronic testing assume a simple dose-response (Figure 1) even though sublethal stimulation is a common effect.

"Ecological" risk assessment amalgamates data derived from single species and other tests (usually tests of environmental fate), most commonly by comparing the responses to a toxicant of putative sensitive species and expected environmental exposures (USEPA, 1989, 1992). The species response is defined statistically as the detectable difference from controls (i.e., the NOEC and LOEC) or as the median lethal tolerance limit (LC50) for certain surrogate species. This response level is compared to the expected environmental concentration. If there is a substantial margin of safety (NOEC >> exposure), then a toxicant is assumed to pose no problem. If the NOEC and exposure concentrations are similar, the decision is more difficult. If the NOEC < exposure, then the compound is clearly hazardous, unless sweeping social benefits are available to counteract the presumption of risk.

Several assumptions underlie the simplistic nature of ecological risk assessment as practiced. First, the standard suite of test species is assumed to be sensitive (Cairns, 1986). Second, the interaction of biological species with the test compound (biodegradation, biotransformation) is assumed to be minimal. Third, apart from survival and growth of nonhuman organisms, there is very little "ecology" in ecological risk assessment. (Note that most of the chapters in this volume evaluate important influences of chemicals at nonecological levels in the biological hierarchy.) The power of statistical tests to detect certain differences in test subjects is not usually estimated (Conquest, 1983; Pratt and Bowers, 1992). Effects on populations, communities, or ecosystems are rarely tested (Cairns, 1984a,b; Kimball and Levin, 1985). Criticisms of simplistic approaches to risk assessment are legion and will not be reviewed further here.

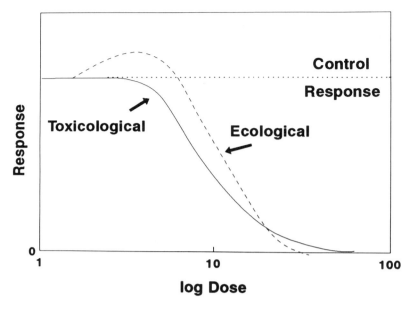

Figure 1 Toxicological and ecological dose responses.

C. ECOLOGICAL PARADIGMS

Although the diversity of paradigms in toxicology and risk assessment is small, the number of ecological interactions and processes that might respond to anthropogenic stressors is large (Cherrett, 1989), as evidenced by the diversity of papers in this volume. Often, communities are portrayed as being in delicate balance, although many ecologists reject ideas based on equilibrium theory (Botkin, 1990). The idea of biological balance is deep-seated and intricately woven into the fabric of biological belief, law, and regulation. Many states retain narrative environmental protection criteria based on "balanced biological communities." Ecological theory has treated ecosystems as predictably stable at some levels (i.e., steady-state process models) and dynamically linked to natural gradients and surrounding ecosystems (e.g., the river continuum concept, Vannote et al., 1980). The equilibrium view of communities and ecosystems is being rapidly replaced.

Most ecologists have studied stressed ecosystems, even systems that are considered to be "pristine." Given the global distribution of pollutants and the commonness of transboundary problems (created when nations differ in their environmental protection goals or capabilities), it is unlikely that ecosystems exist that are unaffected by human activities. Several ecologists have offered insights on the expected trends in ecosystems under stress. Perhaps the most compelling are the predictions of E.P. Odum (1985; Table 2). These predictions have served as hypotheses for ecologists studying ecosystem stress (e.g., Schindler, 1987; Schaeffer et al., 1988; Pratt, 1990).

Table 2 Trends Expected in Stressed Ecosystems

Category	Expected trends
Energetics	Community respiration increases
	P/R (production/respiration) becomes unbalanced (< or >1)
	P/B and R/B (maintenance:biomass) ratios increase
	Importance of auxilliary energy increases
	Exported or unused primary production increases
Nutrient cycling	Nutrient turnover increases
	Horizontal transport increases and vertical cycling decreases
	Nutrient loss increases
Community structure	Proportion of r-strategists increases
	Size of organisms decreases
	Life-spans of organisms or parts decrease
	Food chains shorten
	Species diversity decreases, dominance increases
General trends	Ecosystem becomes more open
	Autogenic successional trends reverse
	Efficiency of resource use decreases
	Parasitism and other negative interactions increase
	Functional properties are more robust than species composition

Based on Odum (1985).

Ecotoxicologists have learned that ecologists expect ecosystems to change predictably under stress. Changes have been expected in both the structure and function of systems. In fact, the importance of various measures has been a matter of considerable controversy concerning the adequacy of measurements needed to ascertain ecosystem "health." The following illustrate the divergence of thinking.

"Theoretically, small changes in water quality should lead to alterations in the structure of a community which is already in delicate balance." Metcalfe (1989), commenting on the use of macroinvertebrate community structure to detect pollution.

"Symptomatic of this schism is the almost complete dependence of the biocriteria development on taxonomic procedures, while basic researchers have moved into questions of process and function." Cummins (1991), on functional views of community organization.

Structurally, the distribution of individuals among taxa has been adequately described by truncate log normal distributions (Preston, 1948), which seem to be characteristic of a large collection of objects subject to certain design constraints (e.g., Gaston et al., 1993). These ecological paradigms have been drawn on in assessing biological diversity as a measure of ecosystem health (e.g., Shannon and Weaver, 1948; Wilhm and Doris, 1968; Weber, 1973), even though the relationship between diversity and ecological "stability" remains elusive (Pimm, 1993). Predictions and empirical evidence suggest that species richness is a useful measure of biological diversity that responds predictably to a number of stressors (Odum, 1985; Karr et al., 1986; Schindler, 1987; Schaeffer et al., 1988; Pratt, 1990).

Ecological processes have been a less common target of ecotoxicological investigations. There is only limited agreement about what constitutes ecological function (Table 3) beyond processes such as primary production and

Table 3   Comparison of Processes Considered to be
           Ecological Functions

Karr (1993)	Lawton and Brown (1993)
Nutrient cycling	Major biogeochemical cycles
Photosynthesis	Primary production, consumption, secondary production
Water cycling	Food web assembly
Speciation	Ecological stability, resilience
Competition, predation	Succession
Mutualism	

nutrient cycling. Allochthonous primary production drives many small stream ecosystems, but not larger ones (Vannote et al., 1980). Because of the importance of imported organic material in small streams, detrital processing — processes that unite several biogeochemical cycles — have been used to infer stressor-related effects (Lugthart and Wallace, 1992; Reice and Wollenberg, 1993). Basic ecological processes may not change predictably with increasing stress because many processes are substrate limited and other participants in these processes may have underutilized capacity so that the effects attributable to anthropogenic stressors may not be observed unless the stressor also alters the substrate supply (Levine, 1989). Ecosystem-level studies, and even highly controlled microcosm studies, have often failed to detect predicted changes in ecological structure and function (e.g., Schindler, 1990; Pratt et al., 1993).

V. LINKING ECOLOGY AND ECOTOXICOLOGY

A. WHAT COULD ECOLOGY LEARN FROM ECOTOXICOLOGY?

Ecotoxicologists have searched for better means to predict the possible adverse effects of human activities on ecosystems. Principally, research has been devoted to testing the effects of chemicals on individual species and to finding means to assess impacts in the field. While much ecotoxicological research has centered on species of interest to humans (i.e., economically important species), ecotoxicologists working at the community and ecosystem levels have studied the implications of chemical stressors, which (usually) lead to biotic impoverishment. Ecologists have long debated the relationship between biological diversity and ecosystem stability. While a strong relationship has seemed intuitive (and is part of the basis for biological criteria; Plafkin et al., 1989), the results of experimentation and modeling studies have been equivocal.

Ecologists have been surprised that ecological theory has provided little guidance in making predictions about the relationship between changing ecological structure and function (Lawton and Brown, 1993). These questions have become important to understanding the repercussions of species loss in relation to wholesale habitat destruction, principally in the tropics. Ecotoxicologists have addressed the structure-function relationship for some time in microcosm, mesocosm, and whole ecosystem studies, but ecologists have eschewed these studies. For many studies, data linking structure and function are qualitative, even anecdotal, but several studies have attempted to link directly changing structure and function.

Ample evidence for the preference of structural measures is available from a number of impact-related studies in both terrestrial and aquatic environments. For example, Bormann (1990) dwelled on structural changes in forests exposed to acidic deposition and noted only that primary production and nutrient cycling were altered. In aquatic ecosystems, the traditional reliance on structure measures of impact is even clearer (Rosenberg and Resh 1993), although aquatic biologists have estimated secondary production and done numerous studies on leaf decomposition (e.g., Lugthart and Wallace, 1992).

Despite the historical preference for structural measures, ecotoxicologists have made many attempts to measure system function in experiments (see Cairns and Pratt, 1989). Given Odum's predictions that ecosystems should be functionally robust, the motivation for repeatedly attempting to measure function is unclear. Several reasons seem plausible. First, ecosystem science has focused on the importance of energetic relationships, and processing and cycling of matter. Second, ecosystem functions are seen as emergent properties integrating the collective functioning of communities. Third, functional measures are "nontaxonomic" and, therefore, do not require laborious "critter counting" to produce species lists. That is, the produced data are univariate rather than multivariate.

B. REDUNDANCY IN TERRESTRIAL SYSTEMS

Manipulative field studies have shown a strong relationship been species richness and ecosystem function. Alteration of species richness by manipulation of nitrogen has been used to examine drought resistance in grassland plants (Tilman and Downing, 1994). These studies showed drought produced a 45% reduction in plant biomass and a 35% reduction in species richness of control plots during a severe drought year. Drought resistance (expressed as biomass deviation, a measure of net production) was strongly related to species richness. The investigators concluded that the results "suggest that each additional species lost from our grasslands had a progressively greater impact on drought resistance."

The Envirotron laboratory studies (Naeem et al., 1994) have also shown a strong link between structure and function in artificial communities consisting of 2, 5, or 16 plant species and additional herbivore and soil organisms (Figure 2). The strong relationship between diversity and productivity led the authors to conclude that their study "demonstrates for the first time under controlled environmental conditions that loss of biodiversity ... may also alter or impair the services that ecosystems provide."

A much earlier, ecotoxicological demonstration of the same effect was provided by van Voris and colleagues (1980) by using the pattern of carbon dioxide flux from cadmium-treated soil core microcosms (large, intact cores from an old field of herbaceous plants). There was a strong statistical link

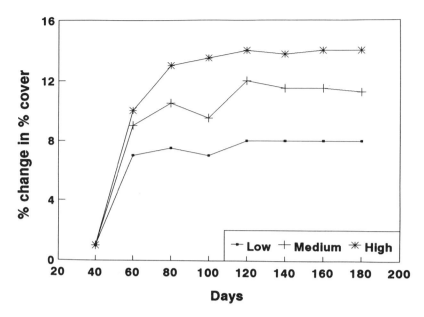

Figure 2   Effect of diversity on the change in plant cover in experimental ecosystems (redrawn from Naeem et al., 1994). Low, medium, and high diversity communities had 2, 5, and 16 plant species, respectively. (Redrawn from Naeem et al., Nature 368:734–737.)

between diversity and the number of peaks in the CO_2 flux power spectrum (Figure 3), leading the investigators to conclude that the number of peaks in the power spectrum (measures of strong repeating frequencies) was a measure of "functional complexity."

C. REDUNDANCY IN AQUATIC SYSTEMS

A number of experiments using model and natural aquatic systems show the linkage between structure and function clearly. We have shown, using a variety of chemical stressors, that statistically detectable changes often occur in structural variables at the same or lower levels of contaminants that produce changes in system function (Pratt et al., 1993). The close link between aquatic insects and detrital processing has been shown in streams treated with insecticide (methoxychlor, Lugthart and Wallace, 1992). Only a small number of illustrative examples will be given here.

Communities exposed to the herbicide atrazine showed altered primary productivity (atrazine interferes with electron transport in photosystem II), nutrient cycling, and biological diversity (de Noyelles et al., 1982; Pratt et al., 1988). Concomitant with changes in biomass and species richness, communities became "leaky" and unable to recover phosphorus. Such experiments produce conflicting evidence of redundancy. In the case of primary production (inferred from oxygen balance), there was no redundancy: structure and function

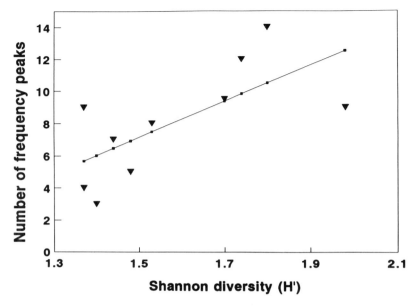

Figure 3 Relationship between diversity (Shannon diversity, based on information theory) and spectral density, based on van Voris et al. (1980). Spectral density was defined as the number of frequency peaks in the power spectrum of CO_2 in soil core microcosms treated with cadmium.

changed at the same levels of stressor (Figure 4a). In the case of phosphate metabolism, the processing rate remained stable as biomass declined (Figure 4a and b). The phosphate supply was constant, and significant changes in organic phosphate processing rate and phosphate loss changed only at levels of the stressor that had already significantly altered community structure.

A more compelling relationship between structure and function is shown in experiments with copper as the stressor (Pratt and Rosenberger, 1993). We supplied copper at concentrations (10, 20, 40, 80, 160 µg/l) near the calculated water quality criterion (~30 µg/l in our hard dilution water) to replicate microcosms and measured a suite of structure and function variables. We examined effects on community function by monitoring the diurnal pattern of microcosm pH and then examined the frequency pattern of pH data using power spectral analysis much like van Voris and colleagues (1980). System pH reflected the daily production-respiration cycle (12L/12D). We collected pH data every 15 min (96 times per day), so we expected to see a repeating cycle with a period of 96 time intervals. We evaluated not only species richness in microcosms, but also community composition (using similarity analysis, Pratt and Smith, 1991). We also measured biomass (chlorophyll, protein), macronutrients (calcium, magnesium, potassium), oxygen concentration, and alkaline phosphatase activity.

Most variables showed dose-related adverse effects. Species richness declined and showed significant changes at concentrations near the water quality criterion (Figure 5), while reduction in community function such as oxygen

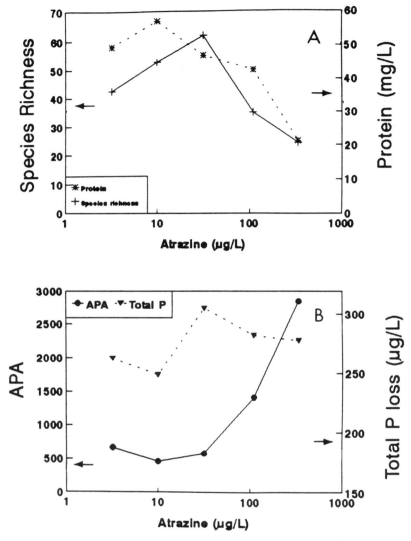

Figure 4 (A). Effect of atrazine on biomass and species richness in experimental microcosms. Microcosms were dosed continuously with atrazine for 21 d. (B). Effect of atrazine on phosphorus metabolism (loss of total phosphorus and biomass-specific activity of alkaline phosphatase enzyme systems [APA]). Units of APA are nmol p-nitrophenol/mg protein/h (based on Pratt et al., 1988). Arrows point to the mean control values on the respective axes.

balance (Figure 5A) and alkaline phosphatase activity (Figure 5B) occurred only at higher doses of the toxicant. We also examined changes in community composition (measured as changes in community similarity) that paralleled declining richness. Significant effects on other variables also occurred at either 20 or 40 µg Cu/l doses. The diurnal pattern of production and respiration (as measured by the pH power spectrum) in control microcosms showed the

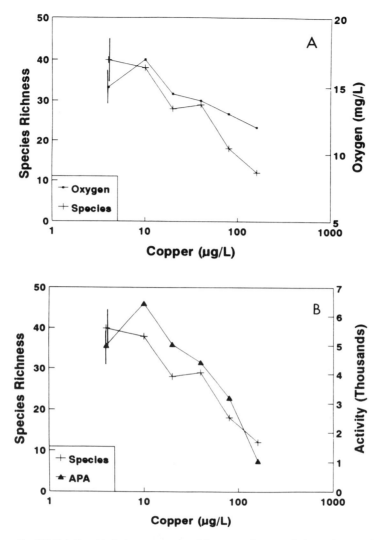

Figure 5 (A) Relationship between species richness and oxygen balance in experimental microcosms dosed with copper. (B) Relationship between species richness and gross alkaline phosphatase activity in experimental microcosms dosed with copper. Units of alkaline phosphatase activity (APA) are nmol *p*-nitrophenol/h. Error bars on controls identify the interval of nonsignificance based on Dunnett's procedure. Treatment means outside this interval are significantly different from controls (*p* <.05).

expected peak frequency (96 time intervals) corresponding to the light-dark cycle (Figure 6). But, this pattern was virtually decomposed at the lowest copper dose (10 ug/l), and there was no discernable pattern at any higher dose, suggesting that community function had been fully compromised at a level of the stressor below which structural effects (species richness, biomass) were detectable. So, while the pattern of daily production and respiration had changed

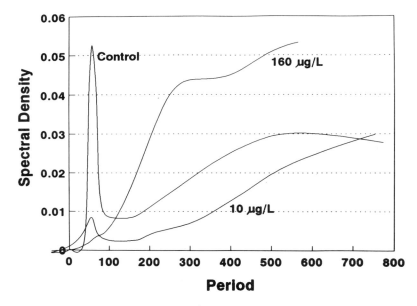

Figure 6   Effects of copper on spectral density of diurnal pH patterns in microcosm
(based on Pratt and Rosenberger, 1993). Microcosm pH was measured every
15 min. The peak of spectral density was expected at 96 time intervals
(period), which was equivalent to a repeating 24-h cycle.

at the lowest level of copper dosing, other structural and functional changes
followed at higher doses.

Experiments such as this reveal a feature of ecological studies often not
discussed: some changes, especially changes in function, are difficult to detect.
Detecting changes in structure is based on generally accepted methods, but
measuring system function is more complex and demanding. In the case of the
pH power spectrum, over 30,000 measurements were made (about 3000 measures on 12 of 18 microcosms). This level of intervention is not generally
feasible in field studies, and we have previously shown that some measures of
function such as enzyme activities are more variable than structural measures
(Pratt and Bowers, 1992). At least part of the problem, then, in claiming
functional redundancy is statistical: the power of analyses of function is
probably lower than that of structural measures.

D. SPECIES TURNOVER

As the size of organisms decreases, the turnover of species reveals much
less predictable community structure, further enhancing ideas of redundancy.
Aquatic insects are at least partially terrestrial, so there is a cycling of populations as nymphal stages mature, pupate, and emerge from streams and lakes
as adults. But, in smaller organisms, there is a perpetual replacement of species

in communities, even though the number of viable taxa at any given time may be relatively constant. The ecological reasons for species replacement are unclear even though the general pattern of tolerances is known for a number of species (Lowe, 1974; Foissner, 1988).

Species turnover in communities is clearly related to organism size (Schoener, 1983), and failure to account for these dynamic processes can lead to gross underestimations of the number of extant taxa. For example, historical surveys of the Flint River (Georgia) showed that the number of protozoan species in the river was about 71 when natural substrata were sampled at a small number of stations (Patrick, 1961). Several years later, a comparable study sampled the same ecosystem over several seasons and utilized sampling of both natural and artificial substrata to collect benthic protozoa. The resulting estimate of species richness for this ecosystem increased by nearly an order of magnitude (485 species; Pratt et al., 1987).

We have monitored protozoa in manipulated agricultural watersheds converted from traditional to best management practices to minimize nutrient and pesticide loss. We observed increasing annual species richness and species persistence over an 8-year period using replicate samples taken twice each month. Had we simply sampled at the expected low flow period when conditions might be worst (i.e., October), we would not have seen the annual community pattern, and we would have concluded that no change had occurred. The annual community pattern of frequency of species occurrence resembled plant community models (so-called "core-satellite models"; Collins et al., 1993) in which a core of species persists and other species are transient (Figure 7). Such a model for protozoan communities was proposed over 20 years ago (Cairns et al., 1971) and now demonstrates a feature of communities that also responds to stressors.

Species turnover, which can be readily observed in microbial communities, also occurs in communities of long-lived species, even though the measurement of this dynamic may be somewhat difficult (Johnson and Mayeaux, 1992). Plant ecologists would regard the assortative processes of species replacement as succession, and, despite one's belief in the presence or absence of an absolute climax, such processes also occur in a variety of post-disturbance ecosystems. Indeed, the differences among ecosystems may relate more to the frequency of disturbance and the processes that lead to recovery than to life history characters of the dominant species, since most ecosystems seem to be regularly disturbed by either natural or anthropogenic forces (Vogl, 1980).

Our point here is not that any individual species may be particularly important but that the character of the sampled communities is dynamic, making the elucidation of patterns problematic unless sampling is extensive. Reliance on a few "indicator" species, such as those used to predict hazards in ecosystems, is particularly problematic if ecosystem structure is in a perpetual state of change. Therefore, quality assurance steps (i.e., biological monitoring) become even more important if biological balance does not exist.

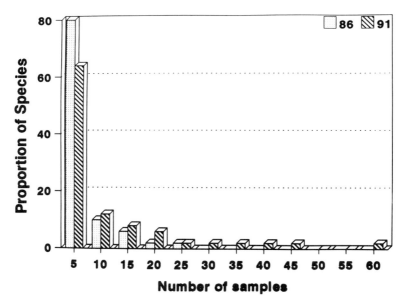

Figure 7   Changing community structure in protozoan communities during recovery of
an agricultural watershed under best management practices. The proportion
of species is shown by the number of samples in which species occurred
(maximum 72) in two different years — 1986 (86), before best management
practices and 1991 (91), after best management practices had been used for
3 years. Before best management practices were used, species rarely re-
occurred for long periods. After best management practices were used, a
"core" community of a few species persisted for much of the year.

VI. TOWARD A NEW ECOTOXICOLOGY: SLAYING SOME MYTHS

A. ECOLOGICAL REDUNDANCY

Communities appear to lack appreciable redundancy. Based on studies in
both terrestrial and aquatic systems, there is currently more support for the
"rivet popper hypothesis" that each structural change decreases system performance. That is, impacts that result in the local extinction of species have
functional consequences, even though these putative changes in function may
be difficult to measure. In fact, measuring some functional changes may be
inherently difficult, leading to the most environmentally dangerous statistical
error — concluding that no impact has occurred when one could not be
adequately measured. It would be incorrect to assume that communities and
ecosystems lack functional redundancy, but empirical evidence contrasts with
conventional wisdom. Invoking ecological or functional redundancy as an
explanation of retention of function in the face of structural change should be
done cautiously, because the very close linkage between structure and function
is now clearer (see Chapter 13). In fact, the concept of ecological redundancy
needs critical testing: the consequences of allowing species to become locally
extinct may be severe even though functional measures suggest no change in

ecosystem function. Further, the scale over which ecosystem structure and function is assessed needs precise definition. In our experiments, we defined ecosystems operationally as microcosms, but in natural landscapes, detection (or failed detection) of a local impact may be an error, and detection may only be possible at other scales of measurement. The failure to detect effects using coarse measuring instruments may be especially problematic.

Concluding that ecosystems lack redundancy bears careful consideration. Many communities, especially those of short-lived species, experience species replacement (species turnover) during the annual cycle (see below). Analysis of these changes must be careful to distinguish the significance of compositional changes. For example, Bartell (1973) found "predictable" smooth seasonal changes in ordination of phytoplankton, while Allen et al. (1977) found discrete clusters (using the same data), not seasonal pattern, using ordination by first difference: the scale of observation (empirical vs. derived) affected interpretation of the same data set. This suggests that understanding the natural pattern of species turnover is necessary to define the normal operating range of an ecosystem.

B. BIOLOGICAL BALANCE AND SPECIES TURNOVER

Biological communities are not delicately balanced. Rather, communities are dynamic and species turnover is the rule. The number of species that appear in an aquatic ecosystem on an annual basis is large (and bewildering). This turnover of species is closely linked to the concept of apparent ecological redundancy: many species with similar "functions" appear throughout the year. The common human frame of reference for terrestrial ecosystems suggests that the world operates seasonally, with little biological activity during colder months and relatively great biological activity during the "growing season." We are aware of the coming and going of certain species, such as migratory birds, and we may even be aware of the sequential flowering of land plants. In general, however, we realize that most of these organisms are present but differentially active during the entire year. This "equilibrium" model can be rejected for many communities. Again, the concept of biological balance needs critical testing, especially in regard to ecological interactions with toxic chemicals. It is problematic to build realistic process models with very many species compartments (see Chapter 10), even though these models can reveal a great deal about our knowledge (or lack of knowledge) of ecological systems and the effects of anthropogenic stressors.

C. ECOLOGICAL RESILIENCE

All communities are not equal in their resistance to, or resilience following, disturbance by anthropogenic stressors. This fact is not accounted for in any meaningful way by simplistic risk assessment procedures. Communities vary and their sensitivity to toxicants also varies. For example, Barreiro and Pratt (1994) manipulated natural periphyton communities with nutrients (nitrogen, phosphorus) to produce communities of different biomass and species composition. After exposure to the herbicide diquat, gross photosynthesis in low nutrient (= low biomass) communities never recovered to *predisturbance* levels, whereas higher biomass systems did (Figure 8). Experiments such as this suggest hazards vary among community types, but there are few studies to draw on for comparisons. The concept of ecological resilience needs further honing to be useful in assessing ecosystem condition. Obviously, a community with great potential for recovery (a resilient system) might be viewed very differently from a community with low resilience.

Ecotoxicological studies provide direct and indirect measurements of ecological resistance, the tendency to be insensitive to toxicant impacts, and this forms the basis for most of the use of ecotoxicological information. Regardless of the comparative sensitivity of structure or function in predicting ecological impacts, ecological resilience — the ability to recover from an impact — may be more important (Cairns and Niederlehner, 1993). Here again, understanding the linkage between ecological structure and function is essential for decision making and for communicating to the general public. Judging the acceptability of risk requires knowing whether a damaged ecosystem can recover to an approximation of its predisturbance condition following episodic or continual natural and anthropogenic stressors.

Ecological functions will probably be most appreciated by the general public (and its representatives in government and industry) if functions are described as ecosystem services (Cairns and Pratt, 1995). These services include maintaining atmospheric gas balance, transformation of societal wastes, storage of flood waters, and the provision of exploitable genetic material, in addition to the expected provision of food and fiber. One of the most persuasive reasons for restoring damaged ecosystems is the reestablishment of services lost to society (e.g., National Research Council, 1992). The clear link between structure and function suggests that structures that mediate functions need to be preserved (or at least conserved for reintroduction). Ecosystem services per capita are declining because the functioning of ecosystems is being reduced (Turner et al., 1990), both by anthropogenic stressors and by human population growth.

D. COMMUNITIES

Communities do not exist. Communities are convenient pieces of the ecological tapestry that are in some way tractable; however, strong ecological interactions among community members characteristic of other levels in the biological hierarchy are the exception, not the rule. Populations interact and the collective interactions of populations form ecosystems. Often the interactions are very specific and complex, but they do not occur at the community level. This does not mean that community ecology is nonscience, just that communities are subsets of ecosystems and these subsets are defined differently by different ecologists.

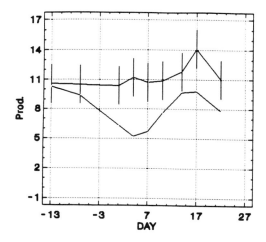

LOW

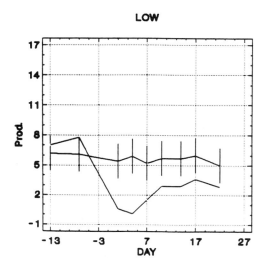

Figure 8 Interaction of nutrients and effects of diquat (0.3 mg/l) on gross primary production (Prod.) measured as oxygen gain in laboratory microcosms and maintained at differing nutrient status. Treatment occurred on day 0. The designations *medium* and *low* refer to nutrient manipulations in the micro-cosms. Nutrients were manipulated so that low microcosms had an average of 0.013 mg nitrate-N/l and 0.009 mg phosphate-P/l. Medium microcosms had an average of 0.54 mg nitrate-N/l and 0.052 mg phosphate-P/l. Gross primary production did not recover to pretreatment levels in low nutrient microcosms, but did recover to pretreatment levels in medium nutrient microcosms but did not attain productivity equal to controls. (Based on Barreiro and Pratt, 1994.)

Over the years, we have worked with "protozoan communities," "periphyton communities," and "microbial communities." When we focused on protozoa, we ignored algae and bacteria in our samples. As we developed our abilities to quantify interactions among bacteria, protozoa, algae, rotifers, and other members of the periphyton biofilm, we changed the definition of our experimental community. Even now, our work with microbial communities generally ignores the importance of fungi: we lack coherent and tractable ways to measure fungi and fungal activity (although it is certainly included in measures of nutrient cycling or biomass). Similar problems are experienced by community ecologists who study bird communities, fish communities, or virtually any other biotic community.

Ecotoxicologists must realize that communities are convenient subsets of ecosystems and share ecosystem properties such as high species richness, succession, nutrient cycling or matter processing, and productivity. As such, communities are good samples of ecosystems, even if communities are ill-defined.

E. A NEW ECOTOXICOLOGY

One definition of ecotoxicology holds that the science is simply toxicology using different test species (i.e., algae, invertebrates, fish, birds). For example, the European Economic Community defines mammalian toxicity tests as *toxicological studies* and tests of fish and crustaceans as *ecotoxicological studies* (Forbes and Forbes, 1994). A contrasting view is that attributed to Forth and colleagues (1980) of "environmental toxicology" which studies effects of chemicals on man and on ecological systems and their repercussions on human health. In our view, ecotoxicology is a science that uses ecological variables to assess the effects of chemicals in the environment.

A new ecotoxicology would embrace the different kinds of information available at differing levels of biological organization, and all of these levels are explored in this volume. Acute and chronic toxicity tests can provide predictions about the range of stressor levels at which organisms respond in the laboratory, but provide little information on toxic mechanisms. Mechanistic information can only be provided by physiological, cellular, and molecular studies. While it is absolutely necessary to understand toxic mechanisms, it is unlikely that mechanistic studies can be extrapolated to predicting effects at the ecosystem level. For example, it is unlikely that knowing the mode of action of a toxicant can allow us to predict which populations will go extinct under toxic stress from that toxicant, nor the structural or functional consequences of such an extinction. It is, in fact, unrealistic to even consider such expectations.

Population studies provide information on organismal groups with age and sex structure. At the community and ecosystem levels, the interactions of the contaminant with the biota and of the biota with the contaminant can be studied. Community and ecosystem studies need to be conducted with full realization of the way ecosystems work. That is, effects on biotic structure and function must be examined, and these data need to be incorporated (along with knowledge about uncertainties) into risk assessment. While ecosystems cannot be created in the laboratory, community-level experimentation is a closer approximation of the ecosystem than independent tests of surrogate species in isolation.

The result of incorporating information from additional levels of the biological hierarchy into risk assessment should not automatically lead to assumptions of even greater limitations on the environmental release of chemicals. Rather, decisions should be based on better information. In some circumstances, effects on untested species are underestimated for persistent chemicals. But, for other chemicals, especially those subject to modification by their interaction with organisms, basing environmental criteria on tests conducted under clean laboratory situations clearly overestimates hazard. Ecotoxicologists have long prided themselves on estimating risk using exposure criteria not simply environmental concentration (Cairns, 1980). In community-level testing, exposure tests (i.e., environmental fate) and biological effects tests can be realistically coupled. Similarly, ecosystems vary in their ability to withstand the effects of anthropogenic stressors, but we do not now know what the characteristics of resistant and vulnerable systems are. Community ecotoxicology can allow us to make the transition to a new, more complete ecotoxicology.

REFERENCES

Abrams, M.D., 1992. Fire and the development of oak forests. BioScience 42, 346–353.

Allen, T.F.H., S.M. Bartell and J.F. Koonce, 1977. Multiple stable configurations in ordination of phytoplankton community change rates. Ecology, 58, 1076–1084.

Barreiro, R. and J.R. Pratt, 1994. Interaction of toxicants and communities: the role of nutrients. Environ. Toxicol. Chem. 13, 361–368.

Bartell, S., 1973. A multivariate statistical approach to the phytoplankton community structure and dynamics of the Lake Wingra ecosystem. M.S. Thesis, Department of Botany, University of Wisconsin, Madison.

Bormann, F., 1990. Air pollution and temperate forests: creeping degradation. In: The Earth in Transition, edited by G.M. Woodwell, Cambridge University Press, Cambridge, pp. 25–44.

Botkin, D.B., 1990. Discordant Harmonies. Oxford University Press, New York.

Cairns, J., Jr., 1980. Estimating hazard. BioScience 30, 101–107.

Cairns, J., Jr., 1984a. Community Toxicity Testing. American Society for Testing and Materials, Philadelphia, PA.

Cairns, J., Jr., 1984b. Multispecies toxicity testing. Environ. Toxicol. Chem. 3, 1–3.

Cairns, J., Jr., 1986. The myth of the most sensitive species. BioScience 36, 670–672.

Cairns, J., Jr. and B.R. Niederlehner, 1993. Ecological function and resiliency: neglected criteria for environmental impact assessment and ecological risk analysis. Environ. Prof. 15, 116–124.

Cairns, J., Jr. and J.R. Pratt, 1986. Commentary: on the relationship between structure and function. Environ. Toxicol. Chem. 5, 785–786.

Cairns, J., Jr. and J.R. Pratt, 1989. Functional Testing of Aquatic Biota for Estimating Hazard of Chemicals. American Society for Testing and Materials, Philadelphia, PA.

Cairns, J., Jr. and J.R. Pratt, 1995. The relationship between ecosystem health and the delivery of ecosystem services. In: Evaluating and Monitoring the Health of Large-Scale Ecosystems, edited by D.J. Rapport and P. Calow, Springer-Verlag, Berlin, pp. 63–76.

Cairns, J. Jr., K.L. Dickson and W.H. Yongue. 1971. The consequences of nonselective periodic removal of portions of fresh-water protozoan communities, Trans. Am. Microsc. Soc. 90, 71–80.

Cherrett, J.M., 1989. Ecological Concepts, the Contributions of Ecology to an Understanding of the Natural World. Blackwell Scientific Publications, Boston, MA.

Collins, S.L., S.M. Glenn and D.W. Roberts, 1993. The hierarchical continuum concept. J. Veg. Sci. 4, 149–156.

Conquest, L.L., 1983. Assessing the statistical effectiveness of ecological experiments: utility of the coefficient of variation. Int. J. Environ. Stud. 13, 141–147.

Cummins, K.W., 1991. Establishing biological criteria: functional views of biotic community organization. In: Biological Criteria: Research and Regulation, EPA-440/5-91-005, Office of Water, Washington, D.C., pp. 3–8.

de Noyelles, F., Jr., W.D. Kettle and D.E. Sinn, 1982. The response of plankton communities in experimental ponds to atrazine, the most heavily used pesticide in the United States. Ecology 63, 1285–1293.

Ehrlich, P.R. and A.H. Ehrlich, 1981. Extinction, the Causes and Consequences of the Disappearance of Species. Random House, New York.

Foissner, W., 1988. Taxonomic and nomenclatural revision of Sladecek's list of ciliates (Protozoa: Ciliophora) as indicators of water quality. Hydrobiologia 166, 1–64.

Forth, W., D. Henschler and W. Rummel, 1980. Allgemeine und spezielle pharmakologie und toxikologie. BI Wissenschaftsverlag, Mannheim.

Forbes, V.E. and T.L. Forbes, 1994. Ecotoxicology in Theory and Practice. Chapman and Hall, London.

Gaston, K.J., T.M. Blackburn and J.H. Lawton, 1993. Comparing animals and automobiles: a vehicle for understanding body size and abundance relationships in species assemblages? Oikos 66, 172–179.

Graney, R.L., J.H. Kennedy and J.H. Rodgers, Jr., 1994. Aquatic Mesocosm Studies in Ecological Risk Assessment. Lewis Publishers, Boca Raton, FL.

Herricks, E.E. and J. Cairns, Jr., 1976. The recovery of stream macrobenthos from low pH stress. Rev. Biol. (Lisb.) 10, 1–11.

Johnson, H.B. and H.S. Mayeaux, 1992. Viewpoint: a view on species additions and deletions and the balance of nature. J. Range Manage. 45, 322–333.

Karr, J.R., 1993. Measuring biological integrity: lessons from streams. In: Ecological Integrity and the Management of Ecosystems, edited by S. Woodley, J. Kay and G. Francis, St. Lucie Press, Delray Beach, FL, pp. 83–104.

Karr, J.R., K.D. Fausch, P.L. Angermeier, P.R. Yant and I.J. Schlosser, 1986. Assessing biological integrity in running waters a method and its rationale. Ill. Nat. Hist. Surv. Spec. Publ. 5, 1–28.

Kimball, K.D. and S.A. Levin, 1985. Limitations of bioassays: the need for ecosystem-level testing. BioScience 35, 165–171.

Lawton, J.H. and V.K. Brown, 1993. Redundancy in ecosystems. In: Biodiversity and Ecosystem Function, edited by E.-D. Schultze and H.A. Mooney, Springer-Verlag, Berlin, pp. 255–270.

Levine, S.N., 1989. Theoretical and methodological reasons for variability in the responses of aquatic ecosystem processes to chemical stresses. In: Ecotoxicology: Problems and Approaches, edited by S.A. Levin, M.A. Harwell, J.R. Kelly and K.D. Kimball, Springer-Verlag, New York, pp. 145–180.

Lowe, R.L., 1974. Environmental requirements and pollution tolerance of freshwater diatoms, EPA 670/4-74-005. U.S. Environmental Protection Agency, Cincinnati, OH.

Lugthart, G.J. and J.B. Wallace, 1992. Effects of disturbance on benthic functional structure and production in mountain streams. J. North Am. Benthol. Soc. 11, 138–164.

Metcalfe, J.L., 1989. Biological water quality assessment of running waters based on macroinvertebrate communities: history and present status in Europe. Environ. Pollut. Ser. A 60, 101–139.

Naeem, S., L.J. Thompson, S.P. Lawler, J.H. Lawton and R.M. Woodfin, 1994. Declining biodiversity can alter the performance of ecosystems. Nature 368, 734–737.

National Research Council (NRC), 1992. Restoration of Aquatic Ecosystems. National Academy Press, Washington, DC.

Odum, E.P., 1985. Trends expected in stressed ecosystems. BioScience 35, 419–422.

Orians, G.H. and W.E. Kunin, 1990. Ecological uniqueness and loss of species. In: The Preservation and Valuation of Biological Resources, edited by G.H. Orians, University of Washington Press, Seattle, WA, pp. 146–184.

Overbeck, J. and R.J. Chrost, 1990. Aquatic Microbial Ecology. Springer-Verlag, New York.

Patrick, R., 1961. A study of the numbers and kinds of species found in rivers in the eastern United States. Proc. Acad. Nat. Sci. Philadelphia 113, 215–258.

Pimm, S.L. 1993. Biodiversity and the balance of nature. In: Biodiversity and Ecosystem Function, edited by E.-D. Schultze and H.A. Mooney, Springer-Verlag, Berlin, pp. 347–359.

Plafkin, J.L., M.T. Barbour, K.D. Porter, S.K. Gross and R.M. Hughes, 1989. Rapid bioassessment protocols for use in streams and rivers: benthic macroinvertebrates and fish, EPA 444/4-89-001. Office of Water, Washington, D.C.

Pratt, J.R., 1990. Aquatic community response to stress: prediction and detection of adverse effects. In: Aquatic Toxicology and Risk Assessment, Vol. 13, edited by W.G. Landis and W.H. van der Schalie, American Society for Testing and Materials, Philadelphia, PA, pp. 16–26.

Pratt, J.R. and N.J. Bowers, 1990. A microcosm procedure for estimating ecological effects of chemicals and mixtures. Toxicol. Assess. 2, 417–436.

Pratt, J.R. and N.J. Bowers, 1992. Variability of community metrics: detecting changes in structure and function. Environ. Toxicol. Chem. 11, 451–457.

Pratt, J.R. and E.P. Smith, 1991. Significance of change in community structure: a new method for testing differences. In: Biological Criteria: Research and Regulation, EPA-440/5-91-005, Office of Water, Washington, D.C., pp. 91–103.

Pratt, J.R. and J.L. Rosenberger, 1993. Community change and ecosystem functional complexity: a microcosm study of copper toxicity. In: Environmental Toxicology and Risk Assessment, Vol. 2, edited by J.W. Gorsuch, F.J. Dwyer, C.G. Ingersoll and T.W. LaPoint. American Society for Testing and Materials, Philadelphia, PA. pp. 88–102.

Pratt, J.R., J. Cairns, Jr. and R. Horowitz, 1987. Protozoan communities of the Flint River (Georgia, USA). Hydrobiologia 148, 159–174.

Pratt, J.R., N.J. Bowers, B.R. Niederlehner, and J. Cairns Jr., 1988. Effects of atrazine on freshwater microbial communities. Arch. Environ. Contam. Toxicol. 17, 449–457.

Pratt, J.R., N.J. Bowers and J.M. Balczon, 1993. A microcosm using naturally-derived communities: comparative ecotoxicology. In: Environmental Toxicology and Risk Assessment, edited by W.G. Landis, J.S. Hughes and M.A. Lewis, American Society for Testing and Materials, Philadelphia, PA. pp. 178–191.

Preston, F., 1948. The commonness and rarity of species. Ecology 29, 254–283.

Reice, S.R. and M. Wollenberg, 1993. Monitoring freshwater benthic macroinvertebrates and benthic processes: measures for assessment of ecosystem health. In: Freshwater Biomonitoring and Benthic Macroinvertebrates, edited by D.M. Rosenberg and V.H. Resh, Chapman and Hall, New York, pp. 287–305.

Rosenberg, D.M. and V.H. Resh (Eds.), 1993. Freshwater Biomonitoring and Benthic Macroinvertebrates. Chapman and Hall, New York.

Schaeffer, D., E.E. Herricks and H.W. Kerster, 1988. Ecosystem health. I. Measuring ecosystem health. Environ. Manage. 12, 445–455.

Schindler, D.W., 1987. Detecting ecosystem responses to anthropogenic stress. Can. J. Fish. Aquat. Sci. 44(Suppl.1), 6–25.

Schindler, D.W., 1990. Experimental perturbations of whole lakes as test of hypotheses concerning ecosystem structure and function. Oikos 57, 25–41.

Schoener, T., 1983. Rate of species turnover decreases from lower to higher organisms: a review of the data. Oikos 41, 372–377.

Schultze, E.D. and H.A. Mooney. 1993. Ecosystem function of biodiversity: a summary. In: Biodiversity and Ecosystem Function, edited by E.-D. Schultze and H.A. Mooney, Springer-Verlag, Berlin, pp. 497–510.

Shannon, C.E. and W. Weaver, 1948. The Mathematical Theory of Communication. University of Illinois Press, Urbana, IL.

Tilman, D. and J.A. Downing, 1994. Biodiversity and stability in grasslands. Nature 367, 363–365.

Turner, M.G., E.P. Odum, R. Costanza, and T.M. Springer, 1990. Market and nonmarket values of the Georgia landscape. Environ. Manage. 12, 209–217.

USEPA (U.S. Environmental Protection Agency), 1989. Short-term methods for estimating the chronic effects of effluents and receiving waters to freshwater organisms, 2nd ed., EPA 600/4-89-001. Environmental Monitoring and Systems Laboratory, Cincinnati, OH.

USEPA (U.S. Environmental Protection Agency), 1992. Framework for ecological risk assessment, EPA/630/R-92/001. Risk Assessment Forum, USEPA, Washington, DC.

Vannote, R.L., G.W. Minshall, K.W. Cummins, J.R. Sedell and C.E. Cushing, 1980. The river continuum concept. Can. J. Fish. Aquat. Sci. 37, 130–137

van Voris, P., R.V. O'Neill, W.R. Emanuel and H.H. Shugart, 1980. Functional complexity and ecosystem stability. Ecology 61, 1352–1360.

Vogl, S., 1980. Perturbation dependent ecosystems. In: The Recovery Process in Damaged Ecosystems, edited by J. Cairns, Jr., Ann Arbor Science, Ann Arbor, MI, pp. 63–94.

Walker, B., 1991. Biodiversity and ecological redundancy. Conserv. Biol. 6, 12–23.

Weber, C.I. (Ed.), 1973. Biological field and laboratory methods for measuring the quality of surface waters and effluents, EPA-670/4-73-001. U.S. Environmental Protection Agency, Cincinnati, OH.

Wilhm, J.L. and T.C. Doris, 1968. Biological parameters for water quality criteria. BioScience 18, 477–481.

CHAPTER 13

Ecosystems and Ecotoxicology: A Personal Perspective

David W. Schindler

I. INTRODUCTION

In my opinion, whether the goal of ecotoxicology is to protect whole ecosystems or important species and processes, it is necessary to place ecotoxicology studies in an ecosystem-level context. Yet, as several authors have pointed out (most recently Cairns et al., 1993; see also Levin et al., 1989), ecosystem criteria have been seldom considered in developing the bioassays that have been widely employed for managing water quality or analogous measures of "ecosystem integrity" or "ecosystem health." This point was recently made by several papers in Levin et al. (1989), and I agree with the conclusions of those works. Consequently, I will devote this chapter to my own thoughts about what new directions might be taken by ecotoxicology at the ecosystem level, rather than performing yet another literature review.

At the ecosystem level, the practice of ecotoxicology has goals that are more or less similar to studies of "ecosystem health," "ecosystem risk," "ecosystem stress," or "ecosystem integrity" (for example, Rapport et al., 1985; Schaeffer et al. 1988, Levin, 1989, Rapport, 1989, Karr, 1991, Cairns et al., 1993, and Suter, 1993). Throughout these papers, there is considerable debate as to whether the degree of human stress on ecosystems can be expressed simply in the form of indices or requires more complex forms of analysis. I intend to avoid this debate, except to remark that the term "quantitative" must be more meaningfully used than to assign numerical values to factors that are not readily amenable to quantification. I shall use the term "ecosystem stresses" to refer to both insults to ecosystems and their effects, as has been done in the past.

It is unrealistic to expect that complex ecosystem-level assessments will ever become a part of routine ecotoxicological procedures. The vast numbers of environmental stresses that must be monitored, the high cost and long duration of ecosystem-scale tests, and the problem of assessing synergisms, antagonisms, and other interactive effects dictate that, for the foreseeable future, the prediction of ecosystem stress must rely heavily on bioassays and monitoring at simpler scales.

This does not downplay the importance of ecosystem-level tests. I believe that a small number of controlled, whole ecosystem studies can provide an invaluable service by allowing us to calibrate and compare lower-level "indicators of stress," whether they be bioassays or monitoring methods, to complex ecosystem responses under controlled stresses, such as experimental lakes (Schindler, 1987, 1988, 1990a, 1991) or simple ecosystem manipulations done for nonexperimental reasons (see examples in Mooney et al., 1991). Such calibrations enhance our ability to correctly interpret what simpler indicators tell us about toxic threats to ecosystems, or to species in whole-ecosystem settings.

The results of several such comparisons done in the Experimental Lakes Area (ELA) have been described, along with results of a number of other settings of comparative simplicity (Schindler, 1987, 1988, 1990a, 1991). Those studies, are updated here, with some speculation about other ecosystem settings and possible courses of action that might be taken to calibrate or test methods used to detect ecosystem stress, and some opinions about indicators of ecosystem-level toxicity that might, when monitored, prove useful as supplements or replacements for more traditional toxicological approaches.

II. A DEFINITION OF ECOSYSTEM ECOLOGY

Despite Tansley's (1935) precise definition of an ecosystem as "... not only the organism-complex, but also the whole complex of physical factors ... the habitat factors in the widest sense" and his admonition that "we cannot separate them (organisms) from their physical environment," it is important to recognize that practitioners of ecology have largely used the term "ecosystem ecology" to refer to studies of ecosystem functions such as photosynthesis, respiration, biogeochemical cycling and, occasionally, general structural attributes. Consideration of direct interactive effects among species has been left as the province of "community ecologists." I have never been able to see the logic in such separation, because most ecosystem functions are complex manifestations of community structure (Schindler, 1985, 1988). I have been able to avoid it by studying lakes, where practical studies of both communities and ecosystems can be done simultaneously, for reasons described briefly below and in more detail elsewhere (Schindler, 1988, 1995). There is now widespread recognition of the need to link community and ecosystem studies (Jones and Lawton, 1995). I propose that in the future, a term like "functional ecology" or "process ecology" be used to refer to studies that focus exclusively on function, and that the term "ecosystem ecology" be reserved for studies of communities and their relation to abiotic components, in line with Tansley's original definition. In this chapter, the term ecosystem is used as Tansley originally defined it.

III. THE CONCEPTUAL BASIS FOR EVALUATING INDICATORS OF STRESS IN ECOSYSTEMS

The scarcity of past ecosystem studies, where both communities and ecosystem functions have been studied simultaneously, must be viewed as one of the tragedies of how ecology has developed. As a result of the dichotomy between community ecology and functional ecology, it is difficult to compare directly the sensitivity of responses of communities, populations, or other "structural" or taxonomic indicators of stress on ecosystems with responses in functions like nutrient cycling or productivity or respiration. It is also difficult to compare responses across different ecosystem types with confidence. Early predictions of general ecosystem-level effects of human-imposed stresses were made by Odum (1985), based on results drawn from a large number of ecosystem types, which were subjected to different types of stresses. The individual studies that Odum reviewed did not cross the community ecology-ecosystem ecology rift mentioned above, i.e., none were truly whole-ecosystem studies.

Despite these limitations, when Odum's predictions were tested by examining their applicability to experimental lakes, where both community structure and several ecosystem functions were monitored simultaneously under deliberate, controlled perturbation regimes, most of them were verified (Odum, 1990; Schindler, 1990a). However, a disturbing number of the responses in experimental lakes were also specific to either particular stresses or particular ecosystems (Schindler, 1990a), indicating that diagnoses must be made with caution.

In the past few years, we have recognized the importance of detecting stresses before ecosystems are harmed. It is clear that such preventive ecosystem management must become the focus of future applied ecology, for our record at restoration and amelioration of damaged ecosystems is so pathetic and expensive that it virtually excludes the possibility of "ecologically sustainable" development. In multiply stressed ecosystems, it is also necessary to consider which ecosystem components are most sensitive to particular stresses, or whether the effects of combinations of stresses can be evaluated. If cost is a consideration, and results are to be compared across many ecosystems and over long time periods, simplicity and intercomparability of methods are also important considerations.

A. ADVANTAGES OF LAKES

Small lakes are unquestionably the easiest of all common ecosystems to study thoroughly, because of the relative ease with which physical, chemical, and biological processes can be monitored, and the relatively short life spans and comparatively well-known taxonomy of most of the important groups of organisms. A small team of investigators, using rather simple, often homemade equipment, is therefore able to conduct studies that are truly ecosystem-scale. These characteristics make lakes ideal pilot ecosystems to compare simple indicators with ecosystem-scale responses to perturbation.

The sediments of lakes also contain historical records that allow us to investigate past changes in many biological and geochemical features, in addition to chronological records of deposition of contaminants. Because lakes generally act as "sewers" for their catchments, many of the changes in terrestrial communities are also recorded in lake sediments, making them excellent sites for investigation of ecological integrity on a landscape scale. Sediments also allow us the luxury of using hindcasting to test ecological models before attempting to use them to predict future events: if the past communities can be predicted by such models, there is considerable confidence that they can be used for predicting future communities. There is a price: we must learn to read the sediments, whose text is more complicated than any contemporary or past human language. As a result, paleoecology has seldom been used as a tool in ecosystem management, and seldom calibrated or interfaced with other types of ecological studies (Schindler, 1987, 1991; Smol, 1990, 1992). Particular examples of how paleoecology can be used as an "ecotoxicological" tool are discussed later in this chapter.

B. MONITORING APPROACHES

Monitoring is often regarded as the thoughtless, pedestrian measurement of a routine selection of parameters. But well-designed monitoring programs, based on measurements known to indicate specific stresses, can be among the most sensitive of ecosystem-level indicators of ecosystem stress. If variables sensitive to perturbation are known to recover quickly after perturbation ceases, sensitive monitoring programs can be used for preventive, as well as retroactive ecosystem management. In past papers (Schindler, 1987, 1990a), I have reviewed population-, community-, and ecosystem-level indicators of ecological stress in lakes, based heavily on whole-ecosystem experiments with acidification and eutrophication at the Experimental Lakes Area (ELA). To summarize my observations briefly, both sensitive individual species and ecological communities have usually, though not always, been more sensitive to stress than most ecosystem functions, such as nutrient cycling and production, or respiration (Table 1). This conclusion was also reached in reviews of aquatic ecology by Ford (1989) and Levine (1989). While there are few cases in terrestrial systems where both community structure and ecosystem functions have been studied simultaneously, Vitousek (1990) and McNaughton (1977) have reached conclusions similar to mine, as have Rapport et al. (1985). Cairns et al. (1993) also review the advantages and disadvantages of various measures of ecosystem stress. As discussed later, some recently devised physiological analyses appear to be more sensitive than even sensitive species or communities.

Table 1   A Comparison of the Sensitivity Indicators of Ecosystem Stress to Acidification of Whole Lakes with H$_2$SO$_4$

	pH				
	6.5	6.0	5.5	5.0	4.5
Sensitive species decline or disappear		--- ♦ ---	*Mysis, Pimephales, Hyalella*		
Fish cease reproducing		♦ Fat-head minnow	♦ Lake trout, Pearl dace	♦ Sucker	
Species diversity declines				♦ Phytoplankton, chironomids	
Major food web disruptions		-------- ♦ --------			
Fish mortalities increase			♦ *Pimephales*	♦ Lake trout	
Conspicuous algal mats				♦ Mougeotia, Zygnema	
Increase in average phytoplankton size				♦ Dinoflagellates	
Phytoplankton production		No decline			
Winter respiratory declines				♦	
Ecosystem P/B		No change			
Ecosystem R/B increases				♦	
Ecosystem P/R decreases				♦	
Periphyton production decreases	♦				
Periphyton respiration increases	♦				
Periphyton P/R decreases	♦				
Σ Nitrogen increases				♦	
Nitrification ceases		♦			
Net internal alkalinity generation ceases					♦

Data from studies in Lakes 223 and 302S; modified from Schindler, 1991

However, it is still not clear how much sensitivity is justified for ecosystem protection, so for the examples described below it is assumed simply that changes in populations of single species, communities, or ecosystem functions are acceptable thresholds for detecting undesirable ecosystem responses to toxic conditions. All are examples from my own experience.

i. Indicator Populations

Indicator species and populations must usually be carefully chosen to be tailored to specific types of stresses. Caswell (Chapter 9) treats general considerations. Some organisms seem to be peculiarly susceptible to a variety of stresses. Where these "born losers" occur as "keystone" species in aquatic communities, their disappearance is of special concern. For example, *Mysis relicta* (Mysidaceae, Crustacea) is among the most sensitive species to acidification, disappearing at pH values of 6 or greater (Nero and Schindler, 1983). It is also very susceptible to high cadmium, high temperature (>16°C or so), and predation by planktivorous fish (Schindler, 1995). It occupies a "niche" where no other species perform the same function. As a voracious zooplanktivore, its elimination from natural habitats, or addition to lakes where it did not occur naturally, cause dramatic changes in the size-structure of zooplankton populations, which in turn reshape the phytoplankton community via trophic cascading (Carpenter et al., 1985). It is a key food for small lake trout in many lakes (Trippel and Beamish, 1993), and its disappearance requires a major reorientation of feeding strategies for that species. Organisms with such characteristics appear to be tailor-made sentinels for early detection of ecosystem stress. However, caution must be exercised, for *Mysis* disappear quickly when its tolerances are exceeded (Nero and Schindler, 1983). Once extirpated, *Mysis* appears to have poor powers of recovery (Schindler et al., 1991). This vulnerability decreases its value as a "sentinel" for preventive action. In this regard, shifts in species with better recovery powers, such as rotifers or daphnids, may be more practical to use in preventive monitoring (Schindler et al., 1991).

ii. Lacustrine Communities as Indicators of Ecosystem Stress

Other chapters in this volume (10, 11, and 12) discuss a number of ways in which communities can be affected by toxicants. The sensitivity of various components of aquatic communities to different stresses is well demonstrated by the success of community analysis in contemporary monitoring (Patrick, 1949; Wiederholm, 1992) and in paleoecology (Warwick, 1980; Smol, 1992). Stresses as diverse as toxic contaminants, acidification, climate change, watershed clearing, and eutrophication have been shown to produce reliable, predictable responses in community composition, both in contemporary and paleoecological studies.

Moreover, many of the community responses appear to be replicable among ecosystem types. For example, the changes in algal and invertebrate communities of acidified lakes are very similar, whether the lakes have been acidified by atmospheric sulfur deposition or deliberate experimentation (Schindler et al., 1991).

iii. Ecosystem Processes and Food Chain Functions as Indicators of Stress

Many of the ecosystem processes in lakes appear to be quite robust, for there is "functional redundancy" built into communities, i.e., species able to tolerate new stresses tend to replace those that are poorly adapted to the new conditions, allowing functions to continue as normal. Examples are seen in the phytoplankton and zooplankton communities of acidified lakes, which maintain normal biomass and production over a manyfold increase in acidity (Schindler, 1987, 1991). There are some exceptions to this general rule. For example, the decreases in production, increases in respiration, and, particularly, declines in production/biomass (P/B) and production/respiration (P/R) in benthic algal communities of acidified lakes were detectable at very early stages of acidification (M. A. Turner, personal communication; Turner et al., 1995). It is noteworthy that this functional change was accompanied by a rapid shift in algal species to domination by filamentous chlorophytes, so that community structure was still as sensitive as ecosystem function as an indicator of stress. There were also a number of functions of intermediate sensitivity, including chironomid emergence, nitrification, and respiration under winter ice. Reduced chironomid emergence was accompanied by a dramatic shift in community composition, so that it was still less sensitive than community structure (I. J. Davies, personal communication). No attempts were made to identify the bacteria involved in nitrification and winter respiration, so that the relative sensitivity of structure and function cannot be compared.

Functional redundancy also applies to community interactions, such as predator-prey relationships. Unless there is some dramatic difference in size, behavior, or palatability of the replacement species, a change in species will probably scarcely affect the functioning of a food chain. For example, lake trout in Lake 223 appeared to switch to pearl dace (*Semotilus margarita*) without any difficulty once acidification had eliminated the original dominant forage fish, the fathead minnow (*Pimephales promelas*). After pearl dace were decimated too, lake trout starved (Schindler et al., 1985), but lake trout in communities with higher diversity of forage fishes would probably have suffered less damage (see below). It follows that while impairment or elimination of ecosystem functions would certainly be cause for alarm, early warning indicators in aquatic populations or communities are probably sensitive enough to allow us to protect most ecosystem functions, if we monitor them closely.

iv. Functional Redundancy and Early Indicators of Stress

Functional redundancy is not a new concept. While not explicitly stated, it was clearly recognized and understood as fundamental to developing effective ecological management strategies. To quote Leopold (1944), in commenting on how humans affected the land: "presumably the greater the losses and alterations (to communities), the greater the risk of impairments and disorganizations." However, demonstrated examples of such links are rare, and probably of varying importance and sensitivity among ecosystem types.

The case of lake trout (*Salvelinus namaycush*) in Lake 223 is described briefly above. In this case, the top predator depended for food upon a small handful of cyprinid and crustacean species, all of which were relatively sensitive to acidification (Schindler et al., 1985).

However, a similar acidification experiment performed a few years later, in Lake 302S, gave a somewhat different response. The food chain was topped by lake whitefish (*Coregonus clupeaformis*). *Mysis relicta* and *S. namaycush* were not present, although *P. promelas*, *S. margarita*, and *Catostomus commersoni* were. Other elements of the plankton, invertebrate, and benthic algal communities of the two lakes were very similar, and responses to acidification closely resembled those observed in Lake 223 (Schindler et al., 1991). However, whitefish seemed less sensitive to changes in forage species than trout. This is to be expected, for they are zooplanktivores and benthivores, and forage fish are not an important part of their diet. Whitefish did not begin to decline until pH values were low enough to affect their reproduction directly. In Little Rock Lake, which had a more diverse, warm water fish assemblage, the general response was similar to that in Lake 302. It is noteworthy that, with the exception of fish, the responses of communities in all three lakes were remarkably similar, with species of algae, zooplankton, and benthic assemblages responding in nearly identical fashion. The responses were also very similar to those observed in lakes acidified by the smelters at Sudbury, Ontario (Schindler et al., 1991). In summary, it appears that, although the responses of lower trophic levels to acidification are remarkably predictable, the responses of upper levels are determined by the community diversity and the sensitivities of particular species to an applied stress. Simple energetic considerations dictate that the most likely trophic level to display low functional redundancy will be that of top predator, where there are seldom more than a few species present. The repercussions of changes in top predators via the "trophic cascade" are well known (Paine, 1966; Carpenter et al., 1985; Power, 1990). It follows that considerable emphasis in "ecotoxicology" should be placed on top predators. I shall come back to this later, to suggest some changes in the way that we have traditionally monitored the "health" of predator populations.

However, ELA acidification experiments show that components of northern aquatic communities other than top predators may also have low functional redundancy. While the low redundancy at the top of the trophic pyramid is the result of energy limitations, at lower trophic levels it appears more likely to occur in simple communities that are limited by harsh environments, poor nutrient supplies, short evolutionary times, or perhaps previous impoverishment by human or other stresses (Schindler, 1990b). It has always astounded me that northern ecosystems, with a few hundred species, appear to carry out the same basic ecosystem and foodweb functions as tropical systems that are orders of magnitude more complex. It follows logically that in northern ecosystems, disruption of food web structures and ecosystem functions might occur with the elimination of relatively few invertebrate or algal species, or probably of a few microbial types (Schindler, 1987, 1990a). If such areas of weakness upon which in community/ecosystem structure can be located, they are obvious points upon which to focus attention in ecotoxicology. In particular, the exceptionally low diversity of lakes of high altitude, latitude, or salinity, and in other harsh environments, renders them more vulnerable to disruption by anthropogenic stressors, simply because of the high likelihood that low functional redundancy will occur at several points in food webs and ecosystem functions. If this hypothesis is correct, it follows that extirpation of a few species from a northern ecosystem may imperil key community interactions and ecosystem functions more than elimination of a large number of species from older, more productive and diverse temperate or tropical communities.

v. How Many Species Do We Really Need?

Recently, it has been suggested that only a few species are needed to maintain key ecosystem functions (Baskin, 1994a, b). A number of ecologists have theorized that other species are simply "along for the ride" (Walker, 1992), suggesting that we could do without much of the diversity that we find in natural ecosystems. My observations in acidified lakes at ELA, and some examples from the literature, leave me very uncomfortable with this notion because, in some cases, ecosystem functions decline even when "protected" by considerable redundancy. For example, the production and respiration of epilithic algal communities in acidified lakes decline dramatically, while 30 to 54 species are still present (M.A. Turner, personal communication). Also, most of the evidence cited (Baskin, 1994a) comes from consideration of simple, artificially manipulated tree communities where photosynthesis was the only function examined. A more broadly based, but still rather simple experiment in controlled environment chambers where 31 species representing 4 trophic levels were combined to test the link between diversity, respiration, and productivity, reached different conclusions. As diversity increased two- to threefold, productivity increased by similar amounts (Naeem et al., 1994). While several other functions were examined, but not affected, the studies clearly indicate that at least some ecosystem processes are dependent on diversity.

Similar examples are provided by the benthic communities of Lake 302S at ELA. At circumneutral pH values, the benthic algal flora (epilithon and metaphyton) contained about 80 species of several taxonomic groups. As the pH was lowered, the number of species that were common declined. Those remaining maintained positive net photosynthesis only in midsummer, and only for the shallower part of the littoral zone (Turner, 1993; Turner et al., 1995). At pH values of 5 and less, and below a depth of 3m in Lake 302, the rock surfaces appeared to have been scrubbed with a wire brush! Photosynthesis was insufficient to meet the respiratory demands of algae, and they disappeared. Even at shallower depths, filamentous blooms survived at great energetic cost, for photosynthesis was sufficient to exceed respiratory costs for only a few months of the year.

Acidification also changed the zoobenthic communities. The most complex group of benthic animals is the chironomids (Diptera), where a typical circumneutral softwater lake at ELA has 60 to 70 species emerging in a year, and where long-term studies have revealed several hundred species for the area (I.J. Davies, personal communication). Acidification from pH 6.5 to 5.0–5.1 reduced the total numbers of species present to 35–40. Moreover, many of the species present at low pH were previously undetectable or rare in the lake. Chironomid emergence shifted from a balance between a number of taxa to domination by a handful of species. At first glance, this appeared to lend support to the conclusion that only a few species are needed to maintain key ecosystem functions. However, upon closer examination, a very disturbing feature emerged: Not only did a few species constitute most of the emerging insects, but almost all were of one genus! In fact, 60 percent of the emergence was by *Cladotanytarsus aeiparthenus*, a species so rare that it was not previously described in the literature (Bilyj and Davies, 1989) and has so far not been found in other lakes at ELA! This suggests that rare species may provide "insurance" for lakes, allowing the integrity of communities to be maintained. Many examples of the importance of such "species banks" can be found in the documentation of the community changes in lakes as they acidify or recover (Schindler et al., 1991), or are exposed to other stressful conditions (Hairston and Munns, 1984; Schindler et al., 1989). The best known of these are the resting spores and eggs of many species of algae and invertebrates (Hairston and Munns, 1984; Hutchinson, 1967). One of the shortcomings of many laboratory toxicological studies is that they do not address the possibility of substitutions, for there are no "species banks" available to respond to perturbations, either within or external to the experimental containers. This problem has been addressed in recent multispecies laboratory studies, where microcosms have been designed to provide a suite of species large enough to allow measurements of shifts in dominance and redundancy (Cowgill and Williams, 1989). Mobile taxa in adjacent, unperturbed ecosystems can also provide stocks of organisms that may reinvade damaged systems once suitable conditions return.

Overall, several ecosystem functions and about half of the original species in the biotic community were lost as Lake 223 was acidified from pH values of 6.5–6.8 to 5.0–5.1. Only about one third of the species eliminated by acidification were replaced by more acid-tolerant analogs. Clearly, functional redundancy in this system is not high. Results from acidification studies of Lakes 302S and Little Rock Lake are similar (Schindler et al., 1991; Webster et al., 1992). Regional models based on these results suggest that hundreds of thousands of populations of fishes and millions of populations of invertebrates have been lost from the lakes of eastern Canada and the northeastern U.S. due to acid precipitation (Schindler et al., 1989; Minns et al., 1990, 1992). (A population is defined as one species in one lake.) In most lakes, the effect on functional redundancy and ecosystem functions is probably great, but on a regional scale (where nearly 700,000 lakes receive acidic precipitation), the loss is unquestionably of significance.

vi. Where Are the Reference Ecosystems?

Elsewhere, we have illustrated that even the communities and ecosystems of very remote arctic and alpine areas can be severely affected by humans (Schindler et al., 1995). For example, several algal taxa have disappeared from remote arctic ponds and lakes since the industrial revolution (Douglas et al., 1994). I am concerned that the disappearance of ecosystems in which full complements of organisms are present, and the increased reliance for the development of ecological principles upon modified or constructed "ecosystems," will lead us to underestimate the full range of resilience of ecosystems. We need studies of communities and ecosystems in near-pristine regions, while a few remain more or less intact, in order to know whether we are unwittingly losing important species gene pools, and functional redundancy.

vii. The Element of Surprise

There will always be hidden changes that, while very important, may be too complex to evaluate by any method except very complete ecosystem studies. Often, the elucidation of these unexpected features is serendipitous. For example, in-lake alkalinity generation (IAG) by microbial processes is an extremely important process in softwater lakes, minimizing the acidifying effect of inputs of strong acids and thus changes in communities and loss of diversity.

Yet it could not be detected until precise mass-balances could be performed for both base cations and strong acid anions, changes that reflect the importance of various geochemical and microbial processes in lakes (Cook et al., 1986; Schindler, 1986; Schindler et al., 1986). In softwater lakes where precipitation contains high concentrations of sulfate, the most important source of IAG is normally the reduction of sulfate to sulfides, followed by their storage in sediments as iron sulfides or organic sulfides. The most important areas of IAG generation are littoral sediments, at depths where oxygen is depleted but sulfate is supplied by diffusion from overlying water, typically within a few millimeters of the mud-water interface (Rudd et al., 1986).

However, extreme stages of acidification of Lake 302 (pH 4.5) brought even more surprises. IAG suddenly ceased! By pure coincidence, the mechanism responsible was discovered by two graduate students conducting coordinated studies of periphyton and sulfate reduction in the littoral zone. They found that the above-mentioned littoral mats of filamentous algae had anoxic conditions in the mats, allowing the anoxic boundary, and thus the site of IAG generation in summer to move above the sediment surface and into the algal mats. When the mats decayed during fall and winter, all of the reduced sulfur stored in algal mats was reoxidized, causing net annual IAG to be greatly reduced (Kelly et al., 1995) (Figure 1).

Such complex surprises seem even more likely to occur at catchment and landscape scales, where terrestrial and aquatic geochemical processes, and communities are inextricably intertwined. Two examples are given below.

Under average conditions, atmospheric inputs of sulfate to terrestrial and wetland ecosystems are partly removed by vegetation and soils (Bayley et al., 1986). However, when droughts occur, stored sulfur is reoxidized, so that subsequent rain showers cause acid pulses in streams draining the catchments (Bayley et al., 1992). Such events appear to delay the recovery of acidified lakes during droughts, even when acid deposition has been dramatically reduced (Dillon and Molot, 1994).

In Lake 302S, the number and biomass of emerging chironomids did not decrease as the result of experimental acidification, although there were changes in the species assemblages. As a result, the quantity of food for tree swallows, which rely heavily on emerging chironomids, was not affected. However, despite the constant food supply, reproduction by tree swallows failed as the result of acidification. Investigations revealed two possible reasons. Firstly, the calcium content of chironomids declined. In addition, the abundance of fish bones and mollusc shells, which are used as calcium supplements by swallows, declined as the species that produced them were eliminated with acidification. Together, these caused the diet of swallows to become calcium-deficient. As a result, eggs were smaller, hatching success was poorer, and nestlings were both smaller and slower growing than those of populations on nearby circumneutral lakes. Overall, the recruitment of swallows was severely impaired (St. Louis, 1992, 1993). Currently, while swallows can be seen feeding over the lake, they have avoided nesting on its shores for several years (V.L. St. Louis, personal communication). Secondly, the higher concentrations of

Normal

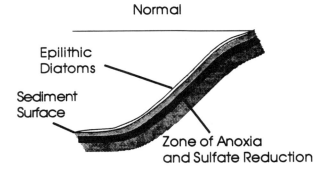

Epilithic
Diatoms

Sediment
Surface

Zone of Anoxia
and Sulfate Reduction

After Acidification to pH 4.5

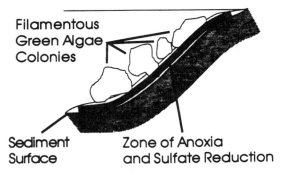

Filamentous
Green Algae
Colonies

Sediment
Surface

Zone of Anoxia
and Sulfate Reduction

Figure 1 A depiction of how the seasonal formation of filamentous algal mats disrupts the permanent generation of alkalinity in an acidified lake, by causing the zone of anoxia and sulfate reduction to move from the sediments into the overlying algal mat. Details are given in Kelly (1995).

toxic metals in the acidified lake caused increased concentrations in chironomids and in swallows (St. Louis et al., 1993).

In summary, we must expect that ecosystems will continue to provide surprises for us. Mother Nature's bag of dirty tricks is far too complicated to permit us to make predictions with 100% accuracy, even when the most cleverly conceived monitoring or ecotoxicological bioassay program is deployed. Good "ecotoxicological" approaches must therefore be flexible enough to deal with such problems as they are detected. Detection of surprise is likely to require community- or ecosystem-scale studies. Some responses may be ecosystem-specific, suggesting that considerable margins of safety should be included in setting regulatory limits, even when ecosystem-scale studies are used for assessment. It is noteworthy that astute field observations were important in discovering the above "surprises." The continuing de-emphasis of field studies and natural history in teaching modern ecology can be expected to significantly erode our ability to detect future surprises.

C. IF WE CANNOT MONITOR, WHAT SHOULD WE MEASURE?

If we are not to monitor all ecosystem processes and community character- istics, we should certainly attempt to select reliable early warning indicators. While there is no single prescription for what constitutes such an indicator, I believe that we can make some intelligent guesses as to what the most impor- tant present and future stresses on aquatic communities might be. From such predictions, astute community and ecosystem ecologists familiar with particu- lar ecosystem types should be able to predict many of the "hot spots" in aquatic ecosystems and communities. In earlier papers (Schindler, 1987, 1988), I have reviewed a number of methods and will not repeat them here. Below, I describe some examples of what I believe to be reliable indicators of various sorts of ecosystem stresses, as shown from retroactive studies. I propose that such indicators could be used proactively to detect impending problems at the ecosystem level.

i. Toxicants in Lake Sediments and Top Carnivores

In our past approach to the measurement of effects of airborne toxicants, much time and technology have been spent in attempting to measure astro- nomically small concentrations in air or water. In the case of some substances, detection limits are in the parts per trillion or even in the parts per quadrillion range! We have all read newspaper articles where various industries have announced with great fanfare that their emissions to air or water now contain "virtually undetectable" amounts of mercury, or dioxin, or whatever pollutant has been cause for past concern. Of course, they hope that the public at large will confuse "virtually undetectable" with "zero" or "not causing any prob- lems," which may or may not be the case. For example, many rivers below pulp mills have "virtually undetectable" concentrations of 2,3,7,8-TCDD (dioxin) in water, while carnivorous fish are highly contaminated.

Some characteristics of aquatic ecosystems and food chains are ready- made to facilitate assessment of such problems. Many pollutants, including several pesticides, dioxins, PCBs, and mercury, are bioaccumulated as they are passed up the food chain, or are deposited in lake sediments. Substances that are nearly impossible to detect in water may be easily detectable in top predators or lake sediments, where concentrations may be several orders of magnitude higher (Schindler et al., 1995) (Figure 2). In extreme cases, "virtually undetectable" concentrations of substances in water are concen- trated to the point where the bodies of slain top predators must be disposed of as toxic waste! It follows that the effects of contaminants on top trophic levels may be many times greater than at lower levels in the food chain, where many ecotoxicology programs are focused. Clearly, sediments and top carnivores are often where we will first see signs of ecosystem contamination.

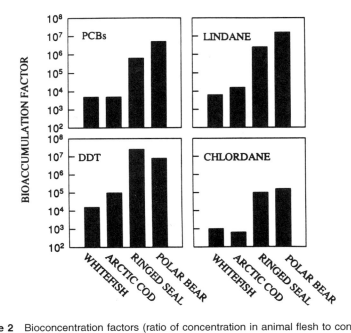

Figure 2 Bioconcentration factors (ratio of concentration in animal flesh to concentration in water) for a number of species harvested as food by aboriginal people in northern Canada. (From Schindler et al., 1995. Sci. Total Environ. 160/161, 1–17. With permission.)

ii. Screening with Stable Isotopes

Fortunately, it appears that toxicants are detectable in sediments and carnivores well before any effects on individual communities or ecosystems are discernible. As such, we should be able to use contaminant concentrations in an early warning capacity. Other studies suggest that there are ways to reduce the cost and increase the efficiency of monitoring top predators. Analyses for many contaminants are difficult, slow, and expensive. In addition, we may not know the true position of a species in a food chain. For example, we have found that lake trout in different lakes of the same area range from largely planktivorous and benthivorous to nearly totally piscivorous (Kidd et al., 1993).

Recently, stable isotopes of nitrogen have been shown to accurately predict an organism's trophic position and to be excellent predictors of contaminant concentrations (Cabana and Rasmussen, 1994; Kidd et al., 1995) (Figure 3). By concentrating our analyses on the highest trophic levels in lakes with the longest food chains, and keeping concentrations of toxicants there to safe levels, we can be reasonably certain that lower trophic levels and the food chains in lakes with shorter food chains will be safe. Stable isotope analyses are both less expensive and less time-consuming than analyses for many of the organic pollutants that are of widespread concern in the northern hemisphere, and they may provide a valuable screening mechanism, allowing us to focus

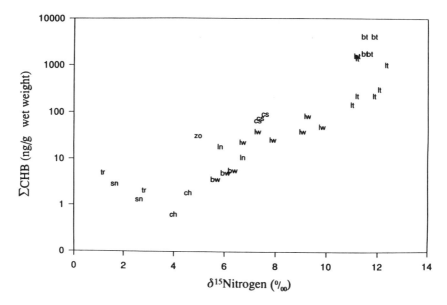

Figure 3   A plot of toxaphene concentration in several species from the food webs of Yukon lakes vs. their stable isotope concentration. Similar relationships are found for mercury and most organochlorines. (From Kidd et al.,1995. Sci. Total Environ. 160/161, 381–390. With permission.)

expensive analyses where they will first detect increasing, or dangerous, levels of pollutants.

iii. Sensitive Physiological Measurements

In recent years, some physiological measurements have also reached the point where they are at least as sensitive as most ecosystem processes. For pollutants that biomagnify, it appears that measurements could be made on predators, providing an early warning of impending problems at the ecosystem level. For example, the fishes, or fish-eating mammals that have been contaminated with PCBs, dioxins, or organochlorine pesticides, all show increasing concentrations of mixed function oxidase (MFO) enzymes in their liver (Lockhart et al., 1992). Concentrations of MFOs rise before any other signs of stress are noticeable, either in predator populations or in ecosystem functions. MFOs are believed to be linked to control of many important functions, such as steroid production and function of the immune system. Indeed, evidence of masculinization of females, feminization of males, and changes in secondary sex characteristics and earlier onset of sexual development in females have been observed in animals where MFOs are elevated (Bortone and Davis, 1994). The term *endocrine disruptor* has been used to denote factors that disrupt ontogenetic development (Colborn and Clement, 1992; Colborn et al., 1993; Bortone and Davis, 1994).

While the exact functions of MFOs are still debated, the long-term trends for sexual characteristics and immune-system functions in humans, wildlife, and populations of test animals suggest that elevated MFOs indicate that something is amiss in the test population (Colborn and Clement, 1992; Colborn et al., 1993). Cairns et al. (1993) also consider MFO to be a valuable early indicator of chemical stress.

An increase in the protein (or group of proteins) referred to as metallotheionine also appear to be a reliable indicator of increased exposure to trace metals in animals of several types (Webb and Cain, 1982), including aquatic species (St. Louis et al., 1993). The protein increases with exposure to a diversity of metals, including cadmium, mercury, and zinc. In an experimental addition of cadmium to Lake 382 over several years, elevated metallotheionine concentrations in many organisms was detectable well before any symptoms of stress to populations, communities, or ecosystem processes occurred (ELA staff, personal communications). The protein also accumulated in tree swallows that ate chironomid larvae from acid lakes, which contained elevated concentrations of several trace metals (St. Louis et al., 1993). It may be a reliable indicator of general exposure to toxic trace metals, which appear to be increasing globally due to human activity (Schindler et al., 1981; Nriagu and Pacyna, 1988).

In brief, knowledge of how food chains and ecosystems operate, how they are structured, and where chemical contaminants are likely to have their earliest effects can enhance our ability to apply simpler toxicological measurements in ways that are meaningful for ecosystem protection.

iv. Interpretation of the Sedimentary Record

As mentioned above, studies of the sedimentary record may provide us with a number of useful features of lakes, and of their catchments or surrounding landscapes. Often, they are our only means of assessing the composition, diversity, and variability of past communities.

In order to evaluate the changes in ecosystems caused by human industrial activities, it is necessary to be able to evaluate a variety of characteristics of ecosystems and how they have changed during the past two centuries. Fortunately, recent developments in coring, dating, and processing techniques have made such determinations possible (Smol, 1992).

Sediments often contain records of the past stresses on ecosystems. Changes in rates of storage of toxic organochlorines, trace metals, nutrients, and sulfur in sediments may indicate changes in toxicant burdens, eutrophication, or internal alkalinity generation (for example, Figure 4). Changes in pollen species, deposition of charcoal, or changes in sedimentation rates can indicate species successions, fire frequency, or erosion patterns in the catchment. Changes in the size, frequency, abundance, and long-term variation of insects, zooplankton, or phytoplankton often indicate changes in the trophic structure of aquatic communities, for example, the addition or deletion of predators.

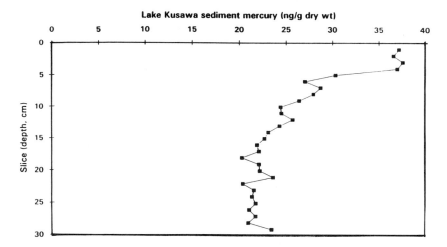

Figure 4   The concentration of mercury in sediments from a lake in the Yukon. Similar increases in concentration in near-surface sediments have been observed in lakes at all latitudes in Canada, reflecting increasing anthropogenic emissions of mercury to the atmosphere.

One of the major problems with evaluating the effects of stress is the lack of ecosystem data before ecosystems are stressed. Fortunately, we are often able to make such evaluations from the sedimentary record. Both presence/absence and natural variability in many taxa can be evaluated, forming a basis for comparison with either contemporary ecological studies or with more recent sedimentary strata, i.e., those deposited during or after stress.

Paleoecology also records some success stories: for example, the deposition of lead in sediments has decreased substantially since leaded gasoline was banned. PCBs and nutrients have also decreased in most lake sediments, at least in temperate regions. However, concentrations are generally still increasing in the arctic, due to a combination of long-range transport in the atmosphere and re-emission from temperate regions contaminated from such substances in earlier years (Schindler et al., 1995).

The fossils of aquatic animals and plants allow us to correlate chemical events with biological effects (or, occasionally, vice versa). Some examples are very clear. For example, the widespread use of toxaphene as a piscicide in the 1950s and 1960s wreaked havoc on invertebrate communities, a feature that was not assessed at the time (Miskimmin and Schindler, 1994) (Figure 5).

Other studies show recent changes in ecosystems, but the causes cannot be identified conclusively. Dramatic changes in the diatom flora of arctic lakes and ponds have occurred since the Industrial Revolution (Douglas et al., 1994). There are widespread increases in the incidence of deformities among fossil chironomids in 20th century lake sediments (Wiederholm, 1984; Warwick, 1985). These changes appear to indicate that communities are extraordinarily sensitivity to stress. But whether changes result from specific stress, are due to

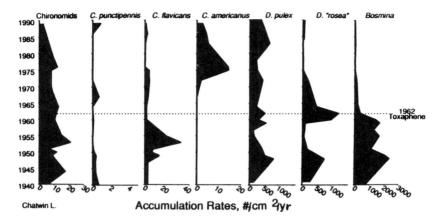

Figure 5 Changes in fossil remains left by several species of dipterans and crustaceans following treatment of Chatwin Lake, Alberta, with toxaphene as a piscicide in 1962. (From Miskimmin and Schindler, 1994. Can. J. Fish. Aquat. Sci. 51, 923–932. With permission.)

a multitude of stressors, or are another of Mother Nature's tricks, has not been conclusively resolved.

Biological stresses on ecosystems also leave their record in sediments. The stocking of fish into previously fishless lakes typically eliminates several large species of invertebrates. Some of these, such as chaoborids and cladocerans, leave fossils in sediments that are easily recognizable to species (Lamontagne and Schindler, 1994; Miskimmin and Schindler, 1994). Typically, changes in algal species or pigments also record such events, as the result of changes in nutrient inputs, grazing pressure, or nutrient recycling when layers are added or removed to food webs (Leavitt et al., 1994, 1995) (Figure 6).

While lake sediments can be used to reconstruct the past history for some taxonomic groups, many of the key organisms do not leave useful fossils. Such diverse groups as the Copepoda, Rotifera, Anostraca, as well as many insect and most fish species leave no record of their past presence, at least that we are able to use. However, new methods are constantly being developed. For example, recent advances in electron microscopy have allowed interpretation of chrysophycean spores (Duff et al., 1995). Copepod spermatophores and fecal pellets are identifiable in sediments. While not presently used, they have the potential to record not only the species of copepods present, but also what algal species they have been eating (H. Kling, personal communication).

One problem in interfacing paleoecological studies with predictions for the future is to ensure that interpretations of sediment records are correct. For example, Warwick (1980) demonstrated that a classical interpretation of changes in the chironomid fossils from the Bay of Quinte would indicate that the Bay had become more oligotrophic after the arrival of Europeans, while other evidence indicated that the Bay had become more eutrophic. The problem was resolved by incorporating historical records for the area, which showed that high rates of erosion caused by deforestation of the catchment had diluted

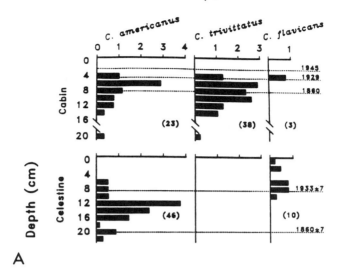

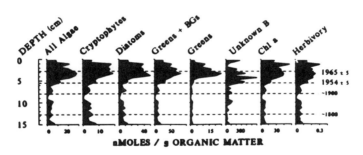

B   **Snowflake Lake**

Figure 6 Examples of changes caused by the introduction of salmonid fish into lakes of the Canadian Rocky Mountains. (A) Changes in the *Chaoborus* communities of two montane lakes in Jasper National Park following stocking with rainbow trout. Both Cabin and Celestine lakes lost *Chaoborus americanus*, a reliable indicator of fishless lakes, following fish stocking. Units are numbers of mandibles per gram of organic matter. Numbers in parenthesis are total numbers of mandibles retrieved per species. (From Lamontagne and Schindler, 1994. Can. J. Fish. Aquat. Sci. 51, 1376–1383. With permission.) (B) Changes in the plant pigments deposited in sediments of an alpine lake in Banff National Park following stocking with rainbow, brook, and cutthroat trout in the 1960s. Pigments used to characterize groups were β-carotene (all algae), alloxanthin (cryptophytes), diatoxanthin (diatoms), lutein-zeaxanthin (greens and cyanobacteria), pheophytin b (greens), unknown derivatives (Unknown B), undegraded chlorophyll *a* (Chl *a*), and total *a*-phorbin pheophorbides (herbivory). Exploratory energetics modeling suggests that increased recycling of phosphorus due to fish excretion probably caused the increase in phytoplankton abundance, resulting in higher pigment deposition. (Reprinted from data contained in Leavitt et al., 1995.) It is noteworthy that a similar relative change in pigment deposition resulted from increasing the phosphorus loading to Lake 227, ELA, by roughly 10-fold (Leavitt et al., 1994).

organic matter in sediments, giving it the characteristics of sediments in a more oligotrophic system. Recently, it has been possible to compare long-term limnological records with paleoecological studies in Lake 227 at ELA. For the most part, the interpretation of changes caused by 25 years of fertilization from sediments accurately reflects the monitoring record (Leavitt et al., 1994).

As mentioned earlier, the sediment records also allow us to identify impending problems, often well in advance. For example, the rapid and relentless increase over time in mercury in sediments of almost any lake in the northern hemisphere (Figure 4) indicates that we should be taking action now to control the mercury problem before it has widespread adverse effects. Due to methylation and bioconcentration, many top aquatic predators already contain concentrations that approach or exceed acceptable limits for human consumption. Increases in the atmosphere over the Atlantic Ocean have averaged 1.5% per year (Slemr and Langer, 1992).

The man-caused sources of mercury are diverse, ranging from incineration of wastes, fossil fuels, and smelting of ores (Schindler et al., 1981; Nriagu and Pacyna, 1988) to increased releases from flooded wetlands (St. Louis et al., 1994), and burning or deforestation (Velga et al., 1994). As a result, it will be difficult to control human-caused releases, for a complexity of control measures are needed.

D. SCALING UP: DOES ECOSYSTEM SIZE MATTER IN MAKING PREDICTIONS?

With the advent of remote sensing techniques, it has become fashionable to speak of the effects of stress at landscape scales, i.e., on ecosystems at the scale of biomes, continents, and planets (Holling et al., 1994). This may be important, for ecotoxicological studies are nearly always done on small systems. However, there are few clear-cut examples of how quantitative scaling to extrapolate ecological or ecotoxicological criteria to larger scales would be done.

i. Effects of Lake Size

In early treatments of the eutrophication problem, I and others assumed that the size of a lake would not change results, other than the general tendency for larger lakes to have longer water renewal times and hence react more slowly (Dillon and Rigler, 1975; Vollenweider, 1976; Schindler, 1978; Schindler et al., 1978).

However, many of the difficulties in applying the results of mesocosm results to lakes appear to result directly from improper or inadequate scaling (Schindler, 1988; Levine and Schindler, 1992). It is logical to expect similar differences caused by size-scales among lakes, for mixing processes, residence time of toxicants, and community complexity, to name a few examples, increase dramatically with lake size.

The Northern Ontario Lake Size Study (NOLSS) was designed to examine the effects of lake size in a manner that would allow results obtained in ELA lakes to be extrapolated to larger lakes (Fee and Hecky, 1992; Fee et al., 1994). A comparative measurement program was set up on lakes in Northwestern Ontario ranging in size over several orders of magnitude, up to and including Lakes Nipigon and Superior. The lakes have similar geological settings, climate, and catchment vegetation. In a given year, they are exposed to more or less similar weather conditions.

Results show that scale has important effects on extrapolation from small freshwater ecosystems to large ones. For example, the fish of small lakes in the same area had higher concentrations of methyl mercury than larger lakes, because the ratio of methylation to demethylation (M:D) is a function of epilimnion temperature (Bodaly et al., 1993). This is because methylation increases with increasing temperature, while demethylation decreases (Ramlal et al., 1993). Because detailed ecological studies, particularly ecosystem experiments, will almost always be carried out on small systems, whereas most management problems focus on large ones, the study of relationship of scales must receive high priority in applied ecology of all ecosystem types. Nutrient supplies and nutrient limitation were also functions of lake size (Fee et al., 1994; Guildford et al., 1994). Because detailed ecological studies, particularly ecosystem experiments, will almost always be carried out on small systems, while most management problems focus on large ones, the study of relationship of scales must receive high priority in applied ecology of all ecosystem types.

ii. Monitoring Landscape-Scale Stresses

It is also important to plan ecological management at landscape scales, where mosaics of several ecosystem types are represented. Satellite imagery and other remote sensing techniques, geographic information systems, and aircraft-mounted analyzers for gases allow us to determine the general shapes, uptakes, and emissions of the overall landscape. While well beyond the current scope of ecotoxicology, indicators of stress at landscape levels would be very useful indeed.

For wild ecosystems, it appears to me that large carnivores might fill such a role. As I have pointed out earlier, bioconcentration of toxicants by passage up food chains generally makes top carnivores the most contaminated members of aquatic communities (Figure 2). Also, as occupants of the top of the food chain, they generally have few competitors that could take over ecological functions if they were eliminated, i.e., changes in their abundance would have dramatic effects on communities and/or ecosystem functions.

Large carnivores generally have ranges of hundreds of square kilometers. Many, such as the large bears, use prey from both aquatic and terrestrial ecosystems. For example, polar bears travel long distances in the arctic, feeding largely on seals and walrus (I. Stirling, personal communication). They may be useful analogs to aboriginal populations who live in the same region, for they have diets that are relatively similar (Kinloch et al., 1992). At present, polar bears contain high concentrations of organochlorines, mercury, and other pollutants (Muir et al., 1992).

Polar bears, beluga whales, cormorants, gulls, ospreys, eagles, large salmon, and other carnivorous fish are among species that also integrate large landscapes into their activities. They contain high concentrations of pollutants throughout the northern hemisphere, including remote arctic regions (Landers and Christie, 1995). In many cases, these pollutant burdens have caused detectable changes in physiology and, in some cases, increases in fetal abnormalities or in cancers of various types. Clearly, these signs of stress in such species should be cause for great concern.

E. HUMANS AS ECOLOGICAL INDICATORS

We now recognize that humans are a part of ecosystems, and that principles of sustainable development must be practiced if we are to avoid disrupting ecological balances. Some have proposed that indices of ecological health are useful measures of human health (Haines et al., 1993; Bortone and Davis, 1994). Indeed, the similarities between the responses of humans and various wildlife species to pollutants or other environmental insults have been of great concern recently (Colborn et al., 1993; Haines et al., 1993). Humans may also serve as useful early-warning indicators of broad-scale ecosystem stress. Aboriginal human populations generally take food from the entire landscapes in which they live. In particular, northern populations tend to feed near the top of food chains, so that aboriginal humans contain high concentrations of lipophilic toxicants (Kinloch et al., 1992). In humans, we have an indicator species that is capable of telling us about its own ailments, as well as about changes in its environment. A number of native health studies are now underway in Canada, to monitor linkages between human health and environmental stress. The ecological and ecotoxicological implications of such studies should not be overlooked.

Many of the compounds of recent concern in both ecotoxicology and human health are endocrine disruptors. Somewhat similar signs of stress are being detected in both human and animal populations (Colborn and Clement, 1992; Colborn et al., 1993; Jeevan and Kripke, 1993; Bortone and Davis, 1994). Clearly, the relationships between human health and ecosystem health must receive closer scrutiny from both ecologists and the medical profession.

F. A PREVENTIVE APPROACH TO ECOSYSTEM MANAGEMENT

Another analog between human health and ecosystem health appears to have some utility. There has been an accelerating movement to replace the old "diagnose and excise" approach to medicine (which has many parallels to the reactive way in which we've managed ecosystems in the past) with "preventive medicine," or "preventive health care" (Mahler, 1986). In brief, most of us now engage in practices that can be shown to have general, statistically defensible positive effects on overall human health, even though detailed, mechanistic linkages to specific health problems cannot be made. Examples include avoidance of fatty and high cholesterol foods, consumption of vitamins, avoidance of stress, and exercise.

In strange contrast, in protecting ecosystem "health," it has generally been accepted that we must prove conclusively that chemical inputs or other stresses are having clear-cut, statistically defensible effects before regulatory action is taken. Certainly, given the human health experience, it is high time that a precautionary or preventive approach was taken to ecosystem management. The implications for ecotoxicology are obvious.

ACKNOWLEDGMENTS

This review was funded by National Sciences and Engineering Research Council of Canada operating grant number 89673 to D.W. Schindler. Review by John Cairns and three anonymous reviewers were greatly appreciated. Brian Parker and Margaret Foxcroft prepared the figures and a final draft of the manuscript.

REFERENCES

Atchison, G.J., M. Sandheinrich and M.D. Bryan, 1995. Effects of environmental stressors on interspecific interactions of aquatic animals. Chapter 11, in this volume.

Baskin, Y., 1994a. Ecologists dare to ask: how much does diversity matter? Science 264, 202-203.

Baskin, Y., 1994b. Ecosystem function of biodiversity. BioScience 44, 657-660.

Bayley, S.E., R.F. Behr and C.A. Kelly, 1986. Retention and release of sulfur from a freshwater wetland. Water Air Soil Pollut. 31, 101-114.

Bayley, S.E., D.W. Schindler, R. Parker, M.P. Stainton and K.G. Beaty, 1992. Effects of forest fire and drought on acidity of a base-poor boreal forest stream: similarities between climatic warming and acidic precipitation. Biogeochemistry 17, 191-204.

Bilyj, B. and I.J. Davies, 1989. Descriptions and ecological notes on seven new species of Cladotanytarsus (Chironomidae: Diptera) collected from an experimentally acidified lake. Can. J. Zool. 67, 948-962.

Bodaly, R.A., J.W.M. Rudd, R.J.P. Fudge and C.A. Kelly, 1993. Mercury concentrations in fish related to size of remote Canadian Shield lakes. Can. J. Fish. Aquat. Sci. 50, 980-987.

Bortone, S.A. and W.P. Davis, W.P., 1994. Fish intersexuality as indicator of environmental stress. Bioscience 44, 165-172.

Cabana, G.A. and J.B. Rasmussen, 1994. Modelling food chain structure and contaminant bioaccumulation using stable nitrogen isotopes. Nature 372, 255-257.

Cairns, J., Jr., P.V. McCormick and R. Niederlehner, 1993. A proposed framework for developing indicators of ecosystem health. Hydrobiologia 263, 1-44.

Carpenter, S.R., J.F. Kitchell and J.R. Hodgson, 1985. Cascading trophic interactions and lake productivity. Bioscience 35, 634-639.

Chesser, R. and D.W. Sugg, 1995. Toxicants as selective agents in population and community dynamics. Chapter 10, in this volume.

Colborn, T. and C. Clement (Eds.), 1992. Chemically-Induced Alterations in Sexual and Functional Development: The Wildlife/Human Connection. Princeton Scientific Publishing, Princeton, NJ.

Colborn, T., F.S. Vom-Saal and A.M. Soto, 1993. Developmental effects of endocrine-disrupting chemicals in wildlife and humans. Environ. Health Perspect. 101, 378-384.

Cook, R., C.A. Kelly, D.W. Schindler and M.A. Turner, 1986. Mechanisms of hydrogen ion neutralization in an experimentally acidified lake. Limnol. Oceanogr. 31, 134-148.

Cowgill, U.M. and L.R. Williams, 1989. Aquatic Toxicology and Hazard Assessment, Vol. 12, STP 1027. American Society for Testing and Materials, Philadelphia, PA., pp. 159-242.

Dillon, P.J. and L.A. Molot, 1994. Sulfate fluxes in lakes and catchments: interactions between reduced atmospheric deposition, reduced runoff and wetlands. Symposium on Regional Assessment of Freshwater Ecosystems and Climate Change in North America. Leesburg, VA, Oct. 24-26, 1994.

Dillon, P.J. and F.H. Rigler, 1975. A simple method for predicting the capacity of a lake for development based on lake trophic status. J. Fish. Res. Board Can. 32, 1519-1531.

Douglas, M.S.V., J.P. Smol and W. Blake, Jr., 1994. Marked post-18th century change in high arctic ecosystems. Science 266, 416-419.

Duff, K.E., A. Zeeb and J.P. Smol, 1995. Atlas of Chrysophycean Stomatocysts. Kluwer Academic Press, Dordrecht, The Netherlands, 189 pp.

Fee, E.J. and R.E. Hecky, 1992. Introduction to the Northwest Ontario lake size series (NOLSS). Can. J. Fish. Aquat. Sci. 49, 2434-2444.

Fee, E.J., R.E. Hecky, G.W. Regehr, L.L. Hendzel and P. Wilkinson, 1994. Effects of lake size on nutrient availability in the mixed layer during summer stratification. Can. J. Fish. Aquat. Sci. 51, 2756–2768.

Fee, E.J., J.A. Shearer, E.R. DeBruyn and E.U. Schindler, 1992. Effects of lake size on phytoplankton photosynthesis. Can. J. Fish. Aquat. Sci. 49, 2445-2459.

Ford, J., 1989. The effects of chemical stress on aquatic species composition and community structure. In: Ecotoxicology: Problems and Approaches, edited by S.A. Levin, M.A. Harwell, J.R. Kelly and K.D. Kimball, Springer Verlag, New York, pp. 100-144.

Frost, T.M., S.R. Carpenter, T.K. Kratz and A. Ives, 1994. Functional redundancy in the linkages between species and ecosystem processes. Sci. Total Environ. (in press).

Guildford, S.J., L.L. Hendzel, H.J. Kling, E.J. Fee, G.G.C. Robinson, R.E. Hecky and S.E.M. Kasian, 1994. Effects of lake size on phytoplankton nutrient status. Can. J. Fish. Aquat. Sci. 51, 2769–2783.

Haines, A., P.R. Epstein and A.J. McMichael, 1993. Global health watch: monitoring impacts of environmental change. Lancet 342, 1464-1469.

Hairston, N.G., Jr. and W.R. Munns, Jr., 1984. The timing of copepod diapause as an evolutionarily stable strategy. Am. Nat. 123, 733-775.

Holling, C.S., D.W. Schindler, W. Walker and J. Roughgarden, 1994. Biodiversity in the functioning of ecosystems: an ecological primer and synthesis. In: Biodiversity Loss: Ecological and Economic Issues, edited by C. Perrings, K.-G. Maler, C. Folke, C. S. Holling and B.-O. Jansson. Cambridge University Press, New York.

Hutchinson, G.E., 1967. A Treatise on Limnology, Vol. 2. John Wiley & Sons, New York.

Jeevan, A. and M.L. Kripke, 1993. Ozone depletion and the immune system. Lancet 342, 1159-1160.

Jones C.G. and J.H. Lawton, 1995. Linking Species and Ecosystems. Chapman and Hall, New York.

Karr, J.R., 1991. Biological integrity: a long-neglected aspect of water resource management. Ecol. Appl. 1, 66-84.

Kelly, C.A., J.A. Amaral, M.A. Turner, J.W.M. Rudd, D.W. Schindler and M.P. Stainton, 1994. Disruption of sulfur cycling and acid neutralization in lakes at low pH. Biogeochemistry 28, 115–130.

Kidd, K.A., J.E. Eamer and D.C.G. Muir, 1993. Spatial variability of chlorinated bornanes (toxaphene) in fishes from Yukon lakes. Chemosphere 27, 1975-1986.

Kidd, K.A., D.W. Schindler, R.H. Hesslein and D.C.G. Muir, 1995. Correlation between stable nitrogen isotope ratios and concentrations in biota from a freshwater food web. Sci. Total Environ. 160/161, 381-390.

Kinloch, D., H. Kuhnlein and D.C.G. Muir, 1992. Inuit foods and diet: a preliminary assessment of benefits and risks. Sci. Total Environ. 122, 247-278.

Lamontagne, S. and D.W. Schindler, 1994. Historical status of fish populations in Canadian Rocky Mountain lakes inferred from subfossil *Chaoborus* mandibles. Can. J. Fish. Aquat. Sci. 51, 1376-1383.

Landers, D.H. and S.J. Christie (Eds.), 1995. Ecological effects of arctic airborne contaminants. Sci. Total Environ. Special issue 160/161.

Leavitt, P.R., J. Hann, J.P. Smol, A. Zeeb, C.C. Christie, B. Wolfe and H.J. Kling, 1994. Analysis of whole-lake experiments with paleolimnology: an overview of results from Lake 227, Experimental Lakes Area, Ontario. Can. J. Fish. Aquat. Sci. 51, 2322-2332.

Leavitt, P.R., D.E. Schindler, A.J. Paul, A.K Hardie, and D.W. Schindler, 1995. Fossil pigment records of phytoplankton in trout stocked alpine lakes. Can. J. Fish. Aquat. Sci. 51, 2411-2423.

Leopold, A., 1944. Conservation: in whole or in part? Reprinted in: The River of the Mother of God and Other Essays by Aldo Leopold, edited by S.L. Flader and J. Callicott, University of Wisconsin Press, Madison, WI, 1991.

Levin, S.A., M.A. Harwell, J.R. Kelly and K.D. Kimball, 1989. Ecotoxicology: Problems and Approaches. Springer-Verlag, New York.

Levine, S.N., 1989. Theoretical and methodological reasons for variability in the responses of aquatic ecosystem processes to chemical stresses. In: Ecotoxicology: Problems and Approaches, edited by S.A. Levin, M.A. Harwell, J.R. Kelly and K.D. Kimball. Springer Verlag, New York, pp.145-179.

Levine, S. N. and D.W. Schindler, 1992. Modification of the N:P ratio in lakes by in situ processes. Limnol. Oceanogr. 37, 917-935.

Lockhart, W.L., R. Wagemann, B. Tracey, D. Sutherland and D.J. Thomas, 1992. Presence and implications of chemical contaminants in the freshwaters of the Canadian arctic. Sci. Total Environ. 122, 165-243.

Mahler, H., 1986. Towards a new public health. Health Promotion 1, 1.

McNaughton, S.J., 1977. Ecology of a grazing ecosystem: the Serengeti. Ecol. Monogr. 55, 259-294.

Minns, C.K., J.E. Moore, D.W. Schindler and M.L. Jones, 1990. Assessing the potential extent of damage to inland lakes in eastern Canada due to acidic deposition. IV. Predicting the response of potential species richness. Can. J. Fish. Aquat. Sci. 47, 821-830.

Minns, C.K., J.E. Moore, D.W. Schindler, P.G.C. Campbell, P.J. Dillon, J.K. Underwood and D.M. Whelpdale, 1992. Expected reduction in damage to Canadian lakes under legislated and proposed decreases in sulphur dioxide emission, Rpt. 92-1. Committee on Acid Deposition, Royal Society of Canada, Ottawa.

Miskimmin, B.M. and D.W. Schindler, 1994. Long-term invertebrate community response to toxaphene treatment in two lakes: 50-yr records reconstructed from lake sediments. Can. J. Fish. Aquat. Sci. 51, 923-932.

Mooney, H.A., E. Medina, D.W. Schindler, E.-D. Schulze and H. Walker, 1991. Ecosystem Experiments, Scope 45. John Wiley & Sons, New York.

Muir, D.C.G., R. Wagemann, T. Hargrave, T. Thomas, D. Peakall and R.J. Norstrom, 1992. Arctic marine ecosystem contamination. Sci. Total Environ. 122, 75-134.

Naeem, S., L.J. Thompson, S.P. Lawler, J.H. Lawton and R.M.Woodfin, 1994. Declining biodiversity can alter the performance of ecosystems. Nature 368, 734-737.

Nero, R.W. and D.W. Schindler, 1983. Decline of *Mysis relicta* during the acidification of Lake 223. Can. J. Fish. Aquat. Sci. 40, 1905-1911.

Nriagu, J.O. and J.M. Pacyna, 1988. Quantitative assessment of worldwide contamination of air, water and soils by trace metals. Nature 333, 134-139.

Odum, E.P., 1985. Trends expected in stressed ecosystems. Bioscience 35, 419-423.

Odum, E.P., 1990. Field experimental tests of ecosystem-level hypotheses. TREE 5, 204-205.

Paine, R.T., 1966. Food web complexity and species diversity. Am. Nat. 100, 65-75.

Patrick, R., 1949. A proposed biological measure of stream conditions, based on a survey of Conestoga Basin, Lancaster County, Pennsylvania. Proc. Acad. Nat. Sci. Philadelphia 101, 277-341.

Power, M.E., 1990. Effects of fish in river food webs. Science 35, 811-814.

Pratt, J.R. and J. Cairns, Jr., 1995. Ecotoxicity and the redundancy problem: understanding community structure and function, Chapter 12, this volume.

Ramlal, P.S., C.M. Kelly, J.W.M. Rudd and A. Furutani, 1993. Sites of methyl mercury production in remote Canadian Shield lakes. Can. J. Fish. Aquat. Sci. 50, 972-979.

Rapport, D. J., 1989. What constitutes ecosystem health? Perspect. Biol. Med. 33, 120-134.

Rapport, D.J., H.A. Regier and T.C. Hutchinson, 1985. Ecosystem behavior under stress. Am. Nat. 125, 617-640.

Rudd, J.W.M., C.A. Kelly and A. Furutani, 1986. The role of sulfate reduction in long-term accumulation of organic and inorganic sulfur in lake sediments. Limnol. Oceanogr. 19, 1281-1291.

Schaeffer, D.J., E.E. Herricks and H.W. Kerster, 1988. Ecosystem health. I. Measuring ecosystem health. Environ. Manage. 12, 445-455.

Schindler, D. W., 1978. Factors regulating phytoplankton production and standing crop in the world's freshwaters. Limnol. Oceanogr. 23, 478-486.

Schindler, D.W., 1985. The coupling of elemental cycles by organisms: evidence from whole-lake chemical perturbations. In: Chemical Processes in Lakes, edited by W. Stumm. Wiley-Interscience, New York, pp. 225-250.

Schindler, D.W., 1986. The significance of in-lake production of alkalinity. Water Air Soil Pollut. 30, 931-944.

Schindler, D.W., 1987. Detecting ecosystem responses to anthropogenic stress. Can. J. Fish. Aquat. Sci. 44 (Suppl. 1), 6-25.

Schindler, D.W., 1988. Experimental studies of chemical stressors on whole lake ecosystems. Verh. Int. Verein. Limnol. 23, 11-41.

Schindler, D.W., 1990a. Experimental perturbations of whole lakes as tests of hypotheses concerning ecosystem structure and function. Oikos 57, 25-41.

Schindler, D.W., 1990b. Natural and anthropogenically imposed limitations to biotic richness in freshwaters. In: The Earth in Transition: Patterns and Processes of Biotic Impoverishment, edited by G. Woodwell. Cambridge University Press, Cambridge, pp. 425-462.

Schindler, D.W., 1991. Whole lake experiments in the Experimental Lakes Area. In: Ecosystem Experiments, Scope 45, edited by H.A. Mooney, E. Medina, D.W. Schindler, E.-D. Schulze and H. Walker. John Wiley & Sons, New York, pp. 108-122.

Schindler, D.W., 1995. Linking species and communities to ecosystem management. In: Linking Species and Ecosystems, edited by C.G. Jones and J.H. Lawton, Chapman and Hall, New York, pp. 313-325.

Schindler, D. W., E.J. Fee and T. Ruszczynski, 1978. Phosphorus input and its consequences for standing crop and production in the Experimental Lakes Area and in similar lakes. J. Fish. Res. Board Can. 35, 190-196.

Schindler, D.W., T.M. Frost, K.H. Mills, et al. 1991. Freshwater acidification, reversibility and recovery: comparisons of experimental and atmospherically-acidified lakes. In: Acidic Deposition: Its Nature and Impacts, edited by F.T. Last and R. Watling, Proc. R. Soc., Edinburgh 97B, 193-226.

Schindler, D.W., S.E.M. Kasian and R.H. Hesslein, 1989. Biological impoverishment in lakes of the midwestern and northeastern United States from acid rain. Environ. Sci. Technol. 23, 573-579.

Schindler, D.W., K.A. Kidd, D.C.G. Muir, W.L. Lockhart, B.M. Miskimmin and P.J. Curtis, 1995. The effects of ecosystem characteristics on contaminant distribution in northern freshwater lakes. Sci. Total Environ. 160/161, 1-17.

Schindler, D.W., K.H. Mills, D.F. Malley, D.L. Findlay, J.A. Shearer, I.J. Davies, M.A. Turner, G.A. Linsey and D.R. Cruikshank, 1985. Long-term ecosystem stress: the effects of years of experimental acidification on a small lake. Science 228, 1395-1401.

Schindler, D.W., M.A. Turner, M.P. Stainton and G.A. Linsey, G.A., 1986. Natural sources of acid neutralizing capacity in low alkalinity lakes of the Precambrian Shield. Science 232, 844-847.

Schindler, D.W., M. Alexander, E.D. Goldberg, et al., 1981. Atmosphere-Biosphere Interactions. Toward a Better Understanding of the Ecological Consequences of Fossil Fuel Combustion. National Academy Press, Washington, D.C.

Slemr, F. and E. Langer, 1992. Increase in global atmospheric concentrations of mercury inferred from measurements of the Atlantic Ocean. Nature 355, 343-437.

Smol, J.P., 1990. Are we building enough bridges between paleolimnology and aquatic ecology? Hydrobiologia 214, 201-216.

Smol, J.P., 1992. Paleolimnology: an important tool for effective ecosystem management. J. Aquat. Ecol. Health 1, 49-58.

St. Louis, V.L., 1992. The effects of experimental lake acidification on the reproductive success of tree swallows, Ph.D. Thesis, University of Toronto, Toronto, Ontario.

St. Louis, V.L., 1993. Element concentrations in Chironomids and their abundance in the littoral zone of acidified lakes in Northwestern Ontario. Can. J. Fish. Aquat. Sci. 50, 953-963.

St. Louis, V.L., L. Breebaart, J.C. Barlow and J.F. Klaverkamp, 1993. Metal accumulation and metallotheionine concentrations in tree swallow nestlings near acidified lakes. Environ. Toxicol. Chem. 12, 1203-1207.

St. Louis, V.L., J.W.M. Rudd, C.A. Kelly, K.G. Beaty, N.S. Bloom and R.J. Flett., 1994. Importance of wetlands as sources of methyl mercury to boreal forest ecosystems. Can. J. Fish. Aquat. Sci. 51, 1065-1076.

Suter, G.W., II, 1993. A critique of ecosystem health concepts and indexes. Environ. Toxicol. Chem. 12, 1533-1539.

Tansley, A.G., 1935. The use and abuse of vegetational concepts and terms. Ecology 16, 284-307.

Trippel, E.A. and F.W.H. Beamish, 1993. Multiple trophic level structuring in *Salvelinus-Coregonus* assemblages in boreal forest lakes. Can. J. Fish. Aquat. Sci. 50, 1442-1455.

Turner, M.A., 1993. The ecological effects of experimental acidification upon littoral algal associations of lakes in the boreal forest. Ph.D. Thesis, University of Manitoba, Winnipeg.

Turner, M.A., D.W. Schindler, D.W. Findlay, M. Jackson and G.G.C. Robinson, 1995. Disruption of littoral algal associations by Experimental Lake acidification to pH 4.5. Can. J. Fish. Aquat. Sci. (in press).

Velga, M.M., J.A. Meech and N. Onate, 1994. Mercury pollution from deforestation. Nature 368, 816-817.

Vitousek, P. M., 1990. Biological invasions and ecosystem processes: towards an integration of population biology and ecosystem studies. Oikos 57, 7-13.

Vollenweider, R.A., 1976. Advances in defining critical loading levels for phosphorus in lake eutrophication. Mem. Ist. Ital. Idrobiol. 33, 53-83.

Walker, B.H., 1992. Biodiversity and ecological redundancy. Conserv. Biol. 6, 18-23.

Warwick, W.F., 1980. Paleoecology of the Bay of Quinte, Lake Ontario: 2800 years of cultural influence. Can. Bull. Fish. Aquat. Sci. 206, 1-117.

Warwick, W.F., 1985. Morphological abnormalities in Chironomidae (Diptera) larvae as measures of toxic stress in freshwater ecosystems: indexing antennal deformities in *Chironomus* Meigen. Can. J. Fish. Aquat. Sci. 42, 1881-1914.

Webb, M. and K. Cain, 1982. Functions of metallotheionine. Biochem. Pharmacol. 31, 137.

Webster, K.E., T.M. Frost, C.J. Watras, W.A. Swenson, M. Gonzalez and P.J. Garrison. 1992. Complex biological responses to the experimental acidification of Little Rock Lake, Wisconsin, U.S.A. Environ. Pollut. 78, 73–78.

Wiederholm, T., 1984. Incidence of deformed chironomid larvae (Diptera: Chironomidae) in Swedish lakes. Hydrobiologia 109, 243-249.

Wiederholm, T., 1992. Freshwater environmental monitoring in Sweden. Proposals from a working group. Naturvardsverket Rapport 4111, Naturvardsverket, Sweden.

CHAPTER 14

Summary

Carl L. Strojan

INTRODUCTION

One of the goals of the Savannah River Symposia on Environmental Science, of which this was the second, is to stimulate new ways of thinking so as to increase our understanding of how ecological systems function. In the case of this symposium, that sort of thinking began with the organization of the meeting itself. This symposium was undoubtedly the first time that ecotoxicology has been addressed in the context of a hierarchical approach encompassing geochemistry, cells, tissues, individuals, populations, communities, and ecosystems. That type of organization and approach brought together talented people with complementary interests who, because of their diverse backgrounds, would not likely cross paths. Those people were then challenged by the thought-provoking points set forth in Chapter 1 to help ecotoxicology develop more as a science that organizes and classifies knowledge based on explanatory principles and away from its current emphasis on regulatory procedures and descriptive data collection. The symposium organizers and participants are to be commended for making the format work.

One of the topics this symposium addressed is how contaminants affect various levels of ecological organization. Several points emerge when ecotoxicology is viewed from a hierarchical perspective. Two are the prevalence of interfaces and the importance of forging links between these interfaces. Ecotoxicology itself is an interface between ecology and toxicology. As one participant put it, "Ecotoxicology challenges scientists to integrate concepts and data across hierarchies." The preceding chapters have provided numerous examples of important interfaces in ecotoxicology. The most obvious are the interfaces between various levels of organization. Other important interfaces discussed in the preceding chapters, not all of which are unique to ecotoxicology, include:

1-56670-1127-9/95/$0.00+$.50

- Contaminants—receptors
- Organizational structure—biological function
- Ecosystem health—human health
- Experimental data—theoretical models
- Laboratory research—field studies
- Normal science—innovative science
- Qualitative descriptions—quantitative understanding
- Science—regulatory policy
- Ideas—technology

These interfaces are important to understanding processes and principles, not only in ecotoxicology, but in other disciplines as well.

Another point about a hierarchical approach is that it forces one to consider ecotoxicology from a broader perspective than one might otherwise do. This has obvious benefits in expanding one's horizons, but the approach creates problems as well by increasing the complexity that must be dealt with in a larger system. All of the complexity that exists at one level of organization is transferred to the next level, and so on through as many levels as exist. As complexity increases, there can be a temptation to focus efforts on topics that are simpler and more tractable, rather than developing new ideas and techniques that can address complex subjects that may be more important.

The hierarchical levels considered in this book range from the atomic and molecular on through to individuals, populations, communities, and ecosystems. One may argue whether these are somewhat arbitrary classifications and whether other categories could be included; nevertheless, the point is that they represent the organization of chemical, biological, and behavioral activities over a broad scale. Can there be a common language or currency that exists, much less a single, unified theory or explanatory principle, that binds together ecotoxicology over such a broad scale of organization? It is highly unlikely, but that does not preclude the opportunity and challenge to develop new insights and a better understanding of linkages among different levels of organization.

Is there a biological attribute or quantitative measure that is most appropriate, cost effective or efficient for detecting, explaining and predicting ecotoxicological effects across broad scales of biological organization? Almost certainly not, as is discussed in Chapters 12 and 13, given the diversity and complexity of systems, and the variety of potential responses to toxic stressors that exists. Again, however, this should not deter the development of new insights and a better understanding of linkages among different levels of organization. Toxic effects, or at least their manifestation, are typically first observed at the level of the individual or higher. Numerous examples of effects on biological structure and function, as well as quantitative methods for detecting them, are given in Chapters 7 to 13. It should be very clear from the information presented in these chapters, however, that toxic effects may not always occur in the form of an observable physiological response or other apparent damage. Effects may be much more subtle, and include changes in life history traits, changes in gene frequencies over time, and behavioral modifications. Chapter 11 is particularly effective in discussing this topic and pointing out that traditional laboratory toxicity tests do not address such effects, even though these effects can be very important in natural settings.

Although ecotoxicological effects on structure and function are usually observed at higher levels of biological organization, such as individuals, populations, and ecosystems, one must usually go to lower levels of organization, such as cellular and molecular, to understand the underlying mechanism for the toxic response that is observed at a higher level. This emphasizes further the value of a broad, hierarchical approach and reinforces the point that all levels of organization have important implications for ecotoxicology. For example, Chapters 3, 4, and 5 discuss cellular structures, molecular responses, and characteristics of toxicants and the environment that allow, or prevent, the movement of contaminants from the external environment into organisms where they may affect not only the organisms themselves but also higher levels of structure and function. Chapter 2 contains a thorough discussion of geochemical aspects of metals and their implications for ecotoxicology. The information in these four chapters is very relevant to what happens at higher levels of organization. In fact, one can easily argue that the interface between the external environment and a biological receptor is the most important boundary in ecotoxicology. If a contaminant is physically or chemically bound to sediments or soil particles, or if its chemical form is altered by environmental conditions so that it cannot cross a cell membrane and become biologically available, there is little or no likelihood of effects at the individual, population, community, or ecosystem level.

The development and application of new techniques are important for progress in any field. This volume discusses many new quantitative approaches that can be applied in ecotoxicology. For example, a virtual revolution has occurred over the last decade in the area of molecular biology. Chapter 5 discusses the promise, as well as the limitations, of a number of new molecular and biochemical approaches that are beginning to be used to address questions in ecotoxicology. The whole of Chapter 8 is a discussion of statistical methods that can be used to estimate toxic effects in individuals. Chapter 6 makes a strong plea for use of more quantitative methods, particularly in ecotoxicology studies at the cell and tissue levels, and discusses methods that could replace the qualitative descriptions that are commonly used now. Chapters 7 and 9 present innovative methods for examining contaminant effects at the population level. Chapter 10 uses a modeling approach to assess the ability of toxicants to act as selective agents in the process of natural selection.

It is important to remember that any technique, old or new, is based on certain assumptions and may have limitations. For example, some of the molecular and biochemical approaches discussed in Chapter 5 may be too sensitive for detecting toxicity if subcellular effects are repaired before damage occurs to the organism. Conversely, histological techniques may be useful only where high levels of contamination are present and, therefore, may not be sensitive enough in many environmental settings. Chapter 12 suggests that techniques based on measuring changes in structure are easier to use than those based on measuring changes in function, but that functional measures may be more sensitive to stressors. The assumptions and limitations of various techniques are important not only for scientific reasons; they also relate to cost effectiveness and have regulatory implications.

Several areas were discussed at the symposium where participants thought attention is needed for ecotoxicology to move forward as a science. These areas and others are mentioned in various chapters of the book. Listed below are some of the major categories, along with chapters that discuss them.

- New ways of thinking about how science is done, particularly ecotoxicology (Chapter 1)
- Defining the boundaries of ecotoxicology more clearly (Chapter 12)
- New techniques, particularly ones that are quantitative (Chapter 6 and most others)
- Better data with which to test hypotheses and verify models (Chapters 7 and 9)

Finally, in his introductory chapter, Newman developed a strong justification for why ecotoxicology needs to change conceptually and suggested ways this can be done. The various contributing authors of this book have made additional suggestions. Since the number of participants at the symposium was limited by design, one measure of the symposium's success will be how well this book is received by the scientific community and others. We hope it will be a beginning that will foster the development and application of new ideas and analytical methods that lead ecotoxicology into the future.

Index

A

Acid-volatile sulfide model, 18–19, 20
Age at maturity, 273
AIC, see Akaike's information criterion
Akaike's information criterion, 236
Alkyl benzene sulfonate, biotoxicity,
    324
Aluminum, biotoxicity, 164, 187
ANOVA model, 351
Anoxic sediment, trace metal
    concentrations, 18–19
Aquatic ecosystems, 350
    redundancy, 356–360
    stresses and, 376
Aquatic organisms
    interspecific interactions,
        319–321
        competition, 331–337
        predation, 321–331
    organic contaminants
        bioaccumulation, 86–117, 383,
            385–386, 391
        bioavailability, 114–116
    trace metals
        bioassay, 26–29
        field studies, 29–34
        interaction with, 20–36
Aquatic toxicology, 85–86
    behavioral ecology, 319–337
    biochemical indicators, 43–49
    ecosystem management,
        392–393
    ecosystem stress, 373–374
        measuring, 383–393
        monitoring, 374–382
    histopathology, 166
    metals
        biochemical indicators, 43–49
        dissolved, 11–36
        particulate, 36–43
    mortality data analysis, 230
    organic contaminants
        bioaccumulation, 86–117, 383,
            385–386, 391
        bioavailability, 114–116
    predation behavior, 321–331
Assimilative capacity, ecosystem, 347
Atrazine, 356
Autecology, 320

AVS model, 18–20
    FIAM bioassays, 26–29
    field studies, 29–34

B

BAF, see Bioaccumulation factor
Basic prey model, 326–327
BCF, see Bioconcentration factor
Behavioral ecology, aquatic ecotoxicology,
    319–337
Benzo[a]pyrene, binding to DOM, 90
BF, see Bioaccumulation factor
Bioaccumulation, 85–86, 116–117
    carnivores, 383–384
    dissolved cadmium, 27, 29
    dissolved trace metals, 27–32
    elimination, 93–94
    estimation method, 105–108
    feeding studies, 101–103
    food web transfer, 103–105, 383,
        391
    kinetic model, 87, 110–114
    linear solvation energy relationship,
        105–108
    organic contaminants
        dissolved organic matter, 91
        in sediment, 95–103
        particle-bound metals, 39–43
        steady-state models, 108–110
Bioaccumulation factor, 99, 108
Bioactivation, 143
Bioassays
    demographics and population models,
        255–288
    trace metal in sediments, toxicity,
        26–29
Bioavailability
    organic contaminants, 114–116
        aqueous exposure, 89–94
        sediments, 98
    trace metals
        dissolved, 20–36
        particulate, 39–43
Biochemical indicators, 44–45
    cytochrome P450 enzymes, 146–148
    metabolites, 149–150
    metallothionein, 45–49, 149, 386
    mixed function oxidase enzymes,
        385–386

nucleic acids, 150
nucleosides, 150
organic hydrocarbons, 385–386
porphyrins, 150
stress proteins, 148
trace metals, 43–49, 386
Bioconcentration
  aqueous exposure
    determining, 86–89
    linear solvation energy relationships,
      105–108
    normalization, 94–95
    pharmacokinetic models, 113–114
Bioconcentration factor, 87, 94
Bioenergetics, 113, 115
Biological balance, species turnover and,
  363
Biological markers, 139–142, 165
  Mysis relicta, 375–376
  pondweed, 36
Biomagnification, 103–104, 109
Biomonitoring
  biochemical indicators, 45–49
  biological markers, 36, 139–142, 165
  molecular markers, 133–157
Biota–sediment accumulation factors, 108
Biotransformation, 143
Bioturbation, 101
BMDP, 231
BPM, see Basic prey model
BSAF, see Biota–sediment accumulation
  factors

C

Cadmium
  bioaccumulation, 27, 29
  biotoxicity
    fish gill studies, 183–185
    foraging behavior, 324
    in sediments, 26–32
    sorption in oxic sediments, 18
    uptake, 75, 76
    of sediment-bound metals, 40–43
Capture efficiency, 324
Carcinogenesis, 164–165
Carnivores, bioaccumulation, 383–384
Carrying capacities, ecosystem dynamics,
  299–300
Cell membrane, 60
  channels, 63–65
  ecotoxicology hierarchy, 78–80
  fluidity, 67

lipids, 61–62, 67–68
parabolic responses, 68
permeability, 60–61
Cells
  material entry into
    aqueous route, 62–65, 68–70
    endocytic route, 38–39, 65–67, 70
    lipid route, 61–62, 67–68
Channel, see Membrane channel
Channel proteins, 63
Channel selectivity, 69–70
Chlorinated hydrocarbons, bioaccumulation,
  98, 102, 104
Chlorophenols, toxicity, 71, 72
Chromium
  assimilation, sediment-bound, 42
  biotoxicity, 183–185
Clearance volume model, 111–112, 115
C-matrix, 296, 300
Cobalt, assimilation, sediment-bound,
  40–43
Communities, 364, 366
Community dynamics, 293–294, 313–315;
    see also Population dynamics;
    Population growth
  ecological resilience, 363–364
  ecosystem stability, 303–308
  immigration, 302–303, 312–313
  methods, 295–302
    carrying capacities, 299–300
    constants, 295–296
    selection, 301–302
    toxicant flow, 300–301
  perturbations, 308–309, 311–312
  species redundancies, 310–312,
    347–367
  species turnover, 360–362, 363
Community ecology, 364
Competition for resources, 331–337
Contaminants, see Bioaccumulation;
    Bioavailability; Bioconcentration;
    Ecotoxicants; Markers; Metals;
    Organic contaminants
Conveyor belt deposit feeders, 101
Copper
  biotoxicity, 75–77
    foraging behavior, 324, 329
    in sediments, 26–29
  ecosystem dynamics, 357
  histopathology, 163–164
Core-satellite models, 361
Cox proportional hazard model, 233–234
CT-TOX, 227
Cytochrome P450 enzymes, 146–148

D

DDT, food chain accumulation, 104
Decision making models, 351
Decomposing regression effects,
    279–280
Decomposition analysis, 264–266, 286
Demographic models, 257, 258–262, 286
Detoxication, 136, 143
Dieldrin, mortality rate, 202–205,
    281–282
7,12-Dimethyl-benz[a]anthracene, toxicity,
    144–145, 152 .
Diquat, ecological resilience, 364, 365
Dissolved organic matter
    FIAM studies, 24
    organic contaminants, 90–91, 100
DNA markers, 150–154
DOM, see Dissolved organic matter
Dose–response concept, 141, 226–244

E

EBA, see Environmental bioavailability
Ecological redundancy, 362–363
Ecological resilience, 363–364
Ecological risk assessment, 350, 351
Ecology
    functional ecology, 372–373
    paleoecology, 387–390
    paradigms, 352–354
    redundancy, 347–367
    toxicology and, 350–367
Ecosystem ecology, 372–373
Ecosystems, 349–350; see also Community
        dynamics; Ecotoxicology;
        Population dynamics
    assimilative capacity, 347
    carrying capacities, 299–300
    extinction, 308, 309, 315
    food chains, 103–105, 116, 376–377,
        391, 393
    immigration, 302–303, 312–313
    management, 392–393
    monitoring, 374–382
    paleoecology, 387–390
    perturbations, 308–309, 311–312
    protection, 347
    redundancy, 310–312, 347–367, 376–378
    reference ecosystems, 380
    remediation, 315
    selection, 315
    species needed for, 378–380

    stability, 303–308
    stress, 373–393
    structure–function relationship, 348–349,
        355–356
Ecotoxicants; see also Bioaccumulation;
        Bioavailability; Bioconcentration;
        Metals; Organic contaminants
    at the cell-membrane barrier, 59–80
    competition for resources, 331–337
    definition, 59–60
    ecosystem, flow, 300–301
    lake sediments, 383–384
    lethal effect in individuals
        dose–response approach, 141,
            226–229, 244
        time–response approach, 226, 229–244
    markers
        biochemical, 44–49, 149
        biological, 36, 134–142
        DNA, 150–154
        molecular, 142–156
    mortality rate, 202–205
    population and community dynamics,
        293–315
    population growth rate, 197–219
    population models, 255–288
    predation, 321–331
    stage specificity, 256, 269–273
    toxicity, mechanisms, 143
Ecotoxicology, 293–294, 350, 399–402;
        see also Aquatic toxicology;
        Ecosystems; Ecotoxicants; Genetic
        ecotoxicology
    behavioral ecology, 319–337
    biochemical indicators, 43–49, 148–150,
        385–386
    biological markers, 36, 139–142, 165,
        375–376
    cell-membrane barrier, 59–80
    definition, 133, 366
    detoxication, 136, 143
    DNA markers, 150–154
    hierarchical approach, 399–402
    histopathology quantitative methods,
        163–189
    individual organisms
        lethal effects, 225–250
        life histories, 197–206
    molecular markers, 133–157
    population
        community dynamics and, 293–315
        growth rates, 197–219, 259, 261, 333
        models, 255–288

redundancy, 310–312, 347–367,
        376–378
regression designs, 276–285
theoretical
        goals, 1–2
        model maturation, 5–6
        multiple working hypotheses, 4–5
        as science, 1–7
        scientific method and, 2–3, 5
        theory maturation, 6–7
toxicokinetics, 87, 110–114
Elimination, bioaccumulation, 93–94
Endocytosis, 38–39, 70
Environmental bioavailability, 114, 116
Environmental stressors
        ecosystems, 373–374
                measuring, 383–393
                monitoring, 374–382
        interspecific interactions, 318–321
                competition, 331–337
                predation, 321–331
        monitoring landscape-scale stress,
                391–392
Environmental toxicology, 366
Equilibrium partitioning, 95–101,
        109–110
        food chain models, 116
        toxicity, 96–97
Euler-Lotka equation, 218, 258
Evolutionary analysis, 213–217
Exotoxicants, regression designs,
        276–285
External energy, 295, 304
External environmental concentrations, 86
Extinction, 308, 309, 315

F

Factorial decomposition, 268
Failure time approach, 229
Feeding studies, 39
        bioaccumulation, 101–103
        predation behavior, 321–331
FIAM, see Free-ion activity model
Fish gill studies, metal toxicity, 77–78,
        181–189
Fixed effects designs, 266–276
Fluid mosaic model, 67
Food chains, 116, 376–377
Food web transfer, contaminant
        bioaccumulation, 103–105, 391,
        393
Foraging, 321–331

Free-ion activity model, 14
        dissolved metals, 23–25
        formulation, 21–23
        sediments, 25–36
Fugacity, 102, 109, 112, 115
Functional ecology, 372–373
Functional redundancy, 347, 376–378

G

Gated channels, 64
Gene frequencies, 313, 314
Generalized additive model, 281
Genetic ecotoxicology, 154–156
        DNA markers, 150–154
        evolutionary analysis, 213–217
        immigration rate, 313
Genotoxicity, 151–154
Genotype fitness, 301
Geochemistry, aquatic environment,
        11–20
        metals in anoxic/suboxic sediments,
                18–19
        metals in oxic sediments, 15–18
GLIM 4, 231
Growth dilution, 110

H

HAI, see Health assessment index
Hazard assessment, see Risk assessment
Health assessment index, 171
Heat shock proteins, 148
Herbicides, ecological resilience, 364,
        365
Histopathology
        qualitative techniques, 166, 167–169
        quantitative techniques, 163–167,
                171–189
        semiquantitative approach, 169–171
Humans, as ecological indicators, 392
Hutchinsonian niche, 310–311
Hydrocarbons
        chlorinated, 98, 102, 104
        polyaromatic, 167
        polycyclic aromatic, 142–143

I

IAG, see In-lake alkalinity generation
Immigration, ecosystem dynamics, 302–303
        312–313
Indicator populations, 375–376

Indicators; see also Biochemical indicators;
    Biological markers; Molecular
    markers
  stress in ecosystems, 373–374
    measuring, 383–393
    monitoring, 374–382
Ingestion
  bioaccumulation of particulate trace
    metals, 39–43
  feeding studies, 101–103
In-lake alkalinity generation, 380–381
Insecticides, biotoxicity, 71, 73,
    328–329
Interspecific interactions, environmental
    stressors, 319–321
  competition, 331–337
  predation, 321–331
Ion regulation, metal ions and, 77–78, 182
Iron, 66, 70
Isotopes, nitrogen, screening for toxicants,
    384–385

K

Kinetic models, see Toxicokinetics

L

Laboratory fish production index, 319
Lacustrine communities, 376
Lakes
  ecosystem studies, 373–383
  in-lake alkalinity generation, 380–381
  paleoecology, 387–390
  size, 390–391
Landscape-scale stresses, monitoring,
    391–392
Lead, biotoxicity, 26–29
Lethal effects
  dose-response approach, 226, 229–244
  time-response approach, 226, 229–244
LFPI, see Laboratory fish production index
Life cycle graph, 273–274
Life history analysis, 197, 201, 213–214,
    216, 230
Life table response experiment, 261–264
Linear solvation energy relationship,
    105–107
Lipid membrane, narcosis and, 67–68
Lipid normalization, 94–95
Lipid solubility
  cell penetration by, 61–62, 67–68
  chlorophenols, 71, 72

insecticides, 71, 73
oil spills, 71, 73–75
Litchfield-Wilcoxon approach, 227
LOEC, see Lowest-observed-effect
    concentration
Logit, 228
Lowest-observed-effect concentration, 320,
    351
LSER, see Linear solvation energy
    relationship
LTRE, see Life table response experiment

M

Mann-Whitney test, 170
Markers; see also Biochemical indicators;
    Biological markers; Molecular
    markers
  DNA, 150–154
  stress in ecosystems, 373–374
    measuring, 383–393
    monitoring, 374–382
MATLAB, 260
Matrix population models, 259–262
Mechanistic physiological models, 257
Membrane channel, 63–65, 68–69
  cadmium uptake, 75, 76
  copper toxicity, 75–77
  selectivity, 69–70
  zinc effects, 77–78
Membrane ecotoxicology
  channel interference, 70
    cadmium uptake, 75, 76
    copper toxicity, 75–77
    zinc effects, 77–78
  lipid membrane, 71
    chlorophenols, 71, 72
    insecticides, 71, 73
    oil spills, 71, 73–75
Membrane fluidity, 67
Mercury, sources, 390
Metabolites, as indicators, 149–150
Metallothionein, 44, 149
  as biochemical indicator, 45–49,
    386
  function, 45
Metals; see also Trace metals
  aluminum, 164, 187
  biochemical indicators, 43–49
  cadmium, 18, 26–32, 40–43, 75, 76,
    183–185, 324
  chromium, 42, 183–185
  cobalt, 40–43

copper, 26–29, 75–77, 163–164, 324,
    329, 357
ion regulation and, 77–78
iron, 66, 70
lead, 26–29
mercury, 390, 391
nickel, 26–29, 183–185
in sediments, 11–12
    concentration, 11–20
    dissolved, 20–36, 50–51
    free ion activity model, 21–36
    particulate, 36–43
silver, 40–43
speciation, 21
tin, 198–199
toxicity, fish gill studies, 77–78,
    181–189
uptake, particulate, 36–43
zinc, 26–29, 40–43, 77–78, 168, 324
Methylcholanthrene, bioavailability, 91
Methyl mercury, biotoxicity, 391
Methyl parathion, biotoxicity, 324, 328
MFOs, see Mixed function oxidase
    enzymes
Microbial communities, model, 361
Mirex, biotoxicity, foraging behavior,
    328
Mixed function oxidase enzymes,
    385–386
Models
    bioaccumulation, steady-state models,
        108–110
    bioconcentration, pharmacokinetic
        models, 113–114
    bioenergetics models, 327–328
    decision making, 351
    demographic, 257, 258–259, 286
        bioassay studies, 255–288
        matrix population models, 259–262
    foraging models, 327–328
    interspecific interactions, 337
    optimal foraging models, 325–328
    population and community dynamics,
        293–315
Molecular connectivity, bioconcentration
    predictions, 105, 106
Molecular markers, 133–157
    biological markers, 36, 139–142, 165
    cytochrome P450 enzymes, 146–148
    DNA markers, 150–154
    metabolites, 149–150
    metallothionein, 45–49, 149, 368
    protein induction, 146–149

stress proteins, 148
Monitoring
    biochemical indicators, 45–49, 144–150,
        385–386
    biological markers, 36, 139–142, 165,
        375–376
    DNA markers, 150–154
    ecosystems, 374–382
    landscape-scale stresses, 391–392
    molecular markers, 133–157
Monod growth curve, 333
Morphometry, 173, 174–179
Mortality rate, 202–205
    lethal effects estimate, 225–250
        dose-response approach, 141,
            226–229, 244
        time-response approach, 226,
            229–244
    nonparametric methods, 232–233
    parametric methods, 234–244
    population density and, 206–207
    predation and, 331
    semiparametric methods, 233–234
Mortality sensitivities, 205
Multiple-response concept, 141–142
Mysis relicta, 375–376

                    N

Naphthalene, bioaccumulation, 90–91
Narcosis, lipid membrane and, 67–68
Net energy intake, 326
Nickel, biotoxicity, 26–29, 183–185
Nitrogen isotopes, screening for toxicants,
    384–385
NOEC, see No-observed-effect
    concentration
Nonparametric regression, 281
No-observed-effect concentration, 320,
    351
Nucleic acids, as biochemical indicators,
    150
Nucleosides, as biochemical indicators,
    150

                    O

Oil spills, 71, 73–75
Optimal foraging theory, 325–328
Organic contaminants
    alkyl benzene sulfonate, 324
    aquatic organisms
        aqueous exposure, 86–95

bioavailability, 114–116
bioconcentration, 86–110
    estimation method, 105–108
    food web transfer, 103–105, 383
    linear solvation energy relationship,
        105–107
    molecular connectivity, 105, 106
    sediment exposure, 87, 95–103
    steady-state models, 108–110
    toxicokinetics, 87, 110–114
atrazine, 356
benzo[a]pyrene, 90
chlorophenols, 71, 72
DDT, 104
dieldrin, 202–205, 281–282
7,12-dimethyl-benz[a]anthracene,
    144–145, 152
diquat, 264, 265
herbicides, 364, 365
hydrocarbons, 167
    chlorinates, 98, 102, 104
    polycyclic aromatic, 142–143
insecticides, 71, 73, 328–329
methylcholanthrene, 91
methyl mercury, 391
methyl parathion, 324, 328
mirex, 328
naphthalene, 90–91
oil spills, 71, 73–75
pentachlorophenol, 91–92
pesticides, 71, 73, 211, 213, 329, 350
phenanthrene, 73
polychlorinated biphenyl, 88–89, 104,
    164
polydimethylsiloxane, 89
pyrene, 90
pyrethroids, 71, 73
2,2',4,4'-tetrachlorobiphenyl, 95
tributyltin, 198–199
Oxic sediment
    metal bioavailability, 35
    trace metal concentration, 15–18

P

Paleoecology, 387–390
Partitioning model, 95–101, 109–110,
    116
Patch switching behavior, 324
Pentachlorophenol, bioaccumulation,
    91–92
Perturbations, ecosystem, 308–309,
    311–312

Pesticides
    biotoxicity, foraging behavior, 329
    ecotoxicity tests, 350
    population ecology and, 211, 213
    toxicity, 71, 73
Phagocytosis, 61, 70
Pharmacokinetic models, 113–115
Phenanthrene, toxicity, 73
Phenomenological population model, 256,
    257
Pollutants, see Contaminants; Ecotoxicants
Polyaromatic hydrocarbons, carcinogenicity,
    167
Polychlorinated biphenyl, 88–89, 104, 164
Polycyclic aromatic hydrocarbons, toxicity,
    142–143
Polydimethylsiloxane, toxicity, 89
Pondweed, as biological marker, 36
Population density, 206–213
Population dynamics, 293–294, 313–315;
        see also Community dynamics
    ecosystem stability, 303–308
    immigration, 302–303, 312–313
    methods, 295–302
        carrying capacities, 299–300
        constants, 295–296
        initial values, 296–299
        selection, 301–302
        toxicant flow, 300–301
    perturbations, 308–309, 311–312
    population growth, 302
    species redundancies, 310–312
    species turnover, 360–362, 363
Population ecology, 206–213
Population growth
    ecosystem dynamics, 302
    immigration, 302–303, 312–313
    sensitivity analysis, 261
Population growth rate, 197–206, 218–219,
    259
    dynamics, 206–213
    evolutionary analysis, 213–217
    food concentration and, 333
    predation, 213
    sensitivity analysis, 261
Population models
    demographic bioassay studies, 255–288
    matrix population models, 259–262
Porins, 63
Porphyrins, as biochemical indicators, 150
Predation
    balancing predation risk and foraging,
        330–331

hunger and motivation to feed, 322
optimal foraging theory, 325–328
population growth, 213
prey
    capture, 324–325
    choice, 323–324
    risks, 328–330
    searching behavior, 322–323
Preference factors, 108
Prey
    basic prey models, 326–327
    capture, 324–325
    choice, 323–324
    encounter rates, 322
    search, 322–323
Probit, 228
Process ecology, 372–373
Production rate, 198
Product-limit methods, 230, 232–233
Proportional hazard model, 233–234
Protein induction, molecular markers,
    146–149
Protozoan communities, model, 361
Pyrene, binding to DOM, 90
Pyrethroids, toxicity, 71, 73

Q

QSAR, see Quantitative structure-activity
    relationships
Qualitative techniques, histopathology, 166,
    167–169
Quantitative structure-activity relationships,
    6, 60
Quantitative techniques, histopathology,
    163–167, 171–189

R

Redundancy
    aquatic ecosystems, 356–360
    ecological redundancy, 362–363
    functional redundancy, 347, 376–378
    predictive approaches, 350–354
    species, 310–312, 360–362, 363
    structure-function relationship, 348–349,
        355–356
    terrestrial systems, 355–356
Reference ecosystems, 380
Regression designs, ecotoxicology, 276–285
Resistance time approach, 229
Risk assessment, 134
    biochemical indicators, 45–49

biological markers, 36, 139–142, 165
ecological studies, 350, 351
kinetic models, 111
molecular markers, 133–157
Rivet popper hypothesis, 347

S

Sampling, quantitative histopathology,
    179–181
SAS, 231, 232–233, 236, 245–250
Scope for growth, 198
Searching behavior, predation, 322–323
Sediments
    contaminant partitioning, 100
    lake sediments, 383–384
    metal concentration, 11–20
        anoxic/suboxic sediments, 18–19
        oxic sediments, 15–18
    organic contaminants, 87, 95–103, 383
    paleoecology, 387–390
    trace metals
        dissolved, 20–36, 50–51
        free ion activity model, 21–36
        particulate, 36–43
Selection, 301–302, 315
SEM, see Simultaneously extracted metals
Semiquantitative techniques, histopathology,
    169–171
Shannon-Wiener index, 308
Silver, assimilation, 40–43
Simulation, population and community
    dynamics, 293–315
Simultaneously extracted metals, 26
Size-efficiency hypothesis, 334
Small body size hypothesis, 335
Sodium pump, 64
Soil, as ecosystem, 350
Sorption, trace metals in sediments, 15–18,
    50–51
Speciation, 21
Species
    extinction, 308, 309, 315
    interspecific interactions, 319–321
        competition, 331–337
        predation, 321–331
    needed to maintain ecosystem,
        378–380
    redundancy, 310–312, 347–367,
        376–378
    turnover, 360–362, 363
Stable stage distribution, 259
Stage specificity, 256, 269–273

Stereology, 173, 174–179
Stress, see Environmental stressors
Stress proteins, 148
Suboxic sediment, trace metal
 concentrations, 18–19
Surface deposit feeders, 102
Survival time approach, 226, 229–244

T

Terrestrial ecosystems, 349, 363
 redundancy, 355–356
2,2',4,4'-Tetrachlorobiphenyl, partitioning,
 95
Theory maturation, 6–7
Threshold food level, 333
Time endpoint approach, 226–229
Time-response approach, 226, 229–244
Time-to-death approach, 229, 230–231,
 233, 238–242
Toxicants, see Ecotoxicants
Toxicity
 equilibrium partitioning, 96–97
 mechanisms, 143
 trace metals in sediment, AVS study,
  26–29
Toxicokinetics, 87, 110–114
Trace metals
 bioaccumulation, 27–32
 biochemical indicators, 43–49, 386
 in sediments, 11–12
  anoxic/suboxic sediments, 18–19
  concentration, 11–20

dissolved, 20–36, 50–51
 free ion activity model, 21–36
 oxic sediments, 15–18
 particulate, 36–43
 speciation, 21
 uptake, particulate, 36–43
Transferrin, specificity, 70
Transformed logit, 228
Tributyltin, biotoxicity, 198–199
TTD, see Time-to-death approach

V

Valinomycin, 62
Vital rates, 255
 as function of treatment, 280–281
 stage-specific, 269–273

W

Waiting time approach, 229
Weibull model, 229, 243

Z

Zinc
 assimilation of sediment-bound metals,
  40–43
 biotoxicity, 168
  foraging behavior, 324
  in sediments, 26–29
 fish gill ion regulation, 77–78
Z-matrix, 296